ENGINEERING ETHICS

Concepts and Cases

FOURTH EDITION

ENGINEERING ETHICS

Concepts and Cases

CHARLES E. HARRIS
Texas A&M University

MICHAEL S. PRITCHARD
Western Michigan University

MICHAEL J. RABINS
Texas A&M University

WADSWORTH
CENGAGE Learning

Australia • Brazil • Japan • Korea • Mexico • Singapore • Spain • United Kingdom • United States

WADSWORTH
CENGAGE Learning™

Engineering Ethics: Concepts and Cases, Fourth Edition
Charles E. Harris, Michael S. Pritchard, and Michael J. Rabins

Acquisitions Editor: Worth Hawes

Assistant Editor: Sarah Perkins

Editorial Assistant: Daniel Vivacqua

Technology Project Manager: Diane Akerman

Marketing Manager: Christina Shea

Marketing Assistant: Mary Anne Payumo

Marketing Communications Manager: Tami Strang

Project Manager, Editorial Production: Matt Ballantyne

Creative Director: Rob Hugel

Art Director: Cate Barr

Print Buyer: Paula Vang

Permissions Editor: Mardell Glinski-Schultz

Production Service: Aaron Downey, Matrix Productions Inc.

Copy Editor: Dan Hays

Cover Designer: RHDG/Tim Heraldo

Cover Image: SuperStock/Henry Beeker

Compositor: International Typesetting and Composition

For product information and technology assistance, contact us at **Cengage Learning Customer & Sales Support, 1-800-354-9706**

For permission to use material from this text or product, submit all requests online at **cengage.com/permissions** Further permissions questions can be e-mailed to **permissionrequest@cengage.com**

Library of Congress Control Number: 2008924940

ISBN-13: 978-0-495-50279-1

ISBN-10: 0-495-50279-0

Wadsworth
10 Davis Drive
Belmont, CA 94002-3098
USA

Cengage Learning is a leading provider of customized learning solutions with office locations around the globe, including Singapore, the United Kingdom, Australia, Mexico, Brazil, and Japan. Locate your local office at **international.cengage.com/region.**

Cengage Learning products are represented in Canada by Nelson Education, Ltd.

For your course and learning solutions, visit **academic.cengage.com.**

Purchase any of our products at your local college store or at our preferred online store **www.ichapters.com.**

Printed in Canada
1 2 3 4 5 6 7 12 11 10 09 08

To
Michael J. Rabins, PE, 1933–2007
coauthor, collaborator, friend

CONTENTS

LIST OF CASES

WE ARE HAPPY TO OFFER the fourth edition of *Engineering Ethics: Concepts and Cases*. This edition has a number of changes, which we believe will enable the book to keep abreast of recent thinking in engineering ethics and to be more useful to students and teachers in the classroom.

The major changes to the fourth edition are as follows:

- Each chapter now begins with a series of bullet items, summarizing the main ideas in the chapter.
- The first chapter explains several approaches to the nature of professionalism and makes it clearer that the subject of the book is professional ethics, not personal ethics or common moral beliefs.
- The first chapter also introduces the student to a new theme in the book, namely the distinction between "preventive ethics" and "aspirational ethics." We believe the latter should have more prominence in engineering ethics.
- The fifth chapter, while incorporating some of the material in the old chapter on computer ethics, also contains our first attempt to introduce some ideas from science and technology studies and the philosophy of technology into the book.
- Most of the other chapters have been reorganized or rewritten with a view to introducing new ideas or making them more accessible to students.
- Finally, the section on cases at the end of the book has been very extensively revised in ways that will be explained soon.

Let us consider these ideas in more detail.

PROFESSIONAL ETHICS

Students sometimes ask why they should take a course in professional ethics, because they consider themselves to be ethical people. It is important for them to understand, therefore, that their personal morality is not being questioned. Personal morality and professional ethics, however, are not always the same. One might have personal objections to working on military projects, but avoiding such work is not required by professional ethics. On the other hand, professional ethics increasingly requires engineers to protect the environment, regardless of their personal moral convictions. We attempt to explore the nature of professionalism and professional ethics more thoroughly than in previous editions.

PREVENTIVE ETHICS AND ASPIRATIONAL ETHICS

During the past few decades, engineering ethics has focused on what we call "preventive ethics." We believe that two influences have determined this orientation: the so-called "disaster cases" (e.g., the *Challenger* and *Columbia* cases and the Hyatt Regency walkway collapse) and the professional codes of ethics. Following the lead of these influences, engineering ethics has tended to have a negative orientation, focusing on preventing harm to the public and preventing professional misconduct. These have been—and will continue to be—important concerns of engineering ethics. We believe, however, that more emphasis should be placed on the more idealistic and aspirational aspects of engineering work, namely the place of technology in improving the lot of humankind. The codes already suggest this goal when they mention concern for human "welfare," but this reference is not easy to interpret. We believe this more positive orientation is important not only in encouraging engineers to do their best professional work but also in encouraging young people to enter and remain in the engineering profession.

SCIENCE AND TECHNOLOGY STUDIES AND THE PHILOSOPHY OF TECHNOLOGY

Scholars in engineering ethics have become increasingly interested in the question, "How can science and technology studies (STS) and the philosophy of technology be integrated into engineering ethics?" The general relevance of these two fields to engineering ethics is obvious: They both deal with the nature of technology and its relationship to society. Determining the precise nature of this relevance, however, has not been easy. STS is a descriptive, empirically oriented field, having its origins in sociology and history. STS researchers have not for the most part explored the ethical implications of their work. The philosophy of technology is more normatively oriented, but the exploration of its implications for engineering ethics has barely begun. In Chapter 5, we suggest some implications of these areas for engineering ethics in a way that we hope will be provocative for instructors and students alike. We especially welcome comments, criticisms, and suggestions on our work here.

REORGANIZATIONS AND ADDITIONS

In addition to the changes indicated previously, every chapter has undergone some degree of reorganization and addition. Chapter 2 on responsibility places more emphasis on engineering standards, including the standard of care and design standards. Chapters 3 and 4 have similar content, but the order of presentation of ideas has been altered in ways we believe provide greater clarity to our ideas about framing and resolving ethical problems. Chapter 6 has relatively little change, but Chapter 7, in addition to some reorganizing, has some new material on the new "capabilities" approach to risk and disaster analysis. In Chapter 8, we place more emphasis on the importance of understanding the culture of an organization in order to know how to effectively make protests or initiate organizational change. We also introduce Michael Davis' account of whistleblowing. Chapter 9 sets ethical decisions within the guidelines of the requirements of environmental law, which we believe gives ethical decisions a more realistic context. Chapter 10 has been reorganized to highlight the way in which various social and cultural differences in countries set the context for what we call "boundary-crossing problems."

THE CASES SECTION: MAJOR REVISIONS

The case section contains not only many new cases but also cases of more widely varying types. We believe the new mix of cases offers a much richer and more stimulating repertoire for students. In addition to cases involving problems for individual engineers—often called "micro cases"—there are also cases that focus on the institutional settings within which engineers work, on general problems within engineering as a profession (e.g., the place of women), and on larger social policy issues related to technology (e.g., global warming). Cases in this last category are sometimes referred to as "macro cases," and their inclusion is part of our larger aim to give increased emphasis to issues of social policy that illustrate the social embeddedness of technology. Some of the cases also carry forward the theme of the more positive, exemplary, and aspirational aspect of engineering work.

THE PASSING OF MICHAEL J. RABINS

It is with regret and sadness that we note the passing of our colleague and coauthor, Michael J. Rabins, to whom this fourth edition is dedicated. It was Mike's interest in bringing philosophers into engineering ethics that contributed to many years of collaboration among the three of us. There were successful grant applications, publications, the development of courses in engineering ethics, and, finally, the various editions of this book. We also express our gratitude to his wife, Joan Rabins. She not only prepared the indexes for previous editions and offered valuable suggestions on the text but also hosted countless meetings of the three of us at their home. We have many happy memories of day-long meetings on the textbook at the beautiful and spacious home that Mike and Joan designed and that they both loved.

ACKNOWLEDGMENTS

Once again, we acknowledge the responses of our students to the previous editions of this book and to our teaching of engineering ethics. We also thank all of our colleagues who have commented on different aspects of previous editions and our latest efforts to improve the book.

For this edition, we offer special thanks to the following people. Peggy DesAutels (Philosophy, University of Dayton) contributed her essay on women in engineering, "Where Are the Women?" to our Cases section. Ryan Pflum (Philosophy graduate student, Western Michigan University) contributed "Big Dig Collapse" and "Bridges" to our Cases section and provided research assistance for other cases.

Colleen Murphy (Philosophy, Texas A & M University) and Paolo Gardoni (Civil Engineering, Texas A & M University) contributed most of the material in the discussion of the capabilities-based approach to assessing harm and risk. Roy Hann (Civil Engineering, Texas A & M) suggested the reorganization of the chapter on the environment to more clearly convey the idea that engineers work in the context of environmental law, although they can (and, we believe, should) go beyond the legal requirements.

A special thanks to Michael Davis (Philosophy, Illinois Institute of Technology) for his many suggestions on Chapter 1 and other ideas contributed throughout the book. Although he does not bear responsibility for the final product, our book is better than it would have been without his suggestions.

For help in preparing the fourth edition, we thank Worth Hawes, Wadsworth Philosophy Editor, and Matt Ballantyne, Wadsworth Production Manager. Merrill Peterson and Aaron Downey at Matrix Productions contributed greatly to the quality of the text. Thanks also to our copy editor, Dan Hays.

ENGINEERING ETHICS
Concepts and Cases

CHAPTER ONE

Why Professional Ethics?

Main Ideas in this Chapter

- This book focuses on professional ethics, not personal ethics or common morality.
- Engineering is a profession by some definitions of professionalism and not as clearly a profession by other definitions.
- Ethical commitment is central to most accounts of professionalism.
- Professional ethics has several characteristics that distinguish it from personal ethics and common morality.
- Possible conflicts between professional ethics, personal ethics, and common morality raise important moral questions.
- Professional engineering ethics can be divided into a negative part, which focuses on preventing disasters and professional misconduct, and a positive part, which is oriented toward producing a better life for mankind through technology.

"WHY SHOULD I STUDY ETHICS? I am an ethical person." Engineers and engineering students often ask this question when the subject of professional ethics is raised, and the short and simple answer to it is not long in coming: "You are not being asked to study ethics in general, but your profession's ethics." We can also anticipate a response to this answer: "Well, what is the difference?" In order to answer this question, we must have an account of the nature of professionalism and then ask whether engineering is a profession according to this account. After this, we can examine more directly professional ethics as it applies to engineering.

1.1 WHAT IS A PROFESSION?

We can begin by looking at the dictionary definition of professionalism. An early meaning of the term *profession* referred to a free act of commitment to a way of life. When associated with the monastic vows of a religious order, it referred to a monk's public promise to enter a distinct way of life with allegiance to high moral ideals. One "professes" to be a certain type of person and to occupy a special social role that carries with it stringent moral requirements. By the late 17th century, the term had been secularized to refer to anyone who professed to be duly qualified.

Thus, *profession* once meant, according to the *Oxford Shorter Dictionary,* the act or fact of "professing." It has come to mean

> the occupation which one professes to be skilled in and to follow....A vocation in which professed knowledge of some branch of learning is used in its application to the affairs of others, or in the practice of an art based upon it.

This brief historical account, however, is not sufficient for our purposes; this account of professionalism provides only limited insight into the nature of professionalism. We can gain deeper insight if we look at the account of professionalism given by sociologists and philosophers. We begin with a sociological account.

A Sociological Analysis of Professionalism

Among the several traditions of sociological analysis of the professions, one of the most influential has a distinctly economic orientation. These sociologists view attaining professional status as a tactic to gain power or advantage in the marketplace. Professions have considerable power in the marketplace to command high salaries, so they conclude that professional status is highly desirable. If we distinguish between an occupation, which is simply a way to make a living, and a profession, the question is how a transition from a "mere" occupation to a profession (or an occupation that has professional status) is accomplished. The answer is to be found in a series of characteristics that are marks of professional status. Although probably no profession has all of these characteristics to the highest degree possible, the more characteristics an occupation has, the more secure it is in its professional status.[1]

1. Extensive training: Entrance into a profession typically requires an extensive period of training, and this training is of an intellectual character. Many occupations require extensive apprenticeship and training, and they often require practical skills, but the training typically required of professionals focuses more on intellectual content than practical skills. Professionals' knowledge and skills are grounded in a body of theory. This theoretical base is obtained through formal education, usually in an academic institution. Today, most professionals have at least a bachelor's degree from a college or university, and many professions require more advanced degrees, which are often conferred by a professional school. Thus, the professions are usually closely allied in our society with universities, especially the larger and more prestigious ones. Although extensive training may be required for professional work, the requirement of university training serves as a barrier to limit the number of professionals and thus to provide them with an economic advantage.

2. Vital knowledge and skills: Professionals' knowledge and skills are vital to the well-being of the larger society. A society that has a sophisticated scientific and technological base is especially dependent on its professional elite. We rely on the knowledge possessed by physicians to protect us from disease and restore us to health. The lawyer has knowledge vital to our welfare if we have been sued or accused of a crime, if our business has been forced into bankruptcy, or if we want to get a divorce or buy a house. The accountant's knowledge is also important for our business successes or when we have to file our tax returns. Likewise, we are dependent on the knowledge and research of scientists and engineers for our safety in an airplane, for many of the technological advances on which our material civilization rests, and for national

defense. Since professional services are vital to the general welfare, citizens are willing to pay any price to get them.

3. Control of services: Professions usually have a monopoly on, or at least considerable control over, the provision of professional services in their area. This control is achieved in two ways. First, the profession convinces the community that only those who have graduated from a professional school should be allowed to hold the professional title. The profession usually also gains considerable control over professional schools by establishing accreditation standards that regulate the quality, curriculum content, and number of such schools. Second, a profession often attempts to persuade the community that there should be a licensing system for those who want to enter the profession. Those who practice without a license are subject to legal penalties. Although it can be argued that monopoly is necessary to protect the public from unqualified practitioners, it also increases the power of professionals in the marketplace.

4. Autonomy in the workplace: Professionals often have an unusual degree of autonomy in the workplace. This is especially true of professionals in private practice, but even professionals who work in large organizations may exercise a large degree of individual judgment and creativity in carrying out their professional responsibilities. Whether in private practice or in an organizational setting, physicians must determine the most appropriate type of medical treatment for their patients, and lawyers must decide the most successful type of defense of their clients. This is one of the most satisfying aspects of professional work. The justification for this unusual degree of autonomy is that only the professional has sufficient knowledge to determine the appropriate professional services in a given situation. Besides providing a more satisfying work environment for professionals, autonomy may also increase the ability of professionals to more easily promote their economic self-interest. For example, a physician might order more tests than necessary because they are performed by a firm in which she has a financial interest.

5. Claim to ethical regulation: Professionals claim to be regulated by ethical standards, many of which are embodied in a code of ethics. The degree of control that professions possess over the services that are vital to the well-being of the rest of the community provides an obvious temptation for abuse, so most professions attempt to limit these abuses by regulating themselves for the public benefit. Professional codes are ordinarily promulgated by professional societies and, in the United States, by state boards that regulate the professions. Sometimes professional societies attempt to punish members who violate their codes, but their powers are limited to expelling errant members. State boards have much stronger legal powers, including the ability to withdraw professional licenses and even institute criminal proceedings. These regulatory agencies are controlled by professionals themselves, and so the claim to genuine ethical regulation is sometimes seen to be suspicious. The claim to self-regulation does, however, tend to prompt the public to allow professionals to charge what they want and to allow professionals considerable autonomy.

According to this sociological analysis, the identifying characteristics of professions may have one or both of two functions: altruistic and self-interest. Arguments can certainly be made that these characteristics of professionalism are necessary in order to protect and better serve the public. For example, professionals must be adequately trained, and they must have a certain amount of freedom to determine what is best for the patient or client. One can also view these characteristics as ways of promoting the economic self-interest of professionals. Thus, there is a certain amount of moral cynicism in this analysis, or perhaps amoralism. Even the claim to be regulated by ethical considerations may be just that—a claim. The claim may be motivated as much by economic self-interest as by genuine concern for the public good.

The next two accounts give ethical commitment a stronger place.

Professions as Social Practices

This account of professionalism begins with an analysis of a concept, not with empirical research. The concept is of a "social practice," which is, as philosopher Alasdair MacIntyre defined it,

> any coherent and complex form of socially established cooperative human activity through which goods internal to that form of activity are realized in the course of trying to achieve those standards of excellence which are appropriate to, and partially definitive of, that form of activity.[2]

A profession is an example of a social practice. Without following the ideas of MacIntyre or others completely, perhaps we can say the following about a social practice. First, every social practice has one or more aims or goods that are especially associated with it or "internal" to it. For example, medicine (along, of course, with nursing, pharmacy, osteopathy, and the like) aims at the health of patients. One of the aims of law is justice. A practice may also produce other goods, such as money, social prestige, and power, but it is these goods especially associated with the practice that interest us here and that are especially related to its moral legitimacy. Second, a social practice is inconceivable without this distinctive aim. We cannot imagine medicine apart from the aim of producing health or law without the aim of producing justice. Third, the aims of a social practice must be morally justifiable aims. Both health and justice are morally praiseworthy aims. Fourth, the distinctive aim of a social practice provides a moral criterion for evaluating the behavior of those who participate in the social practice and for resolving moral issues that might arise in the practice. Although people will differ about how the term is to be defined, if a medical practice does not promote "health," we might wonder about its moral legitimacy as a medical practice.

The advantage of this account of professionalism is that it has a distinctively moral orientation and characterizes the professions as institutions that must be not only morally permissible but also aim at some moral good. There cannot be a profession of thievery or a profession of torturing because these occupations are inconsistent with ordinary morality.

A Socratic Account of Professionalism

Philosopher Michael Davis has proposed a dialogue approach to the issue of defining "professional." Much like the Greek philosopher Socrates, Davis has engaged professionals from various countries as well as other philosophers in conversations about the meaning of "professional." In typical Socratic fashion, a definition of professionalism is not accepted uncritically but, rather, tested against counterexamples until a definition is arrived at which seems to escape criticism. Following this program for approximately two decades, Davis has derived the following definition:

> A profession is a number of individuals in the same occupation voluntarily organized to earn a living by openly serving a moral ideal in a morally permissible way beyond what law, market, morality, and public opinion would otherwise require.[3]

This definition highlights several features that Davis believes are important in the concept of professionalism that he believes many people, including many professionals, hold:

1. A profession cannot be composed of only one person. It is always composed of a number of individuals.
2. A profession involves a public element. One must openly "profess" to be a physician or attorney, much as the dictionary accounts of the term "profession" suggest.
3. A profession is a way people earn a living and is usually something that occupies them during their working hours. A profession is still an occupation (a way of earning a living) even if the occupation enjoys professional status.
4. A profession is something that people enter into voluntarily and that they can leave voluntarily.
5. Much like advocates of the social practice approach, Davis believes that a profession must serve some morally praiseworthy goal, although this goal may not be unique to a given profession. Physicians cure the sick and comfort the dying. Lawyers help people obtain justice within the law.
6. Professionals must pursue a morally praiseworthy goal by morally permissible means. For example, medicine cannot pursue the goal of health by cruel experimentation or by deception or coercion.
7. Ethical standards in a profession should obligate professionals to act in some way that goes beyond what law, market, morality, and public opinion would otherwise require. Physicians have an obligation to help people (their patients) be healthy in a way that nonphysicians do not, and attorneys have an obligation to help people (their clients) achieve justice that the rest of us do not.

This seems like a reasonable approach to take. We believe that it is an acceptable definition of "professional," although one might ask whether Davis' definition has sufficient empirical basis. The evidence for his definition is informal and anecdotal. Although probably based on more observation than the social practice approach, some might wish for a wider body of evidence in support of it. For our purposes, however, it is enough if engineering students and engineers who read this book find that it catches the meaning of profession relevant to them and engineering ethics.

1.2 ENGINEERING AND PROFESSIONALISM

Is engineering a true profession by these criteria? Occupations are probably best viewed as forming a continuum, extending from those occupations that are unquestionably professional to those that clearly are not. The occupations that clearly are professions include medicine, law, veterinary medicine, architecture, accounting (at least certified public accountancy), and dentistry. Using these three accounts of professionalism, to what extent does engineering qualify as a profession?

Looking at the sociological or economic analysis of professionals, engineering seems to qualify only as a borderline profession. Engineers have extensive training and possess knowledge and skills that are vital to the public. However, engineers do not have anything like complete control of engineering services, at least in the United States, because a license is not required to practice many types of engineering. Because they do not have to have a license to practice, a claim by engineers to be regulated by ethical standards—at least by compulsory ethical standards—can be questioned. Only licensed engineers are governed by a compulsory code of ethics. Finally, engineers who work in large organizations and are subject to the authority of managers and employers may have limited autonomy. However, even doctors and lawyers often work in large organizations nowadays. Given that engineers are highly trained and perform services that are vital to the public, that some engineers are registered and thus work under a legally enforced ethical code, and that autonomy in the workplace may be declining for all professionals, engineering qualifies for at least quasi-professional status by the sociological account.

Some might argue that the social practice definition of professionalism also leaves engineering with a questionable professional status. Taking a cue from engineering codes, one might define the goal of engineering as holding paramount the health, safety, and welfare of the public. However, an engineer who ignores human health, safety, and welfare except insofar as these criteria are taken into account by managers who assign him or her a task should probably still be considered an engineer. On the other hand, if one takes the goal or task of engineering to be something like the production of the most sophisticated and useful technology, the ideal is not a moral one at all because technology can be used for moral or immoral ends. Still, it seems to be a useful insight to state that engineering has a goal of producing technology for the welfare of society.

In contrast to the other two accounts of professionalism, Davis' definition allows engineering full professional status. Engineering is a group activity, which openly professes special knowledge, skill, and judgment. It is the occupation by which most engineers earn their living, and it is entered into voluntarily. Engineering serves a morally good end, namely the production of technology for the benefit of mankind, and there is no reason why morally permissible means to that end cannot be used. Finally, engineers have special obligations, including protecting the health and safety of the public, as this is affected by technology.

Although engineering may not, by some definitions, be a paradigmatic profession in the same way that medicine and perhaps law are, it does have professional status by Davis' definition. From the sociological standpoint, a principal factor standing in the way of full professional status is the fact that in the United States a license is not required to practice engineering. From the standpoint of professional ethics, however, one of the crucial issues in professionalism is a genuine commitment to ethical ideals. Ethical ideals must not be merely a smoke screen for getting the public to trust professionals and impose only minimal regulation but also realized in daily practice.

1.3 TWO MODELS OF PROFESSIONALISM

Another way to understand the importance of the ethical element in professionalism is to examine two models of the professional. The contrast between the understanding of the professions as primarily motivated by economic self-interest and as motivated by genuine ethical commitment is made especially clear by the following two models.[4]

The Business Model

According to the business model, an occupation is primarily oriented toward making a profit within the boundaries set by law. Just like any other business, a profession sells a product or service in the marketplace for a profit; the major constraint on this activity is regulation imposed by law. If people ordinarily called professionals, such as doctors, lawyers, or engineers, followed this model, their claim to professionalism would be severely limited. They might choose to adopt the trappings of professionalism, but they would do so primarily as a means to increase their income and protect themselves from governmental regulation. They would use their professional training and specialized knowledge that the layperson does not have to impress upon laypeople that they deserve a high income and preferential treatment. They would take advantage of the fact that they have knowledge that is important to ordinary citizens to gain a monopoly or virtual monopoly over certain services in order to increase profit and to persuade laypeople and governmental regulators that they should be granted a great deal of autonomy in the workplace. They would promote the ideal of self-regulation in order to avoid close governmental supervision by nonprofessionals. They would insist that governmental regulatory boards be composed primarily of other professionals in order to forestall supervision by nonprofessionals.

The major difference between the so-called professionals who adopt the business model and most other occupations, such as sales or manufacturing, is that the latter seek profit primarily by selling a physical product, such as automobiles or refrigerators, whereas professionals seek profit by selling their expertise. Nevertheless, the ultimate goal is the same in both cases: selling something in the marketplace for profit.

The Professional Model

This model offers a quite a different picture of occupations such as medicine, law, and engineering. Crucial to the professional model is the idea that engineers and other professionals have an implicit trust relationship with the larger public. The terms of this trust relationship, sometimes referred to as a "social contract" with the public, are that professionals agree to regulate their practice so that it promotes the public good. In the words of most engineering codes, they agree to hold paramount the safety, health, and welfare of the public. That is, they agree to regulate themselves in accordance with high standards of technical competence and ethical practice so that they do not take unfair advantage of the public. They may agree to governmental regulation, for example, by state regulatory boards, because they believe that it is the most effective and efficient way to preserve this trust relationship between themselves and the larger society. Finally, professionals may seek a monopoly or at least considerable control over the provision of the services in which they are competent, but this is in order to protect the public from incompetent providers. In return, the public confers on professionals a number of benefits. Professionals are accorded high social standing, a better than average income, and considerable autonomy in the workplace. The public also pays for a considerable percentage of professional education, at least at public universities.

It is obvious that neither the business model nor the professional model, taken by themselves, contains the whole truth about the actual practice of professionals.

Most professionals are probably not so cynical and self-interested that they think of their work wholly in terms of a pursuit of profit. However, they may not be so idealistic that they conceive of themselves as concerned primarily with public service. In terms of a description of how professionals actually operate, both models have some validity. Nevertheless, the notion of professionalism, as it is traditionally understood, requires that a professional embrace the professional model to a substantial degree, and in this model ethical commitment is paramount. Engineers can certainly adopt the professional model, and this means that the ethical component is of central importance in engineering professionalism.

1.4 THREE TYPES OF ETHICS OR MORALITY

If ethical commitment is central to professionalism, we must turn more directly to ethics and especially to professional ethics. How does professional ethics differ from other types of ethics—philosophical ethics, business ethics, personal ethics, and so on? In answering this question, it is helpful to distinguish between three types of ethics or morality.[5]

Common Morality

Common morality is the set of moral beliefs shared by almost everyone. It is the basis, or at least the reference point, for the other two types of morality that we shall discuss. When we think of ethics or morality, we usually think of such precepts as that it is wrong to murder, lie, cheat or steal, break promises, harm others physically, and so forth. It would be very difficult for us to question seriously any of these precepts.

We shall expand the notion of common morality in Chapter 3, but three characteristics of common morality must be mentioned here. First, many of the precepts of common morality are negative. According to some moralists, common morality is designed primarily to protect individuals from various types of violations or invasions of their personhood by others. I can violate your personhood by killing you, lying to you, stealing from you, and so forth.

Second, although common morality on what we might call the "ground floor" is primarily negative, it does contain a positive or aspirational component in such precepts as "Prevent killing," "Prevent deceit," "Prevent cheating," and so forth. However, it might also include even more clearly positive precepts, such as "Help the needy," "Promote human happiness," and "Protect the natural environment." This distinction between the positive and negative aspects of common morality will be important in our discussion of professional ethics.

Third, common morality makes a distinction between an evaluation of a person's actions and an evaluation of his intention. An evaluation of action is based on an application of the types of moral precepts we have been considering, but an evaluation of the person himself is based on intention. The easiest way to illustrate this distinction is to take examples from law, where this important common morality distinction also prevails. If a driver kills a pedestrian in his automobile accidentally, he may be charged with manslaughter (or nothing) but not murder. The pedestrian is just as dead as if he had been murdered, but the driver's intention was not to kill him, and the law treats the driver differently, as long as he was not reckless. The result is the same, but the intent is different. To take another example, if you convey false information to another person with the intent to deceive, you are lying.

If you convey the same false information because you do not know any better, you are not lying and not usually as morally culpable. Again, the result is the same (the person is misled), but the intent is different.

Personal Morality

Personal ethics or *personal morality* is the set of moral beliefs that a person holds. For most of us, our personal moral beliefs closely parallel the precepts of common morality. We believe that murder, lying, cheating, and stealing are wrong. However, our personal moral beliefs may differ from common morality in some areas, especially where common morality seems to be unclear or in a state of change. Thus, we may oppose stem cell research, even though common morality may not be clear on the issue. (Common morality may be unclear at least partially because the issue did not arise until scientific advancement made stem cell research possible and ordinary people have yet to identify decisive arguments.)

Professional Ethics

Professional ethics is the set of standards adopted by professionals insofar as they view themselves acting as professionals. Every profession has its professional ethics: medicine, law, architecture, pharmacy, and so forth. Engineering ethics is that set of ethical standards that applies to the profession of engineering. There are several important characteristics of professional ethics.

First, unlike common morality and personal morality, professional ethics is usually stated in a formal code. In fact, there are usually several such codes, promulgated by various components of the profession. Professional societies usually have codes of ethics, referred to as "code of professional responsibility," "code of professional conduct," and the like. The American Medical Association has a code of ethics, as does the American Bar Association. Many engineering societies have a code of ethics, such as the American Society of Civil Engineers or the American Society of Mechanical Engineers. In addition to the professional societies, there are other sources of codes. State boards that regulate the professions have their own codes of ethics, which generally are similar to the codes of the societies. The various codes of ethics do differ in some important ways. In engineering, for example, some of the codes have begun to make reference to the environment, whereas others still do not.

Second, the professional codes of ethics of a given profession focus on the issues that are important in that profession. Professional codes in the legal profession concern themselves with such questions as perjury of clients and the unauthorized practice of law. Perjury is not an issue that is relevant to medicine or dentistry. In engineering, the code of the Association for Computing Machinery sets out regulations for privacy, intellectual property, and copyrights and patents. These are topics not covered in most of the other engineering codes.

Third, when one is in a professional relationship, professional ethics is supposed to take precedence over personal morality—at least ordinarily. This characteristic of professional ethics has an important advantage, but it can also produce complications. The advantage is that a patient or client can justifiably have certain expectations of a professional, even if the patient or client has no knowledge of the personal morality of the professional. When a patient enters a physician's examining room, she can expect the conversations there to be kept confidential, even if she does not know anything about the personal morality of the physician. When a client or employer reveals details of a business relationship to an engineer, he can expect the engineer to keep these details in confidence, even though he knows nothing about the personal morality of the engineer. In both cases, these expectations are based on knowledge of the professional ethics of medicine and engineering, not on knowledge of the professional's personal morality.

A complication occurs when the professional's personal morality and professional ethics conflict. For example, in the past few years, some pharmacists in the United States have objected to filling prescriptions for contraceptives for unmarried women because their moral beliefs hold that sex outside of marriage is wrong. The code of the American Pharmaceutical Association makes no provision for refusing to fill a prescription on the basis of an objection from one's personal moral beliefs. In fact, the code mandates honoring the autonomy of the client. Nevertheless, some pharmacists have put their personal morality ahead of their professional obligations.

Some professions have made provisions for exceptions to professional obligations based on conscience. Physicians who believe that abortion is wrong are not required to perform an abortion, but there is still an obligation to refer the patient to a physician who will perform the abortion. Attorneys may refuse to take a client if they believe the client's cause is immoral, but they have an obligation to refer the prospective client to another attorney. Still, this compromise between personal morality and professional ethics may seem troubling to some professionals. If you believe deeply that abortion is murder, how can it be morally permissible to refer the patient to another physician who would perform the abortion? If you believe what a prospective client wants you to do is immoral, why would you refer him to another attorney who could help him do it? Nevertheless, this compromise is often seen as the best reconciliation between the rights and autonomy of the physician and the rights and autonomy of the patient, client, or employer.

Similar issues can arise in engineering, although engineering codes have not addressed them. Suppose a client asks a civil engineer to design a project that the engineer, who has strong personal environmental commitments, believes imposes unacceptable damage to a wetland. Suppose this damage is not sufficient to be clearly covered by his engineering code. In this case, the engineer probably should refer the client or employer to another engineer who might do the work.

Fourth, professional ethics sometimes differs from personal morality in its degree of restriction of personal conduct. Sometimes professional ethics is more restrictive than personal morality, and sometimes it is less restrictive. Suppose engineer Jane refuses to design military hardware because she believes war is immoral. Engineering codes do not prohibit engineers from designing military hardware, so this refusal is based on personal ethics and not on professional ethics. Here, Jane's personal ethics is more restrictive than her professional ethics. On the other hand, suppose civil engineer Mary refuses to participate in the design of a project that she believes will be contrary to the principles of sustainable development, which are set out in the code of the American Society of Civil Engineers. She may not personally believe these guidelines are correct, but she might (correctly) believe she is obligated to follow them in her professional work because they are stated in her code of ethics. Here, Mary's professional ethics is more restrictive than her personal ethics.

Similar differences in the degree of restriction between personal ethics and professional ethics can occur in other professions. Suppose a physician's personal ethics states that she should tell a woman that her future husband has a serious disease that can be transmitted through sexual intercourse. Medical confidentiality, however, may forbid her from doing so. The physician's professional ethics in this case is more restrictive than her personal ethics. In a famous case in legal ethics, lawyers found themselves defending a decision not to tell a grieving father where his murdered daughter was buried, even though their client had told them where he had buried the bodies of his victims. They argued that this information had been conveyed to them confidentially and that, as lawyers, they could not break this confidentiality. In their defense of themselves, they emphasized that as individual human beings (following their personal ethics) they deeply sympathized with the father, but as lawyers they felt compelled to protect lawyer–client confidentiality.[6] Here, legal ethics was more restrictive than the personal ethics of the lawyers. It would not let them do something that they very much wanted to do from the standpoint of their personal morality.

In these last two cases, the professional ethics of doctors and lawyers probably also differs from common morality. Sometimes the conflicts between professional ethics, personal morality, and common morality are difficult to resolve. It is not always obvious that professional ethics should take priority, and in some cases a professional might simply conclude that her professional ethics is simply wrong and should be changed. In any case, these conflicts can provoke profound moral controversy.

The professional ethics of engineers is probably generally less likely to differ from common morality than the professional ethics of other professions. With regard to confidentiality, we shall see that confidentiality in engineering can be broken if the public interest requires it. As the previous examples show, however, professional ethics in engineering can be different from an engineer's personal ethics. In Chapter 3, we discuss more directly common morality and the ways in which it can differ from professional ethics and personal morality.

Fifth, professional ethics, like ethics generally, has a negative and a positive dimension. Being ethical has two aspects: preventing and avoiding evil and doing or promoting good. Let us call these two dimensions the two "faces" of ethics: the negative face and the positive face. On the one hand, we should not lie, cheat, or steal, and in certain circumstances we may have an obligation to see that others do not do so as well. On the other hand, we have some general obligation to promote human well-being. This general obligation to avoid evil and do good is intensified and made more specific when people occupy special roles and have special relationships with others.

Role morality is the name given to moral obligations based on special roles and relationships. One example of role morality is the set of special obligations of parents to their children. Parents have an obligation not only not to harm their children but also to care for them and promote their flourishing. Another example of role morality is the obligation of political leaders to promote the well-being of citizens.

Professional ethics is another example of role morality. Professionals have both an obligation not to harm their clients, patients, and employers, and an obligation to contribute to their well-being. The negative aspect of professional ethics is oriented toward the prevention of professional malpractice and harm to the public. Let us call this dimension of professional ethics *preventive ethics* because of its focus on preventing professional misconduct and harm to the public. Professionals also have an obligation to use their knowledge and expertise to promote the public good. Let us call this more positive dimension of professional ethics *aspirational ethics* because it encourages aspirations or ideals in professionals to promote the welfare of the public.

The aspirational component has generally received less emphasis in professional ethics than the preventive component. This is true in engineering ethics as well, so it should not be surprising that the aspirational component of professional ethics has received less emphasis in earlier editions of this textbook. In this edition, we have attempted to redress this imbalance to some extent. At least we shall attempt to give more emphasis to the aspirational component of engineering ethics. Next, we discuss in more detail these two faces of professional ethics as they apply to engineering.

1.5   THE NEGATIVE FACE OF ENGINEERING ETHICS: PREVENTIVE ETHICS

During the past few decades, professional ethics for engineers has, as we have said, focused on its negative face, or what we have called preventive ethics. Preventive ethics is commonly formulated in rules, and these rules are usually stated in codes of ethics. A look at engineering codes of ethics will show not only that they are primarily sets of rules but also that these rules are for the most part negative in character. The rules are often in the form of prohibitions, or statements that probably should be understood primarily as prohibitions. For example, by one way of counting, 80 percent of the code of the National Society of Professional Engineers (NSPE) consists of provisions that are, either explicitly or implicitly, negative and prohibitive in character. Many of the provisions are explicitly negative in that they use terms such as "not" or "only." For example, section 1,c under "Rules of Practice" states that "engineers shall not reveal facts, data, or information without the prior consent of the client or employer except as authorized by law or this Code." Section 1,b under "Rules of Practice" states that "engineers shall approve only those engineering documents that are in conformity with applicable standards." This is another way of saying that engineers shall not approve engineering documents that are not in conformity with applicable standards.

Many provisions that are not stated in a negative form nevertheless have an essentially negative force. The rule having to do with undisclosed conflicts of interest is stated in the following way: "Engineers shall disclose all known or potential conflicts of interest that could influence or appear to influence their judgment or the quality of their services." This could also be stated as follows: "Engineers shall not engage in known or potential undisclosed conflicts of interest that could influence or appear to influence their judgment or the quality of their services." Many other provisions of the code, such as the requirement that engineers notify the appropriate professional bodies or public authorities of code violations (II,1,f) are "policing" provisions and thus essentially negative in character. Even the requirement that engineers be "objective and truthful" (II,3,a) is another way of stating that engineers shall not be biased and deceitful in their professional judgments. Similarly, the provision that engineers continue their professional development (III,9,e) is another way of stating that engineers shall not neglect their professional development.

This negative character of the codes is probably entirely appropriate, and it is easy to think of several reasons for this negative orientation. First, as previously discussed, common sense and common morality support the idea that the first duty of moral agents, including professionals, is not to harm others—not to murder, lie, cheat, or steal, for example. Before engineers have an obligation to do good, they have an obligation to do no harm. Second, the codes are formulated in terms of rules that can be enforced, and it is easier to enforce negative rules than positive rules. A rule that states "avoid undisclosed conflicts of interest" is relatively easy to enforce, at least in comparison to a rule that states "hold paramount the welfare of the public."

Another reason for the negative orientation of engineering ethics is the influence of what are often called "disaster cases," which are incidents that resulted, or could have resulted, in loss of life or harm due to technology. The following are examples of disaster cases that have been important in the development of engineering ethics.

The Bay Area Rapid Transit (BART) Case.   BART went into service in 1972. Holger Hjortsvang, a systems engineer, and Max Blankenzee, a programmer analyst, became concerned that there was no systems engineering group to oversee the development of the control and propulsion systems. When they communicated these concerns to management, both orally and in writing, they were told not to make trouble. At approximately the same time, an electrical engineer, Robert Bruder, reported inadequate work on the installation and testing of control and communications equipment. In November of 1971, the three engineers presented their concerns in a confidential way to Daniel Helix, a member of the BART board of directors. When BART managers identified the three engineers, they were fired.

On October 2, 1972, 3 weeks after BART began carrying passengers, one of the BART trains crashed at the Fremont station due to a short circuit in a transistor. Fortunately, there were no deaths and only a few injuries. The three engineers finally won out-of-court settlements, although their careers were disrupted for almost 2 years. The case generated legal precedents that have been used in subsequent cases, and it had a major impact on the development of engineering ethics.[7]

Goodrich A-7 Brake Case.   In 1968, the B. F. Goodrich Corporation won a contract for the design of the brakes for the Navy A-7 aircraft with an innovative four-rotor brake design. Testing showed, however, that the four-rotor system would not function in accordance with government specifications. Managers attempted to show that the brakes did meet government test standards by directing that the brakes should be allowed to coast longer between applications than allowed by military specifications, be cooled by fans between and during test runs, and be remachined between test runs. Upon learning about these gross violations of governmental standards, Searle Lawson, a young, recently graduated engineer, and Kermit Vandivier, a technical writer, informed the FBI, which in turn alerted the Government Accounting Office. Vandivier was fired by Goodrich, and Lawson resigned and went to work for another company.[8]

The DC-10 Case.   The DC-10, a wide-bodied aircraft, was introduced into commercial service in 1972, during a time of intense competition in the aviation industry in the United States. Since the cargo area is pressurized as well as the cabin, it must be able to withstand pressures up to 38 pounds per square inch. During the first year of service, a rear cargo door that was improperly closed blew open over Windsor, Ontario. Luckily, a skilled pilot was able to land the plane successfully. Two weeks after the accident, Convair engineer Dan Applegate expressed doubts about the "Band-Aid" fixes proposed for the cargo door lock and latch system. Managers rejected his expression of concerns because they believed Convair would have to pay for any fixes they proposed, so the prime contractor, McDonnell Douglas, was not notified of Applegate's concerns. On March 3, 1974, soon after takeoff on a flight from Paris to London, the cargo door of a plane broke off, resulting in a crash that killed 346 passengers. At that time, it was the worst aircraft accident in history.[9]

There are common themes in these cases, as well as in the better known *Challenger* and *Columbia* cases that are discussed later: engineers trying to prevent disasters and being thwarted by managers in their attempts, engineers finding that they have to go public or in some way enlist the support of others, and disasters occurring when engineers do not continue to protest (as in the DC-10 case). These are certainly stories that need to be told, and there are lessons to be learned about the importance of, and the risks involved in, protecting the health and safety of the public. We believe that preventive ethics should always be an important part of engineering ethics. However, there is more to being a good professional than avoiding misconduct and preventing harm to the public. We now discuss this more positive and aspirational aspect of engineering.

1.6   THE POSITIVE FACE OF ENGINEERING ETHICS: ASPIRATIONAL ETHICS

It is easy to see the limitations of a professional ethics that is confined to the negative dimension. One of the limitations is the relative absence of the motivational dimension. Engineers do not choose engineering as a career in order to prevent disasters and avoid professional misconduct. To be sure, many engineering students desire the financial rewards and social position that an engineering career promises, and this is legitimate. We have found, however, that engineering students are also attracted by the prospect of making a difference in the world, and doing so in a positive way. They are excited by projects that alleviate human drudgery through labor-saving devices, eliminate disease by providing clean water and sanitation, develop new medical devices that save lives, create automobiles that run on less fuel and are less polluting, and preserve the environment with recyclable products. Most of us probably believe that these activities—and many others—improve the quality of human life.

This more positive aspect of engineering is recognized to some extent in engineering codes of ethics. The first Fundamental Canon of the NSPE code of ethics requires engineers to promote the "welfare" of the public, as well as prevent violations of safety and health. Virtually all of the major engineering codes begin with similar statements. Nevertheless, the positive face of engineering ethics has taken second place to the negative face in most engineering ethics textbooks, including our own. In this edition, we include this more positive or aspirational aspect of engineering ethics.

In addition to us, several other writers on engineering ethics have come to advocate an increased emphasis on the more positive and welfare-promoting aspect of engineering. Mike Martin, author of an important textbook in engineering ethics, opened a recent monograph with the following statement:

> Personal commitments motivate, guide, and give meaning to the work of professionals. Yet these commitments have yet to receive the attention they deserve in thinking about professional ethics....I seek to widen professional ethics to include personal commitment, especially commitments to ideals not mandatory for all members of a profession.[10]

Personal commitments to ideals, Martin believes, can add an important new and positive dimension to engineering ethics.

P. Aarne Veslilind, engineer and writer on engineering ethics, edited the book, *Peace Engineering; When Personal Values and Engineering Careers Converge.* In one of the essays, written by Robert Textor, the following account of "peace" is given:

- Global environmental management
- Sustainable development, especially in the less developed countries
- Tangible, visible steps toward greater economic justice
- Efforts to control and reduce the production and use of weapons, from land-mines and small arms to nuclear and other weapons of mass destruction
- Awareness of cultural differences and skill in finding common ethical ground[11]

Although all engineers might not want to subscribe to some elements of the political agenda suggested here, Textor's statement again highlights the positive aspect of engineering—enhancing human welfare. The book title also makes reference to personal values.

Promoting the welfare of the public can be done in many different ways, ranging from designing a new energy-saving device in the course of one's ordinary employment to using one's vacation time to design and help install a water purification system in an underdeveloped country. Aspirational ethics, then, involves a spectrum of engineering activities. Let us call the more extreme and altruistic examples of aspirational ethics "good works" and the more ordinary and mundane examples "ordinary positive engineering." Although the division between these two categories is not always sharp, we believe the distinction is useful. Let us begin with the category of good works.

Good Works

Good works refers to the more outstanding and altruistic examples of aspirational ethics—those that often involve an element of self-sacrifice. Good works are exemplary actions that may go beyond what is professionally required. A good work is commendable conduct that goes beyond the basic requirements associated with a particular social role, such as the role of a professional. Good works can include outstanding examples of preventive ethics, such as the attempt of engineer Roger Boisjoly to stop the fatal launch of the *Challenger,* but here we are interested in illustrations of good works that fall into the aspirational ethics category. The following are examples.

The Sealed-Beam Headlight. A group of General Electric engineers on their own time in the late 1930s developed the sealed beam headlight, which greatly reduced the number of accidents caused by night driving. There was considerable doubt as to whether the headlight could be developed, but the engineers persisted and finally achieved success.[12]

Air Bags. Carl Clark helped to develop air bags. Even though he was a scientist and not a degreed engineer, his work might well have been done by an engineer. He is now advocating air bags on bumpers, and he has even invented wearable air bags for the elderly to prevent broken hips. He does not get paid for all of his time, and the bumper air bags were even patented by someone else.[13]

Disaster Relief.   Fredrick C. Cuny attended engineering school, but he never received his degree in engineering due to poor grades. In his early twenties, however, he learned how to conduct disaster relief in such a way that the victims could recover enough to help themselves. At age 27, he founded the Interact Relief and Reconstruction Corporation. He was soon working in Biafra helping to organize an airlift to rescue Biafrans after a war. Later, he organized relief efforts, involving engineering work, in Bosnia after the war and in Iraq after Operation Desert Storm. When his work in Iraq was completed, the Kurds held a farewell celebration. Cuny was the only civilian in a parade with the Marines with whom he had worked.[14]

Engineers Without Borders.   Engineers Without Borders is an international organization for engineering professionals and engineering students who want to use their professional expertise to promote human welfare. Engineering students from the University of Arizona chapter are working on a water supply and purification project in the village of Mafi Zongo, Ghana, West Africa. The project will supply 30 or more villages, with approximately 10,000 people, with safe drinking water. In another project, engineering students from the University of Colorado installed a water system in Muramka, a Rwandan village. The system provides villagers with up to 7000 liters of safe water for everyday use. The system consists of a gravity-fed settling tank, rapid sand filters, and a solar-powered sanitation light.[15]

Ordinary Positive Engineering

Most examples of aspirational ethics do not readily fall into the category of good works. They are done in the course of one's job, and they do not involve any heroism or self-sacrifice. One might even say that most of the things an engineer does are examples of ordinary positive engineering, as long as a good argument can be made that they contribute in some way to human welfare. Although this may be true, we are thinking here of actions that usually involve a more conscious and creative attempt to do something that contributes to human welfare. The following are examples, some fictional and some actual.

An Experimental Automobile.   Daniel is a young engineer who is excited about being put on a project to develop an experimental automobile that has as many recyclable parts as possible, is lightweight but safe, and gets at least 60 miles per gallon.

An Auditory Visual Tracker.   Students in a senior design course at Texas A & M decided to build an auditory visual tracker for use in evaluating the training of visual skills in children with disabilities. The engineering students met the children for whom the equipment was being designed, and this encounter so motivated the students that they worked overtime to complete the project. At the end of the project, they got to see the children use the tracker.

Reducing Emissions.   Jane has just been assigned to a project to reduce the emissions of toxic chemicals below the standards set by governmental regulation. Her managers believe that the emission standards will soon be made more restrictive anyway, and that by beginning early the plant will be "ahead of the game." In fact, however, both Jane and her manager are genuinely committed to reducing environmental pollution.

A Solution to "Gilbane Gold." In a well-known videotape in engineering ethics, a young engineer, David Jackson, believes that his plant's emissions should be reduced to comply with a new and more accurate test that has not yet been enacted into law. His manager refuses to cooperate until the standards are legally changed. David's resolution of the problem is to inform the press, an action that will probably cost him his job. Michael Pritchard and chemical engineer Mark Holtzapple suggest an engineering solution that would both further reduce toxic waste and be less costly than the system David's plant is currently using. The solution would probably have helped the environment, changed the manager's position, and saved David's job.[16]

Aspirational Ethics and Professional Character: The Good Engineer

Two features of aspirational ethics are of special importance. First, as Mike Martin noted, the more positive aspect of engineering ethics has a motivational element that is not present in the same way in preventive ethics. Second, as Martin also suggested, there is a discretionary element in aspirational ethics: An engineer has a considerable degree of freedom in how he or she promotes public welfare. Neither of these two features can be conveyed well in rules. Rules are not very effective motivational instruments, especially motivation to positive action. Rules are also inadequate to handle situations in which there is a great deal of discretion. "Hold paramount public welfare" gives little direction for conduct. It does not tell an engineer whether she should devote her time to Engineers Without Borders or to some special project on which she is willing to work overtime, or to simply designing a product that is more energy efficient. These decisions should be left to the individual engineer, given her interest, abilities, and what is possible in her own situation.

For these reasons, we believe that the more appropriate vocabulary for expressing aspirational ethics is that of professional character rather than the vocabulary of rules, which are more appropriate for preventive ethics. Rules do a good job of expressing prohibitions: "Don't violate confidentiality," "Don't have undisclosed conflicts of interest." Rules are less appropriate for capturing and stimulating motivation to do good. Here, the most relevant question is not "What kinds of rules are important in directing the more positive and aspirational elements of engineering work?" Rather, the question is "What type of person, professionally speaking, will be most likely to promote the welfare of the public through his or her engineering work?"

Let us use the term *professional character* to refer to those character traits that serve to define the kind of person one is, professionally speaking. The "good engineer" is the engineer who has those traits of professional character that make him or her the best or ideal engineer. To be sure, the vocabulary of professional character can also be used to describe the engineer who would be a good exponent of preventive ethics. Considering the examples of preventive ethics discussed previously, it is easy to see that the BART engineers displayed courage in attempting to alert management to the problems they found in the BART system. Vandivier also displayed courage in reporting the problems with the four-rotor brake to outside sources. One can think of other character traits that the engineers in the examples of preventive ethics displayed, such as technical expertise and concern for public safety and health. Nevertheless, preventive ethics can be expressed—and has traditionally been expressed—in terms of negative rules.

We can use the term *professional character portrait* to refer to the set of character traits that would make an engineer a good engineer, and especially an effective practitioner of aspirational ethics. We suggest three character traits that might be a part of such a professional character portrait.

The first professional character trait is professional pride, particularly pride in technical excellence. If an engineer wants her work as a professional to contribute to public welfare, the first thing she must do is be sure that her professional expertise is at the highest possible level. Professional expertise in engineering includes not only the obvious proficiencies in mathematics, physics, and engineering science but also those capacities and sensitivities that only come with a certain level of experience.

The second professional character trait is social awareness, which is an awareness of the way in which technology both affects and is affected by the larger social environment. In other words, engineers need an awareness of what we call in Chapter 5 the "social embeddedness" of technology. Engineers as well as the rest of us are sometimes tempted to view technology as isolated from the larger social context. In the extreme version of this view, technology is governed by considerations internal to technology itself and neither influences nor is influenced by social forces and institutions. In a less extreme view, technology powerfully influences social institutions and forces, but there is little, if any, causal effect in the other direction. However, the engineer who is sufficiently aware of the social dimension of technology understands that technology both influences and is influenced by the larger social context. On the one hand, technology can be an instrument of the power elite and can be used for such things as the deskilling of labor. On the other hand, technology can be utilized by grassroots movements, as protesters did in China and bloggers do in the United States. In any case, engineers are often called on to make design decisions that are not socially neutral. This often requires sensitivities and commitments that cannot be incorporated into rules. We believe that such social awareness is an important aspect of a professional character that will take seriously the obligation to promote public welfare through professional work.

A third professional character trait that can support aspirational ethics is an environmental consciousness. Later in this book, we explore this issue more thoroughly, but here it need only be said that the authors believe that environmental issues will increasingly play a crucial role in almost all aspects of engineering. Increasingly, human welfare will be seen as integral to preserving the integrity of the natural environment that supports human and all other forms of life. Eventually, we believe, being environmentally conscious will be recognized as an important element in professional engineering character.

1.7 CASES, CASES, CASES!

In this chapter, we have frequently referred to cases in engineering ethics. Their importance cannot be overemphasized, and they serve several important functions. First, it is through the study of cases that we learn to recognize the presence of ethical problems, even in situations in which we might have thought there are only technical issues. Second, it is by studying cases that we can most easily develop the abilities necessary to engage in constructive ethical analysis. Cases stimulate the moral imagination by challenging us to anticipate the possible alternatives for resolving them and to think about the consequences of those alternatives. Third, a study of cases is the most effective way to understand that the codes cannot provide ready-made answers to many moral questions that professional engineering practice generates and that individual engineers must become responsible agents in moral deliberation. They must both interpret the codes they have and (occasionally) consider how the codes should be revised. Fourth, the study of cases shows us that there may be some irresolvable uncertainties in ethical analysis and that in some situations rational and responsible professionals may disagree about what is right.

Cases appear throughout the text. Each chapter is introduced with a case, which is usually referred to in the chapter. In many chapters, we present our own attempts to resolve ethical problems. We often use brief cases to illustrate various points in our argument.

Cases are of several types. We have already discussed examples of cases that illustrate both preventive and the more positive aspects of professional ethics. Another way to categorize cases is to state that some focus on micro-level issues about the practice of individual engineers, whereas others have to do with questions of social policy regarding technology.[17] Some cases are fictional but realistic, whereas others are actual cases. Sometimes cases are simplified in order to focus on a particular point, but simplification risks distortion. Ideally, most cases would be given a "thick" (i.e., extended) description instead of a "thin" (i.e., abbreviated) description, but this is not possible here. Many thick descriptions of individual cases require a book-length account. Of course, instructors are free to add details as necessary.

Two final points are important with regard to the use of cases. First, the use of cases is especially appropriate in a text on professional ethics. A medical school dean known to one of the authors once said, "Physicians are tied to the post of use." By this he presumably meant that physicians do not have the luxury of thinking indefinitely about moral problems. They must make decisions about what treatment to administer or what advice to give in a specific case.

Engineers, like other professionals, are also tied to the post of use. They must make decisions about particular designs that will affect the lives and financial well-being of many people, give professional advice to individual managers and clients, make decisions about particular purchases, decide whether to protest a decision by a manager, and take other specific actions that have important consequences for themselves and others. Engineers, like other professionals, are case-oriented. They do not work in generalities, and they must make decisions. The study of cases helps students understand that professional ethics is not simply an irrelevant addition to professional education but, rather, is intimately related to the practice of engineering.

Second, the study of cases is especially valuable for engineers who aspire to management positions. Cases have long been at the center of management education. Many, if not most, of the issues faced by managers have ethical dimensions. Some of the methods for resolving ethical problems discussed in Chapter 3—especially finding what we call a "creative middle way" solution—have much in common with the methods employed by managers. Like engineers, managers must make decisions within constraints, and they usually try to make decisions that satisfy as many of those constraints as possible. The kind of creative problem solving necessary to make such decisions is very similar to the deliberation that is helpful in resolving many ethical problems.

1.8 CHAPTER SUMMARY

This book focuses on professional ethics, not one's personal ethics or what is often called common morality. Sociologists and philosophers have come up with several different accounts of professionalism. By some of them, engineering in the United States does not enjoy full professional status, primarily because in the United States an engineer does not have to be licensed to practice engineering. By Michael Davis' Socratic definition of professionalism, however, engineers do have full professional status.

Running through all of the accounts of professionalism is the idea that ethical commitment, or at least a claim to it, is crucial to a claim to be a professional. This means that professional ethics is central to the idea of professionalism. Professional ethics has a number of distinct characteristics, many of which serve to differentiate it from personal ethics and common morality. Professional ethics is usually stated (in part) in a code of ethics, focuses on issues that are important in a given profession, often takes precedence over personal morality when a professional is in his professional capacity, and sometimes differs from personal morality in its degree of restriction of personal conduct. Finally, professional ethics can usefully be divided into those precepts that aim at preventing professional misconduct and engineering disasters (preventive ethics) and those positive ideals oriented toward producing a better life for humankind through technology (aspirational ethics). In elaborating on aspirational ethics, one can think of those professional qualities that enable one to be more effective in promoting human welfare. Cases are a valuable tool in developing the skills necessary for ethical practice.

NOTES

1. These five characteristics are described in Ernest Greenwood, "Attributes of a Profession," *Social Work*, July 1957, pp. 45–55. (For two more extensive sociological accounts that take this economic approach, see Magali Sarfatti Larson, *The Rise of Professionalism* (Berkeley: University of California Press, 1977) and Andrew Abbott, *The System of Professions* (Chicago: University of Chicago Press, 1988). For this entire discussion, we have profited from e-mail comments and two papers by Michael Davis: "Is There a Profession of Engineering?" *Science and Engineering Ethics*, 3, no. 4, 1997, pp. 407–428, and an unpublished paper, used with permission, "Is Engineering in Japan a Profession?"

2. Alasdair MacIntyre, *After Virtue* (Notre Dame, IN: University of Notre Dame Press, 1984), p. 187. For an elaboration of the concept of social practice and another application to professionalism, see Timo Airaksinen, "Service and Science in Professional Life," in Ruth F. Chadwick, ed., *Ethics and the Professions* (Aldershot, UK: Averbury, 1994).

3. Michael Davis, "Is There a Profession of Engineering?" *Science and Engineering Ethics*, 3, no. 4, 1997, p. 417.

4. We are indebted for some aspects of the elaboration of these two models to Professor Ray James, Department of Civil Engineering, Texas A & M University.

5. Often, we use the terms ethics and morality interchangeably because the terms are usually used interchangeably in philosophical ethics. However, there is some difference in usage, in that the term ethics is sometimes used with a more formalized statement of moral precepts, especially as these precepts are stated in ethical codes. Thus, it is more common to refer to "professional ethics" than "professional morality."

6. Reported in several sources, including *The New York Times*, June 20, 1974.

7. *Encyclopedia of Science and Technology Ethics* (Detroit: Thomson, 2005), vol. 1, pp. 170–172.

8. Kermit Vandivier, "Why Should My Conscience Bother Me?" in Robert Heilbroner, ed., *In the Name of Profit* (Garden City, NY: Doubleday, 1972), p. 29.

9. *Encyclopedia of Science and Technology Ethics* (Detroit: Thomson, 2005), vol. 2, pp. 472–473.

10. Mike W. Martin, *Meaningful Work* (New York: Oxford University Press, 2000), p. vii.

11. P. Aarne Vesilind, *Peace Engineering: When Personal Values and Engineering Careers Converge* (Woodsville, NH: Lakeshore Press, 2005), p. 15.

12. This account is based on G. P. E. Meese, "The Sealed Beam Case," *Business & Professional Ethics,* 1, no. 3, Spring 1982, pp. 1–20.

13. See Michael S. Pritchard, "Professional Responsibility: Focusing on the Exemplary," *Science and Engineering Ethics,* 4, 1998, p. 222. This article contains a discussion of good works, which is a concept first introduced by Pritchard.

14. Ibid., pp. 230–233.

15. See the Engineers Without Borders website at http://www.ewb-usa.org.

16. Michael S. Pritchard and Mark Holtzapple, "Responsible Engineering: *Gilbane Gold* Revisited," *Science and Engineering Ethics,* 3, 1997, pp. 217–230.

17. For a discussion of the distinction between micro- and macro-level issues, see Joseph Herkert, "Future Directions in Engineering Ethics Research: Microethics, Macroethics and the Role of Professional Societies," *Science and Engineering Ethics,* 7, no. 3, 2001, pp. 403–414.

Responsibility in Engineering

Main Ideas in this Chapter

- **Responsibility has to do with accountability, both for what one does in the present and future and for what one has done in the past.**
- **The obligation-responsibilities of engineers require, not only adhering to regulatory norms and standard practices of engineering but also satisfying the standard of reasonable care.**
- **Engineers can expect to be held accountable, if not legally liable, for intentionally, negligently, and recklessly caused harms.**
- **Responsible engineering practice requires good judgment, not simply following algorithms.**
- **A good test of engineering responsibility is the question, "What does an engineer do when no one is looking?"**
- **Impediments to responsible practice include self-interest, fear, self-deception, ignorance, egocentric tendencies, narrow vision, uncritical acceptance of authority, and groupthink.**

ON JANUARY 16, 2003, AT 10:39 A.M. Eastern Standard Time, the *Columbia* lifted off at Kennedy Space Center, destined for a 16-day mission in space.[1] The seven-person *Columbia* crew, which included one Israeli pilot, was scheduled to conduct numerous scientific experiments and return to Earth on February 1. Only 81.7 seconds after lift-off, a briefcase-size piece of the brownish-orange insulating foam that covered the large external tank broke off and hit the leading edge of the orbiter's left wing. Unknown to the *Columbia* crew or the ground support staff, the foam knocked a hole in the leading edge of the wing that was approximately 10 inches across.

Cameras recorded the foam impact, but the images provided insufficient detail to determine either the exact point of impact or its effect. Several engineers, including Rodney Rocha, requested that attempts be made to get clearer images. There were even requests that the *Columbia* crew be directed to examine the wing for possible damage. It had become a matter of faith at NASA, however, that foam strikes, although a known problem, could not cause significant damage and were not a safety-of-flight issue, so management rejected this request. The astronauts were not told of the problem until shortly before reentry, when they were informed that the foam strike was inconsequential, but that they should know about it in case they were asked about the strike by the press on return from their mission.

Upon reentry into the earth's atmosphere, a snaking plume of superheated air, probably exceeding 5000 degrees Fahrenheit, entered the breach in the wing and began to consume the wing from the inside. The destruction of the spacecraft began when it was over the Pacific Ocean and grew worse when it entered U.S. airspace. Eventually, the bottom surface of the left wing began to cave upwards into the interior of the wing, finally causing *Columbia* to go out of control and disintegrate, mostly over east Texas. The entire crew, along with the spacecraft, was lost.

2.1 INTRODUCTION

This tragic event, which has many striking similarities with the *Challenger* disaster 17 years earlier, illustrates many of the issues surrounding the notion of responsibility in the engineering profession. Engineers obviously played a central role in making the *Columbia* flight possible and in safeguarding the spaceship and its travelers. From the outset of the launch, engineers had a special eye out for possible problems. Rodney Rocha and other engineers became concerned about flying debris. Noticing and assessing such details was their responsibility. If they did not handle this well, things could go very badly. Even if they did handle this well, things could go very badly. The stakes were high.

The concept of responsibility is many-faceted. As a notion of accountability, it may be applied to individual engineers, teams of engineers, divisions or units within organizations, or even organizations themselves. It may focus primarily on legal liabilities, job-defined roles, or moral accountability. Our focus in this chapter is mainly on the moral accountability of individual engineers, but this will require attending to these other facets of responsibility as well.

As professionals, engineers are expected to commit themselves to high standards of conduct.[2] The preamble of the code of ethics of the National Society for Professional Engineers (NSPE) states the following:

> Engineering is an important and learned profession. As members of this profession, engineers are expected to exhibit the highest standards of honesty and integrity. Engineering has a direct and vital impact on the quality of life for all people. Accordingly, the services provided by engineers require honesty, impartiality, fairness, and equity, and must be dedicated to the protection of the public health, safety, and welfare. Engineers must perform under a standard of professional behavior that requires adherence to the highest principles of ethical conduct.

Although this preamble insists that such conduct is expected of engineers, this is not a predictive statement about how engineers, in fact, conduct themselves. By and large, it is hoped, engineers do adhere to high principles of ethical conduct. However, the preamble is a normative statement, a statement about how engineers should conduct themselves. This is based on the impact that engineering has on our quality of life. This impact is the result of the exercise of expertise that is the province of those with engineering training and experience. Such expertise carries with it professional responsibility.

William F. May points out the seriousness of the responsibility that comes with professional expertise. Noting our increasing reliance on the services of professionals whose knowledge and expertise is not widely shared or understood, May comments,[3]

> [The professional] had better be virtuous. Few may be in a position to discredit him. The knowledge explosion is also an ignorance explosion; if knowledge is power, then ignorance is powerlessness.

The knowledge that comes with expanding professional expertise is largely confined to specialists. Those outside these circles of expertise experience the ignorance explosion to which May refers. This includes the general public, as well as other professionals who do not share that expertise. May states, "One test of character and virtue is what a person does when no one else is watching. A society that rests on expertise needs more people who can pass that test."[4]

May's observations apply as much to engineers as to accountants, lawyers, doctors, and other professionals. What this means is that in its ignorance, the public must place its trust in the reliable performance of engineers, both as individuals and as members of teams of engineers who work together. In turn, if they are to be given opportunities to provide services to others, it is important for engineers to conduct themselves in ways that do not generate distrust. However, given what May calls the "ignorance explosion," placing trust in the reliable performance of engineers may sometimes provide unscrupulous or less than fully committed engineers with opportunities to fall short of the mark without being noticed.

May concludes, "Important to professional ethics is the moral disposition the professional brings to the structure in which he operates, and that shapes his approach to problems."[5] This is a matter of professional character. This has important implications for a professional's approach to his or her responsibilities. We might think of possible approaches to responsibility along a spectrum. At one end of the spectrum is the minimalist approach of doing as little as one can get away with and still stay out of trouble, keep one's job, and the like. At the other end of the spectrum are attitudes and dispositions that may take one "above and beyond the call of duty." This does not mean that one self-consciously aims at doing more than duty requires. Rather, it involves a thoroughgoing commitment to a level of excellence that others regard as supererogatory, or "going the extra mile." The professional's attitude might be one of "just doing my job," but the dedication to an extraordinarily high level of performance is evident.

Most engineers typically fall somewhere in between these two ends of the spectrum most of the time. We can ask what sorts of attitudes and dispositions employers might look for if they were hoping to hire a highly responsible engineer.[6] We would expect integrity, honesty, civic-mindedness, and a willingness to make some self-sacrifice to make the list. In addition to displaying basic engineering competence, a highly responsible engineer would be expected to exhibit imaginativeness and perseverance, to communicate clearly and informatively, to be committed to objectivity, to be open to acknowledging and correcting mistakes, to work well with others, to be committed to quality, and to be able to see the "big picture" as well as more minute details. No doubt there are other items that could be added to the list. What all these characteristics have in common is that they contribute to the reliability and trustworthiness of engineers.

2.2   ENGINEERING STANDARDS

One way in which engineers can try to gain the trust of those they serve and with whom they work is to commit themselves to a code of ethics that endorses high standards of performance. Standards of responsibility expressed in engineering codes typically call for engineers to approach their work with much more than the minimalist dispositions mentioned previously. At the same time, satisfying the standards that the codes endorse does not require that they operate at a supererogatory level. Nevertheless, as we shall see, if taken seriously, the standards are quite demanding.

Like other engineering codes of ethics, the NSPE code requires that the work of engineers conform with "applicable engineering standards." These may be regulatory standards that specify technical requirements for specific kinds of engineering design—for example, that certain standards of safety be met by bridges or buildings. As such, they focus primarily on the results of engineering practice—on whether the work satisfies certain standards of quality or safety. Engineering standards may also require that certain procedures be undertaken to ascertain that specific, measurable levels of quality or safety are met, or they may require that whatever procedures are used be documented, along with their results.

Equally important, engineering codes of ethics typically insist that engineers conform to standards of competence—standards that have evolved through engineering practice and presumably are commonly accepted, even if only implicitly, in ordinary engineering training and practice.[7] Regulatory standards and standards of competence are intended to provide some assurance of quality, safety, and efficiency in engineering. It is important to realize, however, that they also leave considerable room for professional discretion in engineering design and its implementation. There are few algorithms for engineers to follow here. Therefore, the need for engineering judgment should not be overlooked.[8]

The NSPE code of ethics is the product of the collective reflection of members of one particular professional society of engineers. However, it seems intended to address the ethical responsibilities of all practicing engineers. Given this, the standards endorsed by the code should be supportable by reasons other than the fact that NSPE members publicly endorse and commit themselves to those standards. That is, the standards should be supportable by reasons that are binding on even those engineers who are not members of NSPE. Are they?

In answering this question, it is important to note that the preamble makes no reference to its members creating or committing themselves to the code. Instead, it attempts to depict the role that engineering plays in society, along with the standards of conduct that are required in order for engineers to fulfill this role responsibly. Presumably, this depiction is apt regardless of whether engineers are members of NSPE.

Engineers and nonengineers alike can readily agree that engineers do play the sort of vital societal role depicted by the preamble. It suggests that, first and foremost, engineers have a responsibility to use their specialized knowledge and skills in ways that benefit clients and the public and do not violate the trust placed in them. We make reference to this type of responsibility when we say that professionals should "be responsible" or "act responsibly." We can refer to this as a generally "positive" and forward-looking conception of responsibility. Let us call it *obligation-responsibility*.

Obligation-responsibility sometimes refers to a person who occupies a position or role of supervision. We sometimes say that an engineer is in "responsible charge" of a design or some other engineering project. A person in responsible charge has an obligation to see to it that the engineering project is performed in accordance with professional standards, both technical and ethical.

Related to forward-looking conceptions of responsibility are judgments about how well we think obligation-responsibilities have been handled. Backward-looking, these are judgments of praise and blame. Unfortunately, we have a tendency to focus on the blaming end of this evaluative spectrum. We seem more readily to notice shortcomings and failures than the everyday competent, if not exceptional, performance of engineers. (We expect our cars to start, the elevators and trains to run, and the traffic lights to work.) In any case, we speak of an engineer as "being responsible" for a mistake or as being one of those "responsible" for an accident. This is a fundamentally negative and backward-looking concept of responsibility. Let us refer to it as *blame-responsibility*.

In the first part of this chapter, we develop the notion of the obligation-responsibilities of engineers. Then we turn to the negative notion of responsibility, or blame-responsibility. We consider the relationship of the causing of harm to being responsible for harm. We can speak of physical causes of harm, such as a malfunctioning part that causes an accident. Whether organizations can be moral agents responsible for harm or whether they are best thought of as causes of harm is more controversial. In either case, the importance of organizations in understanding accidents is crucial, as the investigation of the *Columbia* accident has shown. There is no doubt, however, that we can speak of human beings as being responsible for harm.

We conclude the chapter with a consideration of impediments to responsibility. These impediments are factors that keep people from being responsible in the positive or "obligation" sense of responsibility, but they can also be grounds for attribution of blame or responsibility in the negative sense. An engineer who, for example, is guilty of self-deception or ignorance can be held morally responsible if these factors lead to harm.

2.3 THE STANDARD OF CARE

Engineers have a professional obligation to conform to the standard operating procedures and regulations that apply to their profession and to fulfill the basic responsibilities of their job as defined by the terms of their employment. Sometimes, however, it is not enough to follow standard operating procedures and regulations. Unexpected problems can arise that standard operating procedures and current regulations are not well equipped to handle. In light of this, engineers are expected to satisfy a more demanding norm, the *standard of care*. To explain this idea, we can first turn to codes of ethics.

Codes of ethics of professional engineering societies are the result of efforts of their members to organize in a structured way the standards that they believe should govern the conduct of all engineers. However, because particular situations cannot be anticipated in all their relevant nuances, applying these standards requires professional judgment. For example, although sometimes it is clear what would constitute a failure to protect public health and safety, often it is not. Not actively protecting public safety will fail to satisfy the public safety standard only if there is a responsibility to provide that level of safety. However, since no engineering product can be expected to be "absolutely" safe (at least, not if it is to be a useful product) and there are economic costs associated with safety improvements, there can be considerable controversy about what is a reasonable standard of safety.

Rather than leave the determination of what counts as safe solely in the hands of individual engineers, safety standards are set by government agencies (such as the National Institute of Standards and Technology, the Occupational Safety and Health Administration, and the Environmental Protection Agency) or nongovernmental organizations (such as professional engineering societies and the International Organization for Standardization). Nevertheless, standards of safety, as well as standards of quality in general, still leave room for considerable engineering discretion. Although some standards have a high degree of specificity (e.g., minimal requirements regarding the ability of a structure to withstand winds of a certain velocity striking that structure at a 90-degree angle), some simply require that unspecified standard processes be developed, followed, and documented.[9]

Engineering codes of ethics typically make statements about engineers being required to conform to accepted standards of engineering practice. What such standards translate to in actual practice depends, of course, on the area of engineering practice in question, along with whatever formal regulatory standards may be in place. However, underlying all of this is a broader standard of care in engineering practice—a standard appealed to in law and about which experienced, respected engineers can be called on to testify in the courts in particular cases.

Joshua B. Kardon characterizes this standard of care in the following way.[10] Although some errors in engineering judgment and practice can be expected to occur as a matter of course, not all errors are acceptable:

> An engineer is not liable, or responsible, for damages for every error. Society has decided, through case law, that when you hire an engineer, you buy the engineer's normal errors. However, if the error is shown to have been worse than a certain level of error, the engineer is liable. That level, the line between non-negligent and negligent error, is the "standard of care."

How is this line determined in particular cases? It is not up to engineers alone to determine this, but they do play a crucial role in assisting judges and juries in their deliberations:

> A trier of fact, a judge or jury, has to determine what the standard of care is and whether an engineer has failed to achieve that level of performance. They do so by hearing expert testimony. People who are qualified as experts express opinions as to the standard of care and as to the defendant engineer's performance relative to that standard.

For this legal process to be practicable and reasonably fair to engineers, it is necessary that there be an operative notion of accepted practice in engineering that is well understood by competent engineers in the areas of engineering under question. As Kardon notes,[11]

> A good working definition of the standard of care of a professional is: that level or quality of service ordinarily provided by other normally competent practitioners of good standing in that field, contemporaneously providing similar services in the same locality and under the same circumstances.

Given this, we should not expect to find a formal statement of what specifically satisfies the standard. Rather, an appeal is being made to what is commonly and ordinarily done (or not done) by competent engineers.

Engineers who have responsible charge for a project are expected to exercise careful oversight before putting their official stamp of approval on the project. However, what careful oversight requires will vary with the project in question in ways that resist an algorithmic articulation of the precise steps to be taken and the criteria to be used.

Two well-known cases are instructive. In the first instance, those in charge of the construction of the Kansas City Hyatt Regency hotel were charged with professional negligence in regard to the catastrophic walkway collapse in 1981.[12] Although those in charge did not authorize the fatal departure from the original design of the walkway support, it was determined that responsible monitoring on their part would have made them aware of the proposed change. Had it come to their attention, a few simple calculations could have made it evident to them that the resulting structure would be unsafe.

In this case, it was determined that the engineers in charge fell seriously short of accepted engineering practice, resulting in a failure to meet the standard of care. Satisfying the standard of care cannot guarantee that failure will not occur. However, failure to satisfy the standard of care itself is not acceptable. In any particular case, there may be several acceptable ways of meeting the standard. Much depends on the kind of project in question, its specific context, and the particular variables that (sometimes unpredictably) come into play.

The second case also involved a departure from the original design not noted by the chief structural engineer of Manhattan's 59-story Citicorp Center.[13] In contrast to the Hyatt Regency walkway, this was not regarded to be a matter of negligence. William LeMessurier was surprised to learn that Citicorp Center's major structural joints were bolted rather than deep-welded together, as called for in the original design. However, he was confident that the building more than adequately satisfied the New York City building code's requirement that winds striking the structure from a 90-degree angle would pose no serious danger. Assuming he was correct, it is fair to conclude that either deep welds or bolts were regarded to be consistent with accepted engineering practice. The code did not specify which should be chosen, only that the result must satisfy the 90-degree wind test.

Fortunately, LeMessurier did not rest content with the thought that the structure satisfied the city building code. Given the unusual features of the Citicorp structure, he wondered what would happen if winds struck the building diagonally at a 45-degree angle. This question seemed sensible because the first floor of the building is actually several stories above ground, with the ground support of the building being four pillars placed in between the four corners of the structure rather than at the corners. Further calculations by LeMessurier determined that bolted joints rendered the structure much more vulnerable to high winds than had been anticipated. Despite satisfying the city code, the building was unsafe. LeMessurier concluded that corrections must be made. The standard set by the city building code was flawed. The code could not be relied on to set reliable criteria for the standard of care in all cases.

From this it should not be concluded that there is only one acceptable solution to the joint problem. LeMessurier's plan for reinforcing the bolted joints worked. However, the original plan for deep welds apparently would have worked as well. Many other acceptable solutions may have been possible. Therefore, a variety of designs for a particular structure could be consistent with professional engineering standards.

The Hyatt Regency case is a clear illustration of culpable failure. The original design failed to meet building code requirements. The design change made matters worse. The Citicorp case is a clear illustration of how the standard engineering practice of meeting code requirements may not be enough. It is to LeMessurier's credit that he discovered the problem. Not doing so would not have been negligence, even though the structure was flawed. Once the flaw was discovered, however, the standard of care required LeMessurier to do something about it, as he clearly realized.

No doubt William LeMessurier was disappointed to discover a serious fault in Citicorp Center. However, there was much about the structure in which he could take pride. A particularly innovative feature was a 400-ton concrete damper on ball bearings placed near the top of the building. LeMessurier introduced this feature not to improve safety but, rather, to reduce the sway of the building—a matter of comfort to residents, not safety. Of course, this does not mean that the damper has no effect on safety. Although designed for comfort, it is possible that it also enhances safety. Also, especially since its movement needs to be both facilitated and constrained, it is possible that without other controls, it could have a negative effect on safety. In any case, the effect that a 400-ton damper near the top of a 59-story structure might have on the building's ability to handle heavy winds is something that requires careful attention.

Supporting the structure on four pillars midway between the corners of the building is another innovation—one that might explain why it occurred to LeMessurier that it was worthwhile to try to determine what effect 45-degree winds might have on the structure's stability. Both innovations fall within the range of accepted engineering practice, provided that well-conceived efforts are made to determine what effect they might have on the overall integrity and utility of the structure. The risk of relying exclusively on the particular directives of a building code is that its framers are unlikely to be able in advance to take into account all of the relevant effects of innovations in design. That is, it is quite possible for regulations to fail to keep pace with technological innovation.

2.4   BLAME-RESPONSIBILITY AND CAUSATION

Now let us turn to the negative concept of responsibility for harm. We can begin by considering the relationship of responsibility for harm to the causation of harm. When the *Columbia* Accident Investigation Board examined the *Columbia* tragedy, it focused on what it called the "causes" of the accident. It identified two principal causes: the "physical cause" and the "organizational causes." The physical cause was the damage to the leading edge of the left wing by the foam that broke loose from the external tank. The organizational causes were the defects in the organization and culture of NASA that led to an inadequate concern for safety.[14] It also made reference to individuals who were "responsible and accountable" for the accident. The board, however, did not consider its primary mission to be the identification of individuals who should be held responsible and perhaps punished.[15] Thus, it identified three types of explanations of the accident: the physical cause, organizational causes, and individuals responsible or accountable for the accident.

The concept of cause is related in an interesting way to that of responsibility. Generally, the more we are inclined to speak of the physical cause of something, the less we are inclined to speak of responsibility—and the more we are inclined to speak of responsibility, the less inclined we are to focus on physical causes. When we refer only to the physical cause of the accident—namely, the damage produced by the breach in the leading edge of the orbiter's left wing—it is inappropriate to speak of responsibility. Physical causes, as such, cannot be responsible agents. The place of responsibility with respect to organizations and individuals raises more complex issues. Let us turn first to organizations.

The relationship of organizations to the concepts of causation and responsibility is controversial. The *Columbia* Accident Investigation Board preferred to speak of the organization and culture of NASA as a cause of the accident. With respect to the physical cause, the board said,[16]

> The physical cause of the loss of the *Columbia* and its crew was a breach in the Thermal Protection System on the leading edge of the left wing, caused by a piece of insulating foam which separated from the left bipod ramp section of the External Fuel Tank at 81.7 seconds after launch, and struck the wing in the vicinity of the lower half of Reinforced Carbon-Carbon panel number 8.

With respect to the organizational causes of the accident, the board said,[17]

> The organizational causes of this accident are rooted in the Space Shuttle Program's history and culture, including the original compromises that were required to gain approval for the Shuttle, subsequent years of resource constraints, fluctuating priorities, schedule pressures, mischaracterization of the Shuttle as operational rather than developmental, and lack of an agreed national vision for human space flight. Cultural traits and organizational practices detrimental to safety were allowed to develop, including: reliance on past successes as a substitute for sound engineering practices (such as testing to understand why systems were not performing in accordance with requirements); organizational barriers that prevented effective communication of critical safety information and stifled professional differences of opinion; lack of integrated management across program elements; and the evolution of an informal chain of command and decision-making processes that operated outside the organization's rules.

With respect to the relative importance of these two causes, the board concluded,[18]

> In the Board's view, NASA's organizational culture and structure had as much to do with this accident as the External Tank foam. Organizational culture refers to the values, norms, beliefs, and practices that govern how an institution functions. At the most basic level, organizational culture defines the assumptions that employees make as they carry out their work. It is a powerful force that can persist through reorganizations and reassignments of key personnel.

If organizations can be causes, can they also be morally responsible agents, much as humans can be? Some theorists believe it makes no sense to say that organizations (such as General Motors or NASA) can be morally responsible agents.[19] An organization is not, after all, a human person in the ordinary sense. Unlike human persons, corporations do not have a body, cannot be sent to jail, and have an indefinite life. On the other hand, corporations are described as "artificial persons" in the law. According to *Black's Law Dictionary*, "the law treats the corporation itself as a person which can sue and be sued. The corporation is distinct from the individuals who comprise it (shareholders)."[20] Corporations, like persons, can also come into being and pass away and can also be fined.

Philosopher Peter French argues that corporations can, in a significant sense, be morally responsible agents.[21] Although French focuses on corporations, his arguments can also be applied to governmental organizations such as NASA. Corporations have three characteristics that can be said to make them very similar to moral agents. First, corporations, like people, have a decision-making mechanism. People can deliberate and then carry out their decisions. Similarly, corporations have boards of directors and executives who make decisions for the corporation, and these decisions are then carried out by subordinate members of the corporate hierarchy. Second, corporations, like people, have policies that guide their decision making. People have moral rules and other considerations that guide their conduct. Similarly, corporations have corporate policies, including, in many cases, a corporate code of ethics. In addition to policies that guide conduct, corporations also have a "corporate culture" that tends to shape their behavior, much as personality and character shape the actions of individuals. Third, corporations, like people, can be said to have "interests" that are not necessarily the same as those of the executives, employees, and others who make up the corporation. Corporate interests include making a profit, maintaining a good public image, and staying out of legal trouble.

Consider an example of a corporate decision. Suppose an oil corporation is considering beginning a drilling operation in Africa. A mountain of paperwork will be forwarded to the CEO, other top executives, and probably the board of directors. When a decision is made, according to the decision-making procedure established by the corporation, it can properly be called a "corporate decision." It was made for "corporate reasons," presumably in accordance with "corporate policy," to satisfy "corporate interests," hopefully guided by "corporate ethics." Because it is a corporate decision, the corporation can be held responsible for it, both morally and legally.

Whether organizations can be morally responsible agents is, of course, still a matter of debate. The answer to the question depends on the strength of the analogies between organizations and moral agents. Although there are disanalogies between organizations and persons, we find the analogies more convincing. Regardless of whether organizations are seen as moral agents or merely causes of harm, however, organizations can be held responsible in at least three senses.[22] First, they can be criticized for harms, just as the *Columbia* Accident Investigation Board criticized NASA. Second, an organization that harms others can be asked to make reparations for wrong done. Finally, an organization that has harmed others is in need of reform, just as the board believed NASA needs reform.

One worry about treating organizations as morally responsible agents is that individual responsibility might be lost. Instead of holding individuals responsible, it is feared, their organizations will be. However, there need be no incompatibility in holding both organizations and the individuals within them morally accountable for what they do. We now turn to the responsibilities of individuals.

2.5 LIABILITY

Although engineers and their employers might try to excuse failure to provide safety and quality by pointing out that they have met existing regulatory standards, it is evident that the courts will not necessarily agree. The standard of care in tort law (which is concerned with wrongful injury) is not restricted to regulatory standards.

The expectation is that engineers will meet the standard of care as expressed in *Coombs v. Beede:* [23]

> The responsibility resting on an architect is essentially the same as that which rests upon the lawyer to his client, or upon the physician to his patient, or which rests upon anyone to another where such person pretends to possess some special skill and ability in some special employment, and offers his services to the public on account of his fitness to act in the line of business for which he may be employed. The undertaking of an architect implies that he possesses skill and ability, including taste, sufficient enough to enable him to perform the required services at least ordinarily and reasonably well; and that he will exercise and apply, in the given case, his skill and ability, his judgment and taste reasonably and without neglect.

As Joshua B. Kardon points out, this standard does not hold that all failure to provide satisfying services is wrongful injury. However, it does insist that the services provide evidence that reasonable care was taken. What counts as reasonable care is a function of both what the public can reasonably expect and what experienced, competent engineers regard as acceptable practice. Given the desirability of innovative engineering design, it is unrealistic for the public to regard all failures and mishaps to be culpable; at the same time, it is incumbent on engineers to do their best to anticipate and avoid failures and mishaps.

It should also be noted that *Coombs v. Beede* does not say that professionals need only conform to the established standards and practices of their field of expertise. Those standards and practices may be in a state of change, and they may not be able to keep pace with advancing knowledge of risks in particular areas. Furthermore, as many liability cases have shown, reasonable people often disagree about precisely what those standards and practices should be taken to be.

A practical way of examining moral responsibility is to consider the related concept of legal liability for causing harm. Legal liability in many ways parallels moral responsibility, although there are important differences. To be legally liable for causing harm is to warrant punishment for, or to be obligated to make restitution for, harms. Liability for harm ordinarily implies that the person caused the harm, but it also implies something about the conditions under which the harm was caused. These conditions are ordinarily "mental" in nature and can involve such things as malicious intent, recklessness, and negligence. Let us discuss these concepts of liability and moral responsibility for harm in more detail, noting that each connotes a weaker sense of liability than the other.[24] We shall also see that, although the concept of causing harm is present, the notions of liability and responsibility are the focus of attention.

First, a person can *intentionally* or knowingly and deliberately cause harm. If I stab you in the back in order to take your money, I am both morally responsible and legally liable for your death. The causal component in this case is killing you, and the mental component is intending to do you serious harm.

Second, a person can *recklessly* cause harm by not aiming to cause harm but by being aware that harm is likely to result. If I recklessly cause you harm, the causal factor is present, so I am responsible for your harm. In reckless behavior, although there is not an intent to harm, there is an intent to engage in behavior that is known to place others at risk of harm. Furthermore, the person may have what we could call a reckless attitude, in which the well-being of others, and perhaps even of himself, is not uppermost in his mind. The reckless attitude may cause harm, as in the case of a person who drives twice the speed limit and causes an accident.

He may not intend to do harm or even to cause an accident, but he does intend to drive fast, and he may not be thinking about his own safety or that of others. If his reckless action causes harm, then he is responsible for the harm and should be held legally liable for it.

Third, a still weaker kind of legal liability and moral responsibility is usually associated with *negligently* causing harm. Unlike recklessness, where an element of deliberateness or intent is involved (such as a decision to drive fast), in negligent behavior the person may simply overlook something or not even be aware of the factors that could cause harm. Furthermore, there may not be any direct causal component. The person is responsible because she has failed to exercise due care, which is the care that would be expected of a reasonable person in the circumstances. In law, a successful charge of negligence must meet four conditions:

1. A legal obligation to conform to certain standards of conduct is present.
2. The person accused of negligence fails to conform to the standards.
3. There is a reasonably close causal connection between the conduct and the resulting harm.
4. Actual loss or damage to the interests of another results.

These four elements are also present in moral responsibility, except that in the first condition, we must substitute "moral obligation" for "legal obligation." Professions such as engineering have recognized standards of professional practice, with regard to both technical and ethical practice. Professional negligence, therefore, is the failure to perform duties that professionals have implicitly or explicitly assumed by virtue of being professionals. If an engineer does not exercise standard care, according to the recognized standards of his or her profession, and is therefore negligent, then he or she can be held responsible for the harm done.

One concept of legal liability has no exact parallel in moral responsibility. In some areas of the law, there is *strict liability* for harms caused; there is no attribution of fault or blame, but there is a legal responsibility to provide compensation, make repairs, or the like. Strict liability is directed at corporations rather than individual engineers within the organization. However, insofar as they have a duty to be faithful and loyal employees, and perhaps even as a matter of specifically assigned duties, engineers can have responsibilities to their employer to help minimize the likelihood that strict liability will be imposed on the organization. So even strict liability at the corporate level can have moral implications for individual engineers.

A common complaint is that court determinations, particularly those involving juries, are often excessive. However valid this complaint may be, two points should not be lost. First, the fact that these determinations are made, however fair or unfair they may be, has important implications for engineers. As consultants who are themselves subject to liability, they have self-interested reasons for striving to take the standard of care seriously. As corporate employees, they have a responsibility to be concerned about areas of corporate liability that involve their expertise.

Second, the standard of care has a moral basis, regardless of how it plays out in courts of law. From a moral standpoint, intentionally, negligently, or recklessly causing harm to others is to fail to exercise reasonable care. What, if any, legal redress is due is another matter.

Although the standard of care plays a prominent role in law, it is important to realize that it reflects a broader notion of moral responsibility as well. Dwelling on its role in law alone may suggest to some a more calculative, "legalistic" consideration of reasonable care. In calculating the case for or against making a full effort to meet the standard of care, the cost of doing so can be weighed against the chances of facing a tort claim. This involves estimating the likelihood that harm will actually occur—and, if it does, that anyone will take it to court (and that they will be successful). Liability insurance is already an expense, and those whose aim is simply to maximize gains and minimize overall costs might calculate that a less than full commitment to the standard of care is worth the risk. From this perspective, care is not so much a matter of reasonable care as it is taking care not to get caught.

2.6 DESIGN STANDARDS

As previously noted, most engineering codes of ethics insist that, in designing products, engineers are expected to hold considerations of public safety paramount. However, there is likely more than one way to satisfy safety standards, especially when stated broadly. But if there is more than one way to satisfy safety standards, how are designers to proceed?

If we are talking about the overall safety of a product, there may be much latitude—a latitude that, of course, provides space for considerations other than safety (e.g., overall quality, usability, and cost). For example, in the late 1960s, operating under the constraints of developing an appealing automobile that weighed less than 2000 pounds and that would cost consumers no more than $2000, Ford engineers decided to make more trunk space by putting the Pinto's gas tank in an unusual place.[25] This raised a safety question regarding rear-end collisions. Ford claimed that the vehicle passed the current standards. However, some Ford engineers urged that a protective buffer should be inserted between the gas tank and protruding bolts. This, they contended, would enable the Pinto to pass a more demanding standard that it was known would soon be imposed on newer vehicles. They warned that without the buffer, the Pinto would fail to satisfy the new standard, a standard that they believed would come much closer to meeting the standard of care enforced in tort law.

Ford decided not to put in the buffer. It might have been thought that satisfying the current safety standard ensured that courts and their juries would agree that reasonable care was exercised. However, this turned out to be a mistaken view. As noted previously, the courts can determine that existing technical standards are not adequate, and engineers are sometimes called upon to testify to that effect.

Given the bad publicity Ford received regarding the Pinto and its history of subsequent litigation, Ford might regret not having heeded the advice of those engineers who argued for the protective buffer. This could have been included in the original design, or perhaps there were other feasible alternatives during the early design phases. However, even after the car was put on the market, a change could have been made. This would have involved an expensive recall, but this would not have been an unprecedented move in the automotive industry.

These possibilities illustrate a basic point about regulatory standards, accepted standards of engineering practice, and engineering design. Professional standards for engineers underdetermine design. In principle, if not in practice, there will be more than one way to satisfy the standards. This does not mean that professional standards have no effect on practice. As Stuart Shapiro notes,[26]

> Standards are one of the principal mechanisms for managing complexity of any sort, including technological complexity. Standardized terminology, physical properties, and procedures all play a role in constraining the size of the universe in which the practitioner must make decisions.

For a profession, the establishment of standards of practice is typically regarded as contributing to professionalism, thereby enhancing the profession in the eyes of those who receive its services. At the same time, standards of practice can contribute to both the quality and the safety of products in industry. However, standards of practice have to be applied in particular contexts that are not themselves specified in the standards. Shapiro notes,[27]

> There are many degrees of freedom available to the designer and builder of machines and processes. In this context, standards of practice provide a means of mapping the universal onto the local. All one has to do is think of the great variety of local circumstances for which bridges are designed and the equally great variety of designs that result.... Local contingencies must govern the design and construction of any particular bridge within the frame of relative universals embodied in the standards.

Shapiro's observation focuses on how standards of practice allow engineers freedom to adapt their designs to local, variable circumstances. This often brings surprises not only in design but also in regard to the adequacy of formal standards of practice. As Louis L. Bucciarelli points out, standards of practice are based on the previous experience and testing of engineers. Design operates on the edge of "the new and the untried, the unexperienced, the ahistorical."[28] Thus, as engineers develop innovative designs (such as LeMessurier's Citicorp structure), we should expect formal standards of practice sometimes to be challenged and found to be in need of change—all the more reason why courts of law are unwilling simply to equate the standard of care with current formal standards of practice.

2.7 THE RANGE OF STANDARDS OF PRACTICE

Some standards of practice are clearly only local in their scope. The New York City building code requirement that high-rise structures be tested for wind resistance at 90-degree angles applied only within a limited geographic region. Such specific code requirements are local in their origin and applicability. Of course, one would expect somewhat similar requirements to be in place in comparable locales in the United States, as well as in other high-rise locales throughout the world. This suggests that local codes, particularly those that attempt to ensure quality and safety, reflect more general standards of safety and good engineering practice.

One test of whether we can meaningfully talk of more general standards is to ask whether the criteria for engineering competence are only local (e.g., those of New York City civil engineers or Chicago civil engineers). The answer seems clearly to be "no" within the boundaries of the United States, especially for graduates of accredited engineering programs at U.S. colleges and universities.

However, as Vivian Weil has argued, there is good reason to believe that professional standards of engineering practice can cross national boundaries.[29] She offers the example of early 20th-century Russian engineer, Peter Palchinsky. Critical of major engineering projects in Russia, Palchinsky was nevertheless regarded to be a highly competent engineer in his homeland. He also was a highly regarded consultant in Germany, France, England, The Netherlands, and Italy. Although he was regarded as politically dangerous by Russian leaders at the time, no one doubted his engineering abilities—in Russia or elsewhere.

Weil also reminds readers of two fundamental principles of engineering that Palchinsky applied wherever he practiced:[30]

> Recall that the first principle was: gather full and reliable information about the specific situation. The second was: view engineering plans and projects in context, taking into account impacts on workers, the needs of workers, systems of transportation and communication, resources needed, resource accessibility, economic feasibility, impacts on users and on other affected parties, such as people who live downward.

Weil goes on to point out that underlying Palchinsky's two principles are principles of common morality, particularly respect for the well-being of workers—a principle that Palchinsky argued was repeatedly violated by Lenin's favored engineering projects.

We have noted that the codes of ethics of engineering societies typically endorse principles that seem intended to apply to engineers in general rather than only to members of those particular societies. Common morality was suggested as providing the ground for basic provisions of those codes (for example, concern for the safety, health, and welfare of the public). Whether engineers who are not members of professional engineering societies actually do, either explicitly or implicitly, accept the principles articulated in a particular society's code of ethics is, of course, another matter. However, even if some do not, it could be argued that they should. Weil's point is that there is no reason, in principle, to believe that supportable international standards cannot be formulated and adopted. Furthermore, this need not be restricted to abstract statements of ethical principle. As technological developments and their resulting products show up across the globe, they can be expected to be accompanied by global concerns about quality, safety, efficiency, cost-effectiveness, and sustainability. This, in turn, can result in uniform standards in many areas regarding acceptable and unacceptable engineering design, practice, and products. In any case, in the context of an emerging global economy, constructive discussions of these concerns should not be expected to be only local.

2.8   THE PROBLEM OF MANY HANDS

Individuals often attempt to evade personal responsibility for wrongdoing. Perhaps the most common way this is done, especially by individuals in large organizations, is by pointing out that many individuals had a hand in causing the harm. The argument goes as follows: "So many people are responsible for the tragedy that it is irrational and unfair to pin the responsibility on any individual person, including me." Let us call this the *problem of fractured responsibility* or (preferably) the *problem of many hands*.[31] In response to this argument, philosopher Larry May has proposed the following principle to apply to the responsibility of individuals in a situation in which many people are involved: "[I]f a harm has resulted from collective inaction, the degree of individual responsibility of each member of a putative group for the harm should vary based on the role each member could, counterfactually, have played in preventing the inaction."[32] Let us call this the *principle of responsibility for inaction in groups.* Our slightly modified version of this principle reads as follows: In a situation in which a harm has been produced by collective inaction, the degree of responsibility of each member of the group depends on the extent to which the member could reasonably be expected to have tried to prevent the action. The qualification "the extent to which each member could reasonably be expected to have tried to prevent the action" is necessary because there are limits to reasonable expectation here. If a person could have prevented an undesirable action only by taking his own life, sacrificing his legs, or harming someone else, then we cannot reasonably expect him to do it.

A similar principle can apply to collective action. Let us call it the *principle of responsibility for action in groups:* In a situation in which harm has been produced by collective action, the degree of responsibility of each member of the group depends on the extent to which the member caused the action by some action reasonably avoidable on his part. Again, the reason for the qualification is that if an action causing harm can only be avoided by extreme or heroic action on the individual's part (such as taking his own life, sacrificing his legs, or harming someone else), then we may find reason for not holding the person responsible, or at least holding him less responsible.

2.9 IMPEDIMENTS TO RESPONSIBLE ACTION

What attitudes and frames of mind can contribute to less than fully responsible action, whether it be intentional, reckless, or merely negligent? In this section, we discuss some impediments to responsible action.

Self-Interest

Engineers are not simply engineers. They are, like everyone else, people with personal hopes and ambitions that are not restricted to professional ideals. Sometimes concern for our own interests tempts us to act contrary to the interests of others, perhaps even contrary to what others expect from us as professionals. Sometimes concern for self-interest blocks us from seeing or fully understanding our professional responsibilities. As discussed later, this is a major worry about conflicts of interest—a problem standardly addressed in engineering codes of ethics.

Taken to an extreme, concern for self-interest is a form of *egoism*—an exclusive concern to satisfy one's own interests, even at the possible expense of others. This is popularly characterized as "looking out for number one."

Whether a thoroughgoing egoist would act at the expense of others very much depends on the circumstances. All of us depend to some extent on others to get what we want; some degree of mutual support is necessary. However, opportunities for personal gain at the expense of others do arise—or so it seems to most of us. Egoists are prepared to take advantage of this, unless they believe it is likely to work to their long-term disadvantage. But it is not just egoists who are tempted by such opportunities: All of us are, at least occasionally.

Self-interest may have been partly an impediment to responsible action in the case of many NASA managers in the *Columbia* accident. Managers advance their careers by being associated with successful and on-schedule flights. They may have sometimes pursued these goals at the expense of the safety of the crew. Many

NASA managers had positions that involved them in conflicts of interest that may have compromised their professional integrity.[33] NASA contractors also had reasons of self-interest not to place any obstacles in the way of NASA's desire to keep the flights on schedule. This was a powerful consideration in getting Morton Thiokol to reverse its original recommendation not to fly in the case of the *Challenger;* it may have influenced contractors not to press issues of safety in the *Columbia* flight as well.[34]

Self-Deception

One way to resist the temptations of self-interest is to confront ourselves honestly and ask if we would approve of others treating us in the same way we are contemplating treating them. This can have a powerful psychological effect on us. However, for such an approach to work, we must truly recognize what we are contemplating doing. Rationalization often gets in the way of this recognition. Some rationalizations show greater self-awareness than others, particularly those that exhibit self-defensiveness or excuse making. ("I'm not really doing this just for myself." "Everyone takes shortcuts once in a while—it's the only way one can survive.") Other rationalizations seem to betray a willful lack of self-understanding. This is called *self-deception,* an intentional avoidance of truths we would find it painful to confront self-consciously.[35] Because of the nature of self-deception, it is particularly difficult to discover it in oneself. However, open communication with colleagues can help correct biases to which we are susceptible—unless, of course, our colleagues share the same biases (an illustration of groupthink, discussed later).

Self-deception seems to have been pervasive in the NASA space flight program. Rodney Rocha accused NASA managers of "acting like an ostrich with its head in the sand."[36] NASA managers seem to have convinced themselves that past successes are an indication that a known defect would not cause problems, instead of deciding the issue on the basis of testing and sound engineering analysis. Often, instead of attempting to remedy the problem, they simply engaged in a practice that has come to be called "normalizing deviance," in which the boundaries of acceptable risk are enlarged without a sound engineering basis. Instead of attempting to eliminate foam strikes or do extensive testing to determine whether the strikes posed a safety-of-flight issue, managers "increasingly accepted less-than-specification performance of various components and systems, on the grounds that such deviations had not interfered with the success of previous flights."[37] Enlarging on the issue, the board observed: "With each successful landing, it appears that NASA engineers and managers increasingly regarded the foam-shredding as inevitable, and as either unlikely to jeopardize safety or simply an acceptable risk."[38] We consider the normalization of deviance again in our discussion of risk.

There were other aspects of self-deception in the space flight program, such as classifying the shuttle as an operational vehicle rather than one still in the process of development.[39] With "operational" technology, management considerations of economy and scheduling are much more important than they are with a technology that is in the development stage, where quality and safety must be the primary considerations.

Finally, there was a subtle shift in the burden of proof with respect to the shuttle. Instead of requiring engineers to show that the shuttle was safe to fly or that the foam strike did not pose a safety-of-flight issue, which was the appropriate position, managers appear to have required engineers to show that the foam strike was a safety-of-flight issue. Engineers could not meet this standard of proof, especially in the absence of the images of the area of the foam strike. This crucially important shift may have occurred without full awareness by the NASA staff. In any event, the shift had profound consequences, just as a similar shift in the burden of proof had in the *Challenger* accident. Referring to the plight of the Debris Assessment Team that was assigned the task of evaluating the significance of the foam strike, the *Columbia* Accident Investigation Board remarked,[40]

> In the face of Mission managers' low level of concern and desire to get on with the mission, Debris Assessment Team members had to prove unequivocally that a safety-of-flight issue existed before Shuttle Program management would move to obtain images of the left wing. The engineers found themselves in the unusual position of having to prove that the situation was unsafe—a reversal of the usual requirement to prove that a situation is safe.

As the board observed, "Imagine the difference if any Shuttle manager had simply asked, 'Prove to me that *Columbia* has not been harmed.' "[41]

Fear

Even when we are not tempted to take advantage of others for personal gain, we may be moved by various kinds of *fear*—fear of acknowledging our mistakes, of losing our jobs, or of some sort of punishment or other bad consequences. Fears of these sorts can make it difficult for us to act responsibly.

Most well-known whistle-blowing cases are instances in which it is alleged that others have made serious mistakes or engaged in wrongdoing. It is also well-known that whistleblowers commonly endure considerable hardship and suffering as a result of their open opposition. This may involve being shunned by colleagues and others, demotion or the loss of one's job, or serious difficulties in finding new employment (especially in one's profession). Although the circumstances that call for whistle-blowing are extreme, they do occur. Given the typical fate of whistle-blowers, it takes considerable courage to step forward even when it is evident that this is the morally responsible thing to do.

Here there is strength in numbers. Group resistance within an organization is more likely to bring about changes without the need for going outside the organization. When this fails, a group of whistleblowers may be less likely than a single whistleblower to be perceived as simply disloyal or trying to get back at the organization for some grievance. However, the difficulty of finding others with whom to join a cause can itself increase one's fears. Thus, there seems to be no substitute for courage and determination in such circumstances.

One form of fear is the fear of retribution for objecting to actions that violate professional standards. The *Columbia* Accident Investigation Board observed that "fear of retribution" can be a factor inhibiting the expression of minority opinions.[42] As such, it can be a powerful impediment to responsible professional behavior.

Ignorance

An obvious barrier to responsible action is *ignorance* of vital information. If an engineer does not realize that a design poses a safety problem, for example, then he or she will not be in a position to do anything about it. Sometimes such a lack of awareness is willful avoidance—a turning away from information in order to avoid having to deal with the challenges it may pose. However, often it results from a lack of imagination, from not looking in the right places for necessary information, from a failure to persist, or from the pressure of deadlines. Although there are limits to what engineers can be expected to know, these examples suggest that ignorance is not always a good excuse.

NASA managers were often ignorant of serious problems associated with the shuttle. One of the reasons for this is that as information made its way up the organizational hierarchy, increasingly more of the dissenting viewpoints were filtered out, resulting in an excessively sanitized version of the facts. According to the *Columbia* Accident Investigation Board, there was a kind of "cultural fence" between engineers and managers. This resulted in high-level managerial decisions that were based on insufficient knowledge of the facts.[43]

Egocentric Tendencies

A common feature of human experience is that we tend to interpret situations from very limited perspectives and it takes special efforts to acquire a more objective viewpoint. This is what psychologists call *egocentricity*. It is especially prevalent in us as young children, and it never completely leaves us. Although egocentric thinking is sometimes egoistic (self-interested), it need not be. It is actually a special form of ignorance.

It is not just self-interest that interferes with our ability to understand things from other perspectives. We may have good intentions for others but fail to realize that their perspectives are different from ours in important ways. For example, some people may not want to hear bad news about their health. They may also assume that others are like them in this respect. So, if they withhold bad news from others, this is done with the best of intentions—even if others would prefer hearing the bad news. Similarly, an engineer may want to design a useful product but fail to realize how different the average consumer's understanding of how to use it is from those who design it. This is why test runs with typical consumers are needed.

NASA managers probably exhibited egocentric thinking when they made decisions from an exclusively management perspective, concentrating on such factors as schedule, political ramifications, and cost. These were not necessarily self-interested motivations, and in most cases they surely had the well-being of the organization and the astronauts at heart. Nevertheless, making decisions from this exclusively management perspective led to many mistakes.

Microscopic Vision

Like egocentric thinking, *microscopic vision* embraces a limited perspective.[44] However, whereas egocentric thinking tends to be inaccurate (failing to understand the perspectives of others), microscopic vision may be highly accurate and precise but our field of vision is greatly limited. When we look into a microscope, we see things that we could not see before—but only in the narrow field of resolution on which the microscope focuses. We gain accurate, detailed knowledge—at a microscopic level. At the same time, we cease to see things at the more ordinary level. This is the price of seeing things microscopically. Only when we lift our eyes from the microscope will we see what is obvious at the everyday level.

Every skill, says Michael Davis, involves microscopic vision to some extent:

> A shoemaker, for example, can tell more about a shoe in a few seconds than I could tell if I had a week to examine it. He can see that the shoe is well or poorly made, that the materials are good or bad, and so on. I can't see any of that. But the shoemaker's insight has its price. While he is paying attention to people's shoes, he may be missing what the people in them are saying or doing.[45]

Just as shoemakers need to raise their eyes and listen to their customers, engineers sometimes need to raise their eyes from their world of scientific and technical expertise and look around them in order to understand the larger implications of what they are doing.

Large organizations tend to foster microscopic thinking. Each person has his or her own specialized job to do, and he or she is not responsible, from the organizational standpoint, for the work of others. This was evidently generally true of the NASA organizational structure. It may also have been a contributing factor to the *Columbia* accident.

Uncritical Acceptance of Authority

Engineering codes of ethics emphasize the importance of engineers exercising independent, objective judgment in performing their functions. This is sometimes called professional *autonomy*. At the same time, the codes of ethics insist that engineers have a duty of fidelity to their employers and clients. Independent consulting engineers may have an easier time maintaining professional autonomy than the vast majority of engineers, who work in large, hierarchical organizations. Most engineers are not their own bosses, and they are expected to defer to authority in their organizations.

An important finding of the research of social psychologist Stanley Milgram is that a surprisingly high percentage of people are inclined to defer uncritically to authority.[46] In his famous obedience experiments during the 1960s, Milgram asked volunteers to administer electric shocks to "learners" whenever they made a mistake in repeating word pairs (e.g., nice/day and rich/food) that volunteers presented to them earlier. He told volunteers that this was an experiment designed to determine the effects of punishment on learning. No shocks were actually administered, however. Milgram was really testing to determine the extent to which volunteers would continue to follow the orders of the experimenter to administer what they believed were increasingly painful shocks. Surprisingly (even to Milgram), nearly two-thirds of the volunteers continued to follow orders all the way up to what they thought were 450-volt shocks—even when shouts and screams of agony were heard from the adjacent room of the "learner." The experiment was replicated many times to make sure that the original volunteers were a good representation of ordinary people rather than especially cruel or insensitive people.

There is little reason to think that engineers are different from others in regard to obeying authority. In the Milgram experiments, the volunteers were told that the "learners" would experience pain but no permanent harm or injury. Perhaps engineers would have had doubts about this as the apparent shock level moved toward the 450-volt level. This would mean only that the numbers need to be altered for engineers, not that they would be unwilling to administer what they thought were extremely painful shocks.

One of the interesting variables in the Milgram experiments was the respective locations of volunteers and "learners." The greatest compliance occurred when "learners" were not in the same room with the volunteers. Volunteers tended to accept the authority figure's reassurances that he would take all the responsibility for any unfortunate consequences. However, when volunteers and "learners" were in the same room and in full view of one another, volunteers found it much more difficult to divest themselves of responsibility.

Milgram's studies seem to have special implications for engineers. As previously noted, engineers tend to work in large organizations in which the division of labor often makes it difficult to trace responsibility to specific individuals. The combination of the hierarchical structure of large organizations and the division of work into specialized tasks contributes to the sort of "distancing" of an engineer's work from its consequences for the public. This tends to decrease the engineer's sense of personal accountability for those consequences. However, even though such distancing might make it easier psychologically to be indifferent to the ultimate consequences of one's work, this does not really relieve one from at least partial responsibility for those consequences.

One further interesting feature of Milgram's experiments is that volunteers were less likely to continue to administer what they took to be shocks when they were in the presence of other volunteers. Apparently, they reinforced each other's discomfort at continuing, and this made it easier to disobey the experiment. However, as discussed in the next section, group dynamics do not always support critical response. Often quite the opposite occurs, and only concerted effort can overcome the kind of uncritical conformity that so often characterizes cohesive groups.

Groupthink

A noteworthy feature of the organizational settings within which engineers work is that individuals tend to work and deliberate in groups. This means that an engineer will often participate in group decision making rather than function as an individual decision maker. Although this may contribute to better decisions ("two heads are better than one"), it also creates well-known but commonly overlooked tendencies to engage in what Irving Janis calls *groupthink*—situations in which groups come to agreement at the expense of critical thinking.[47] Janis documents instances of groupthink in a variety of settings, including a number of historical fiascos (e.g., the bombing of Pearl Harbor, the Bay of Pigs invasion, and the decision to cross the 38th parallel in the Korean War).[48]

Concentrating on groups that are characterized by high cohesiveness, solidarity, and loyalty (all of which are prized in organizations), Janis identifies eight symptoms of groupthink:[49]

- an *illusion of invulnerability* of the group to failure;
- a strong "we-feeling" that views outsiders as adversaries or enemies and encourages *shared stereotypes* of others;
- *rationalizations* that tend to shift responsibility to others;
- an *illusion of morality* that assumes the inherent morality of the group and thereby discourages careful examination of the moral implications of what the group is doing;
- a tendency of individual members toward *self-censorship,* resulting from a desire not to "rock the boat";
- an *illusion of unanimity,* construing silence of a group member as consent;

- an application of *direct pressure* on those who show signs of disagreement, often exercised by the group leader who intervenes in an effort to keep the group unified; and
- *mindguarding,* or protecting the group from dissenting views by preventing their introduction (by, for example, outsiders who wish to present their views to the group).

Traditionally, engineers have prided themselves on being good team players, which compounds the potential difficulties with groupthink. How can the problem of groupthink be minimized for engineers? Much depends on the attitudes of group leaders, whether they are managers or engineers (or both). Janis suggests that leaders need to be aware of the tendency of groups toward groupthink and take constructive steps to resist it. Janis notes that after the ill-advised Bay of Pigs invasion of Cuba, President John F. Kennedy began to assign each member of his advisory group the role of critic. He also invited outsiders to some of the meetings, and he often absented himself from meetings to avoid influencing unduly its deliberations.

NASA engineers and managers apparently were often affected with the group-think mentality. Commenting on management's decision not to seek clearer images of the leading edge of the left wing of the shuttle in order to determine whether the foam strike had caused damage, one employee said, "I'm not going to be Chicken Little about this."[50] The *Columbia* Accident Investigation Board described an organizational culture in which "people find it intimidating to contradict a leader's strategy or a group consensus," evidently finding this characteristic of the NASA organization.[51] The general absence of a culture of dissent that the board found at NASA would have encouraged the groupthink mentality.

To overcome the problems associated with the uncritical acceptance of authority, organizations must establish a culture in which dissent is accepted and even encouraged. The *Columbia* Accident Investigation Board cites organizations in which dissent is encouraged, including the U.S. Navy Submarine Flooding Prevention and Recovery program and the Naval Nuclear Propulsion programs. In these programs, managers have the responsibility not only of encouraging dissent but also of coming up with dissenting opinions themselves if such opinions are not offered by their subordinates. According to the board, "program managers [at NASA] created huge barriers against dissenting opinions by stating preconceived conclusions based on subjective knowledge and experience, rather than on solid data." Toleration and encouragement of dissent, then, was noticeably absent in the NASA organization. If dissent is absent, then critical thinking is absent.

2.10 CHAPTER SUMMARY

Obligation-responsibility requires that one exercise a standard of care in one's professional work. Engineers need to be concerned with complying with the law, adhering to standard norms and practices, and avoiding wrongful behavior. However, this may not be good enough. The standard of care view insists that existing regulatory standards may be inadequate because these standards may fail to address problems that have yet to be taken adequately into account.

We might wish for some sort of algorithm for determining what our responsibilities are in particular circumstances. However, this is an idle wish. Even the most detailed codes of ethics of professional engineering societies can provide only general guidance. The determination of responsibilities in particular circumstances depends on discernment and judgment on the part of engineers.

Blame-responsibility can be applied to individuals and perhaps to organizations. If we believe organizations can be morally responsible agents, it is because we believe the analogies between undisputed moral agents (people) and organizations are stronger than the disanalogies. In any case, organizations can be criticized for the harms they cause, asked to make reparations for harm done, and assessed as needing to be reformed.

Individuals can be responsible for harm by intentionally, recklessly, or negligently causing harm. Some argue that individuals cannot be responsible for harm in situations in which many individuals have contributed to the harm, but we can proportion responsibility to the degree to which an individual's action or inaction is responsible for the harm.

There are many impediments to the kind of discernment and judgment that responsible engineering practice requires. Self-interest, fear, self-deception, ignorance, egocentric tendencies, microscopic vision, uncritical acceptance of authority, and groupthink are commonplace and require special vigilance if engineers are to resist them.

NOTES

1. This account is based on three sources: *Columbia* Accident Investigation Board, vol. 1 (Washington, DC: National Aeronautics and Space Administration, 2003); "Dogged Engineer's Effort to Assess Shuttle Damage," *New York Times,* Sept. 26, 2003, p. A1; and William Langewiesche, "*Columbia's* Last Flight," *Atlantic Monthly,* Nov. 2003, pp. 58–87.

2. The next several paragraphs, as well as some later segments of this chapter, are drawn from Michael S. Pritchard, "Professional Standards for Engineers," forthcoming in Anthonie Meijers, ser. ed., *Handbook Philosophy of Technology and Engineering Sciences,* Part V, "Normativity and values in technology," Ibo van de Poel, ed. (New York: Elsevier, forthcoming).

3. William F. May, "Professional Virtue and Self-Regulation," in *Ethical Issues in Professional Life* (Oxford: Oxford University Press, 1988), p. 408.

4. Ibid.

5. Ibid.

6. The list that follows is based on interviews of engineers and managers conducted by James Jaksa and Michael S. Pritchard and reported in Michael S. Pritchard, "Responsible Engineering: The Importance of Character and Imagination," *Science and Engineering Ethics,* 7, no. 3, 2001, pp. 394–395.

7. See, for example, the Association for Computing Machinery, *ACM Code of Ethics and Professional Conduct,* section 2.2. Acquire and maintain professional competence.

8. This is a major theme in Stuart Shapiro, "Degrees of Freedom: The Interaction of Standards of Practice and Engineering Judgment," *Science, Technology, & Human Values,* 22, no. 3, Summer 1997.

9. Shapiro, p. 290.

10. Joshua B. Kardon, "The Structural Engineer's Standard of Care," paper presented at the OEC International Conference on Ethics in Engineering and Computer Science, March 1999. This article is available at onlineethics.org.

11. Ibid. Kardon bases this characterization on *Paxton v. County of Alameda* (1953) 119c.C.A. 2d 393, 398, 259P 2d 934.

12. For further discussion of this case, see C. E. Harris, Michael S. Pritchard, and Michael Rabins, "Case 64: Walkway Disaster," in *Engineering Ethics: Concepts and Cases,* 3rd ed. (Belmont, CA: Wadsworth, 2005), pp. 348–349. See also Shapiro, p. 287.

13. For further details, see Harris et al., "Case 11: Citicorp," pp. 307–308. See also Joe Morgenstern, "The Fifty-Nine Story Crisis," *New Yorker,* May 29, 1995, pp. 49–53.
14. *Columbia* Accident Investigation Board, p. 6.
15. Nevertheless, the investigation eventually resulted in the displacement of no less than a dozen key people at NASA, as well as a public vindication of Rocha for doing the right thing.
16. Ibid., p. 9.
17. Ibid.
18. Ibid., p. 177.
19. For discussions of this issue, see Peter French, *Collective and Corporate Responsibility* (New York: Columbia University Press, 1984); Kenneth E. Goodpaster and John B. Matthews, Jr., "Can a Corporation Have a Conscience?" *Harvard Business Review,* 60, Jan.–Feb. 1982, pp. 132–141; and Manuel Velasquez, "Why Corporations Are Not Morally Responsible for Anything They Do," *Business and Professional Ethics Journal,* 2, no. 3, Spring 1983, pp. 1–18.
20. *Black's Law Dictionary,* 6th ed. (St. Paul, MN: West, 1990), p. 340.
21. See Peter French, "Corporate Moral Agency" and "What Is Hamlet to McDonnell-Douglas or McDonnell-Douglas to Hamlet: DC-10," in Joan C. Callahan, ed., *Ethical Issues in Professional Life* (New York: Oxford University Press, 1988), pp. 265–269, 274–281. The following discussion has been suggested by French's ideas, but it diverges from them in several ways.
22. These three senses all fall on the blame-responsibility side. A less explored possibility is that corporations can be morally responsible agents in positive ways.
23. *Coombs v. Beede,* 89 Me. 187, 188, 36 A. 104 (1896). This is cited and discussed in Margaret N. Strand and Kevin Golden, "Consulting Scientist and Engineer Liability: A Survey of Relevant Law," *Science and Engineering Ethics,* 3, no. 4, Oct. 1997, pp. 362–363.
24. We are indebted to Martin Curd and Larry May for outlining parallels between legal and moral notions of responsibility for harms and their possible applications to engineering. See Martin Curd and Larry May, *Professional Responsibility for Harmful Actions,* Module Series in Applied Ethics, Center for the Study of Ethics in the Professions, Illinois Institute of Technology (Dubuque, IA: Kendall/Hunt, 1984).
25. Information on the Ford Pinto is based on a case study prepared by Manuel Velasquez, "The Ford Motor Car," in *Business Ethics: Concepts and Cases,* 3rd ed. (Englewood Cliffs, NJ: Prentice Hall, 1992), pp. 110–113.
26. Shapiro, p. 290.
27. Ibid., p. 293.
28. Louis L. Bucciarelli, *Designing Engineers* (Cambridge, MA: MIT Press, 1994), p. 135.
29. Vivian Weil, "Professional Standards: Can They Shape Practice in an International Context?" *Science and Engineering Ethics,* 4, no. 3, 1998, pp. 303–314.
30. Ibid., p. 306. Similar principles are endorsed by disaster relief specialist Frederick Cuny and his Dallas, Texas, engineering relief agency. Renowned for his relief efforts throughout the world, Cuny's principles of effective and responsible disaster relief are articulated in his *Disasters and Development* (New York: Oxford University Press, 1983).
31. The term "the problem of many hands" is suggested by Helen Nissenbaum in "Computing and Accountability," in Deborah G. Johnson and Helen Nissenbaum, eds., *Computers, Ethics, and Social Values* (Upper Saddle River, NJ: Prentice Hall, 1995), p. 529.
32. Larry May, *Sharing Responsibility* (Chicago: University of Chicago Press, 1992), p. 106.
33. *Columbia* Accident Investigation Board, pp. 186, 200.
34. Ibid., pp. 10, 202.
35. This is Mike Martin's characterization of self-deception. See his *Self-Deception and Morality* (Lawrence: University Press of Kansas, 1986) for an extended analysis of self-deception and its significance for morality.

36. "Dogged Engineer's Effort to Assess Shuttle Damage," p. A1.
37. *Columbia* Accident Investigation Board, p. 24.
38. Ibid., p. 122.
39. Ibid., p. 198.
40. Ibid., p. 169.
41. Ibid., p. 192.
42. Ibid., p. 192.
43. Ibid., pp. 168, 170, 198.
44. This expression was introduced into engineering ethics literature by Michael Davis. See his "Explaining Wrongdoing," *Journal of Social Philosophy,* XX, nos. 1 & 2, Spring–Fall 1989, pp. 74–90. Davis applies this notion to the *Challenger* disaster, especially when Robert Lund was asked to take off his engineer's hat and put on his manager's hat.
45. Ibid., p. 74.
46. Stanley Milgram, *Obedience to Authority* (New York: Harper & Row, 1974).
47. Irving Janis, *Groupthink,* 2nd ed. (Boston: Houghton Mifflin, 1982).
48. The most recent edition of the McGraw-Hill video *Groupthink* features the *Challenger* disaster as illustrating Janis's symptoms of groupthink.
49. Ibid., pp. 174–175.
50. "Dogged Engineer's Effort to Assess Shuttle Damage," p. A1.
51. *Columbia* Accident Investigation Board, p. 203.

CHAPTER THREE

Framing the Problem

Main Ideas in this Chapter

- To a large extent, moral disagreement occurs against the background of widespread moral agreement.
- Disagreement about moral matters is often more a matter of disagreement about facts than moral values.
- Disagreement is also sometimes about conceptual matters—what concepts mean and whether they apply in particular circumstances.
- Much of the content of engineering codes of ethics is based on the application of ideas of our common morality to the context of engineering practice.
- Two general moral perspectives that can be helpful in framing moral problems in engineering are the utilitarian ideal of promoting the greatest good and that of respect for persons.

IN 1977, THE OCCUPATIONAL SAFETY and Health Administration (OSHA) issued an emergency temporary standard requiring that the level of air exposure to benzene in the workplace not exceed 1 part per million (ppm).[1] This was a departure from the then current standard of 10 ppm. OSHA wanted to make this change permanent because of a recent report to the National Institutes of Health of links between leukemia deaths and exposure to benzene. However, the reported deaths were in workplaces with benzene exposure levels above 10 ppm, and there were no animal or human test data for lower levels of exposure. Nevertheless, because of evidence that benzene is carcinogenic, OSHA advocated changing the standard to the lowest level that can be easily monitored (1 ppm).

OSHA's authority seemed clear in the Occupational Safety and Health Act, which provides that "no employee will suffer material impairment of health or functional capacity even if such employee has regular exposure to the hazard dealt with by such standard for the period of his working life."[2] The law went on to state that "other considerations shall be the latest available scientific data in the field, the feasibility of the standards, and experience gained under this and other health and safety laws."[3]

On July 2, 1980, the U.S. Supreme Court ruled that OSHA's proposed 1 ppm standard was too strict. The law, said the Court, does not "give OSHA the unbridled discretion to adopt standards designed to create absolutely risk-free workplaces regardless of the costs."[4] According to the Court, although the current limit is

10 ppm, the actual exposures are often considerably lower. It pointed out that a study by the petrochemical industry reported that out of a total of 496 employees exposed to benzene, only 53 percent were exposed to levels between 1 and 5 ppm, and 7 percent were exposed to between 5 and 10 ppm.[5] But most of the scientific evidence involved exposure well above 10 ppm.

The Court held that a safe work environment need not be risk-free. OSHA, it ruled, bears the burden of proof that reducing the exposure level to 1 ppm will result in substantial health benefits. OSHA, however, believed that in the face of scientific uncertainty and when lives are at risk, it should be able to enforce stricter standards. OSHA officials objected to shifting to them the burden of proof that chemicals such as benzene are dangerous, when it seemed to them that formerly, with support of the law, the burden lay with those who were willing to expose workers to possibly dangerous chemicals.

3.1 INTRODUCTION

The conflicting approaches of OSHA and the Supreme Court illustrate legal and possibly moral disagreement. OSHA officials were concerned about protecting workers, despite the heavy costs in doing so. The Supreme Court justices apparently believed that OSHA officials had not sufficiently taken into account the small number of workers affected, the technological problems involved in implementing the new regulations, and the impact of regulations on employers and the economy.

Despite this disagreement, OSHA officials and the justices probably agreed on many of their basic moral beliefs: that it is wrong to murder, that it is wrong to fail to meet obligations and responsibilities that one has accepted, that it is in general wrong to endanger the well-being and safety of others, and that one should not impose responsibilities on others that are greater than they can legitimately be expected to bear.

These observations point out the important fact that *we usually experience moral disagreement and controversy within a context of agreement.* When we disagree, this is often because we still are not clear enough about important matters that bear on the issue. In this chapter, we consider the importance of getting clear about the fundamental facts and concepts relevant to the case at hand. Then, we discuss the common moral ground that can help us frame the ethical issues facing engineers. In the next chapter, we suggest useful ways of attempting to resolve those issues.

3.2 DETERMINING THE FACTS

We cannot discuss moral issues intelligently apart from a knowledge of the facts that bear on those issues. So we must begin by considering what those facts are. In any given case, many facts will be obvious to all, and they should be taken into account. However, sometimes people come to different moral conclusions because they do not view the facts in the same way. Sometimes they disagree about *what* the facts are. Sometimes they disagree about the *relevance* or relative importance of certain facts. Therefore, close examination of our take on the facts is critical.

To understand the importance of facts in a moral controversy, we propose the following three theses about factual issues:

1. *Often, moral disagreements turn out to be disagreements over the relevant facts.* Imagine a conversation between two engineers, Tom and Jim, that might have taken place shortly before OSHA issued its May 1977 directive that worker exposure to benzene emissions be reduced from 10 to 1 ppm. Their conversation might have proceeded as follows:

Tom: I hear OSHA is about to issue stricter regulations regarding worker exposure to benzene. Oh, boy, here we go again. Complying with the new regulations is going to cost our company several million dollars. It's all well and good for the bureaucrats in Washington to make rules, as long as they don't have to pay the bills. I think OSHA is just irresponsible!

Jim: But Tom, human life is at stake. You know the dangers of benzene. Would you want to be out in the area where benzene exposure is an issue? Would you want your son or your daughter to be subjected to exposures higher than 1 ppm?

Tom: I wouldn't have any problem at all. There is just no scientific evidence that exposure to benzene below 10 ppm has any harmful effect.

2. *Factual issues are sometimes very difficult to resolve.* It is particularly important for engineering students to understand that many apparent moral disagreements are reducible to disagreements over factual (in many cases technical) matters. The dispute between Tom and Jim could be easy to resolve. If Jim reads the literature that has convinced Tom that there is no scientific evidence that exposure to benzene below 10 ppm has harmful effects, they might agree that OSHA plans go too far. Often, however, factual issues are not easily resolved. Sometimes, after a debate over issues in professional ethics, students come away with an attitude that might be stated as follows: "Well, here was another dispute about ethics in which nobody could agree. I'm glad that I'm in engineering, where everything depends on the facts that everybody can agree on. Ethics is just too subjective." But the dispute may pivot more around the difficulty of determining factual matters than any disagreement about moral values as such. Sometimes the information we need is simply not available now, and it is difficult to imagine how it could be available soon, if at all.

3. *Once the factual issues are clearly isolated, disagreement can reemerge on another and often more clearly defined level.*

Suppose Jim replies to Tom's conclusion that exposure to benzene below 10 ppm is not harmful in this way:

Jim: Well, Tom, the literature you've shared with me convinces me that we don't have any convincing evidence yet that exposure to benzene below 10 ppm is harmful. But, as we've so often learned to our regret, in the long run things we thought were harmless turned out to be harmful. That's what happened with asbestos in the workplace. For years the asbestos industry scoffed at any evidence that asbestos might be harmful, and it simply assumed that it wasn't. Maybe OSHA is going beyond what our current data can show, but 1 ppm can be easily monitored. It may cost a bit more to monitor at that level, but isn't it better to be safe than sorry when we're dealing with carcinogenic materials?

Tom: It is better to be safe than sorry, but we need to have positive evidence that taking stronger measures makes us safer. Of course, there are risks in the face of the unknown—but that doesn't mean that we should act now as if we know something we don't know.

Jim: But if we assume that something like benzene is safe at certain levels simply because we can't show right now that it isn't, that's like playing ostrich—burying our heads in the sand until we're hit from behind.

Tom: Well, it seems to me that your view is more like Chicken Little's worry that the sky is falling—jumping to the worst conclusion on the basis of the least evidence.

What this discussion between Jim and Tom reveals is that sometimes our best factual information is much less complete than we would like. In the arena of risk, we must consider probabilities and not certainties. This means that we need to develop standards of *acceptable* risk; and disagreements about such standards are not simply disagreements about facts. They reflect value judgments regarding what levels of risk it is reasonable to expect people to accept.

Known and Unknown Facts

It should not be surprising to find two people disagreeing in their conclusions when they are reasoning from different factual premises. Sometimes these disagreements are very difficult to resolve, especially if it is difficult to obtain the information needed to resolve them. In regard to the benzene issue, Tom and Jim had an initial disagreement about the facts. After examining the evidence available to them at the time, evidently the Supreme Court sided with Tom. However, it is important to realize that all along Tom and Jim apparently agreed that if it were shown that lower levels of exposure to benzene are harmful, stronger regulations would be needed. Both agreed with the general moral rule against harming others.

Frequently, important facts are not known, thereby making it difficult to resolve disagreement. Some of the facts we may want to have at our disposal relate to something that has already happened (e.g., what caused the accident). But we also want to know what consequences are likely to result from the various options before us, and there can be much uncertainty about this. Thus, it is important to distinguish not only between relevant and irrelevant facts but also between known facts and unknown facts. Here, the number of unknown facts is less important than the degree of their relevance or importance. Even a single unknown relevant fact might make a crucial difference to what should be done. In any case, we have a special responsibility to seek answers to unanswered factual questions.

Weighing the Importance of Facts

Even if two or more people agree on which facts are relevant, they might nevertheless disagree about their relative importance. In the automotive industry, for example, two engineers might agree that the evidence indicates that introducing another safety feature in the new model would most likely result in saving a few lives during the next 5 years. One engineer might oppose the feature because of the additional cost, whereas the other thinks the additional cost is well worth the added safety. This raises questions about acceptable risk in relation to cost. One engineer might oppose the feature because he thinks that the burden of responsibility should be shifted to the consumer, whereas the other thinks that it is appropriate to protect consumers from their own negligence.

3.3 CLARIFYING CONCEPTS

Good moral thinking requires not only attending carefully to facts but also having a good grasp of the key concepts we need to use. That is, we need to get as clear as we can about the meanings of key terms. For example, "public health, safety, and welfare," "conflict of interest," "bribery," "extortion," "confidentiality," "trade secret," and "loyalty" are key terms for ethics in engineering.

It would be nice to have precise definitions of all these terms; but like most terms in ethics, their meanings are somewhat open-ended. In many cases, it is sufficient to clarify our meaning by thinking of *paradigms,* or clear-cut examples, of what we have in mind. In less straightforward cases, it is often useful to compare and contrast the case in question with paradigms. Suppose a firm signs a contract with a customer that specifies that all parts of the product will be made in the United States, but the product has a special $1/4$-inch staple hidden from view that was made in England. Is the firm dishonest if it does not tell its customer about this staple? In order to settle this question it is important, first, to get clearer about we mean by "dishonesty" as a basic concept.

A clear-cut case of dishonesty would be if Mark, the firm's representative, answers "No" to the customer asking, "Is there *anything* in this product that wasn't made in the U.S.A.?" Suppose, instead, the customer asks, "Does this product have any parts not made in the U.S.A.?" and Mark replies, "No," silently thinking, "After all, that little staple isn't a *part;* it simply holds parts together." Of course, this raises the question of what is meant by "part." But given the contract's specifications, honesty in this case would seem to call for full disclosure. Then the customer can decide whether the English staple is acceptable. Better yet would be for the firm to contact the customer before using the staple, explaining why it is needed and asking whether using it would be acceptable.

Although in this case we may question the firm's motives (and therefore its honesty), sometimes apparent moral disagreement turns out to rest on conceptual differences where no one's motives are in question. These are issues about the general definitions, or meanings, of concepts. In regard to risk, an obvious conceptual issue of meaning has to do with the proper definition of "safe." If we are talking about risks to health, in addition to the question of what we should mean by "health," we might be concerned about what we should mean by a "substantial" health risk or what is a "material impairment" of health. Finally, the definition of "burden of proof" can be a point of controversy, especially if we are considering the issue from a moral and not merely a legal standpoint, where the term may be more clearly defined.

We can imagine a continuation of the conversation between Tom and Jim that illustrates the importance of some of the conceptual considerations that can arise in the context of apparent moral disagreement:

Jim: Tom, I admit that the evidence that exposures to benzene between 1 and 10 ppm are harmful is weak at best, but this doesn't really end the matter. I'll go back to one of my original points: Human life is involved. I just don't believe we should take a chance on harming people when we aren't certain about the facts. I think we ought to provide a safe environment for our workers, and I wouldn't call an environment "safe" when there is even a chance that the disputed benzene levels are harmful.

Tom: Here we go again on that old saw, "How safe is safe?" How can you say that something is not safe when you don't have any evidence to back up your claim?

Jim: I think something is unsafe when there is any kind of substantial health risk.

Tom: But how can you say there is any substantial health risk when, in fact, the evidence that is available seems to point in the other direction?

Jim: Well, I would say that there is a substantial health risk when there is some reason to suspect that there might be a problem, at least when something like carcinogens are involved. The burden of proof should rest on anyone who wants to expose a worker to even a possible danger.

Tom: I'll agree with you that workers shouldn't be exposed to substantial health risks, but I think you have a strange understanding of "substantial." Let me put the question another way. Suppose the risk of dying from cancer due to benzene exposure in the plant over a period of 30 years is no greater than the risk over the same period of time of dying from an automobile accident while driving home from the plant. Would you consider the health risk from benzene exposure in this case to be "substantial"?

Jim: Yes, I would. The conditions are different. I believe we have made highways about as safe as we can. We have not made health conditions for workers in plants as safe as we can. We can lower the level of benzene in the plant, and with a relatively moderate expenditure. Furthermore, everyone accepts the risks involved in auto travel. Many of the workers don't understand the risk from benzene exposure. They aren't acting as free agents with informed consent.

Tom: Look, suppose at the lower levels of benzene exposure—I mean under 10 ppm—the risk of cancer is virtually nil, but some workers find that the exposure causes the skin on their faces, hands, and arms to be drier than usual. They can treat this with skin lotion. Would you consider this a health problem?

Jim: Yes, I would. I think it would be what some people call a "material impairment" of health, and I would agree. Workers should not have to endure changes in their health or bodily well-being as a result of working at our plant. People are selling their time to the company, but not their bodies and their health. And dry skin is certainly unhealthy. Besides, there's still the problem of tomorrow. We don't really know the long-range effects of lower levels of exposure to benzene. But given the evidence of problems above 10 ppm, we have reason to be concerned about lower levels as well.

Tom: Well, this just seems too strict. I guess we really do disagree. We don't even seem to be able to agree over what we mean by the words we use.

Here, genuine disagreement over moral issues has reappeared, but this time in the form of disagreement over the definitions of crucial terms. Concepts such as "safe," "substantial," "health," and "material impairment" are a blend of factual elements and value elements. Tom and Jim might agree on the effects of exposure to benzene at various levels and still disagree as to what is "safe" or "healthy" and what is not. To know whether benzene is safe, we must have some notion of what the risks are at various exposure levels, but we also must have a notion of what we mean by "acceptable risk." The use of the term *acceptable* should be sufficient to alert us that there is a value element here that cannot be determined by the facts alone.

When disagreements about the meanings of words arise, it may be tempting to say, "We're just quibbling about words" or "It's just a semantic question." Insofar as the choice of meanings we make affects our chosen course of action, this understates the significance of the disagreement. Disputants might interpret regulatory standards differently based on their different understandings of "safe." The different meanings they give "safe" also reflect different levels of risk to which they are willing to give their approval. Although disputants might never resolve their differences, it is desirable for them to try. This might enable them to see more clearly what these differences are. If they can agree that "safe" is best understood in terms of "acceptable risk" rather than "absolutely risk-free" (a standard that is virtually unattainable), they can then proceed to discuss reasonable standards of acceptability.

3.4 APPLICATION ISSUES

So far, we have emphasized that when engaging in ethical reflection, it is important to get as clear as we can about both the relevant facts and the basic meanings of key concepts. However, even when we are reasonably clear about what our concepts mean, disagreement about their applications in particular cases can also arise. If those who disagree are operating from different factual premises, there might well be disagreement about whether certain concepts apply in particular circumstances. For example, a disagreement about bribery might pivot around the question of whether an offer of a free weekend at an exclusive golf resort in exchange for a vendor's business was actually made. It might be agreed that if such an offer were made, this would be an attempt to bribe. However, whether or not such an offer was actually made may be at issue.

If the issue is only over whether or not a certain offer was made, the possible ways of resolving it may be readily apparent. If there were no witnesses and neither party is willing to admit that the offer was made, the issue may remain unresolved for others, but at least we can say, "Look, either the offer was made or it wasn't—there's a fact of the matter."

There is another kind of application issue, one that rests on a common feature of concepts. Attempts to specify the meanings of terms ahead of time can never anticipate all of the cases to which they do and do not apply. No matter how precisely we attempt to define a concept, it will always remain insufficiently specified so that some of its applications to particular circumstances will remain problematic.

We can clarify this further in a somewhat more formal way. If we let "X" refer to a concept, such as "keeping confidentiality" or "proprietary information," a conceptual question can be raised regarding what, in general, are the defining features of X. A question regarding a concept's *application* in a particular situation can also be raised. It is one thing to determine what we *mean* by "safe" and another to determine whether a given situation should *count* as safe, considering the definition.

Asking what we mean by "safe" is a conceptual question. Asking whether a particular situation should count as safe is an application question. Answering this second question may require only determining what the facts are. However, sometimes it requires us to reexamine the concept. In many situations, a clear definition of a term can make its application unproblematic. Often, the concept either clearly does or does not apply to a particular situation. Sometimes, however, this is not the case. This is because definitions cannot possibly be so clear and complete that every possible situation clearly does or does not count as an instance of the concept. This inherent limitation of definitions and explanations of concepts gives rise to problems in applying concepts and calls for further reflection.

One way of dealing with these problems is to change or modify our definitions of crucial concepts in the face of experience. Sometimes an experience may not appear to exemplify the concept as we have defined it, but we believe it should count as an instance of the concept anyway. In such a case, the experience prompts us to modify the definition. When this happens in analyzing a case, it is a good idea to revisit the initial depiction of the case and reassess the relevant facts and ethical considerations before attempting its final resolution.

3.5 COMMON GROUND

An ethics case study describes a set of circumstances that calls for ethical reflection. It is helpful to begin an analysis with two questions: What are the relevant facts? and What are the relevant kinds of ethical considerations? These two questions are interconnected; they cannot be answered independently of one another. Let's see why.

First, let's consider the facts. Which facts? Those that have some bearing on what is ethically at stake. That is, we need to have our eye on what is ethically important in order to know which of the many facts available to us we should be considering. On the one hand, it may be a fact that engineer Joe Smith was wearing a yellow tie on the day he was deciding whether to accept an expensive gift from a vendor. But it is not obvious that this fact is relevant to the question of whether he should accept or refuse the gift. On the other hand, the fact that accepting the gift might incline him to favor the vendor's product regardless of its quality is relevant.

However, we also have to decide what sorts of ethical considerations are relevant. Here, we need to draw on our ethical principles, rules, and concepts. However, again, the key term *relevant* comes into play. Which ethical principles, rules, and concepts are relevant? This depends on the facts of the case. For example, *conflict of interest* may be an ethically important concept to consider—but only when the facts of a case suggest that there might actually be a conflict of interest.

Unfortunately, the relevant facts in a case do not come with labels ("Here I am, an ethically relevant fact"). To determine what facts are relevant, as well as what facts would be useful to know, it is helpful to bear in mind the kinds of moral resources we have available that could help us think through the case. These include the ideas of common morality, professional codes of ethics, and our personal morality. All of these may be helpful in determining what facts are relevant in any given case. To this we should add our ability to evaluate critically all of these resources, including our personal morality.

We can call the stock of common moral beliefs *common morality*. The term is used by analogy with the term *common sense*. Just as most of us share a common body of beliefs about the world and about what we must do in order to survive—a body of beliefs that we call common sense—we share a common stock of basic beliefs about moral standards, rules, and principles we believe should guide our lives. If asked, we may offer different grounds for holding these beliefs. Many of us will appeal to our religious commitments, others to more secular commitments. Nevertheless, there is a surprising degree of agreement about the content of common morality.

We also agree in many specific moral judgments, both general and particular. We not only agree with the general idea that murder is wrong but also commonly agree in particular instances that a murder has occurred—and that this is wrong. We not only agree with the general idea that for engineers not to disclose conflicts of interest is wrong but also commonly agree in particular instances that an engineer has failed to disclose a conflict of interest—and that this is wrong.

Of course, people do differ to some extent in their moral beliefs because of such factors as family background and religious upbringing, but most of these differences occur with respect to beliefs about specific practices—such as abortion, euthanasia, sexual morality, and capital punishment—or with respect to specific moral judgments about, for example, whether this particular person should or should not have an abortion. Differences are not as prevalent at the level on which we are now focusing, our more general moral beliefs.

To examine these general moral beliefs more closely, we must formulate them, which is no easy matter. Fortunately, there are common features of human life that suggest the sorts of general moral beliefs we share. First, we are *vulnerable*. We are susceptible to pain, suffering, unhappiness, disability, and, ultimately, death. Second, we value *autonomy,* our capacity to think for ourselves and make our own decisions. Third, we are *interdependent.* We depend on others to assist us in getting what we want through cooperative endeavors and the division of labor. Our well-being also depends on others refraining from harming us. Fourth, we have *shared expectations and goals.* Beyond wanting things for ourselves as individuals, we may want things together—that is, as groups working toward shared ends. Groups may range from two or more individuals who care for each other to larger groups, such as particular professions, religious institutions, nations, or even international organizations such as the United Nations or the World Health Organization. Finally, we have *common moral traits.* Fair-mindedness, self-respect, respect for others, compassion, and benevolence toward others are common traits. Despite individual differences in their strength, scope, and constancy, these traits can be found to some degree in virtually all human beings.

Without suggesting that this list is complete, it does seem to provide a reasonable basis for understanding why common morality would include general moral rules or principles about how we should treat each other. We briefly discuss attempts by two philosophers to formulate these general considerations.

The first, W. D. Ross, constructed a list of basic duties or obligations, which he called "prima facie" or "conditional" duties.[6] In using these terms, Ross intended to convey the idea that although these duties are generally obligatory, they can be overridden in special circumstances. He disclaimed finality for his list, but he believed that it was reasonably complete. His list of prima facie duties can be summarized as follows:

R1. Duties resting on our previous acts
  (a) Duties of fidelity (to keep promises and not to tell lies)
  (b) Duties of reparation for wrong done
R2. Duties of gratitude (e.g., to parents and benefactors)

R3. Duties of justice (e.g., to support happiness in proportion to merit)

R4. Duties of beneficence (to improve the condition of others)

R5. Duties of self-improvement

R6. Duties not to injure others

Engineers, like others, probably share these moral beliefs, and this is reflected in many engineering codes of ethics. Most codes enjoin engineers to be faithful agents for their employers, and this injunction can be seen to follow from the duties of fidelity (R1) and gratitude (R2). Most codes require engineers to act in ways that protect the health, safety, and welfare of the public, and this obligation follows from the duties of justice (R3) and beneficence (R4), and especially from the duty not to injure others (R6). Finally, most codes encourage engineers to improve their professional skills, a duty reflected in R5.

Bernard Gert formulated a list of 10 "moral rules" that he believes capture the basic elements of common morality:[7]

G1. Don't kill.

G2. Don't cause pain.

G3. Don't disable.

G4. Don't deprive of freedom.

G5. Don't deprive of pleasure.

G6. Don't deceive.

G7. Keep your promise (or don't break your promise).

G8. Don't cheat.

G9. Obey the law (or don't disobey the law).

G10. Do your duty (or don't fail to do your duty).

Ross's prima facie duties and Gert's moral rules can be seen to overlap each other considerably. G1–G9, for example, might be seen as specifications of Ross's duty not to injure others. The wrongness of lying and promise breaking appear on both lists. R2–R5 seem to be of a more positive nature than Gert's moral rules, which focus on not causing harm. However, Gert also has a list of 10 "moral ideals," which focus on preventing harm. In fact, the moral ideals can be formulated by introducing the word "prevent" and changing the wording of the rules slightly. Thus, the moral ideal corresponding to "Don't kill" is "Prevent killing." For Gert, the moral rules specify moral requirements, whereas the moral ideals are aspirational.

Like Ross's prima facie duties, Gert's moral rules are not "absolute." That is, each allows exceptions, but only if a justification is provided. Gert says,[8]

> The claim that there are moral rules prohibiting such actions as killing and deceiving means only that these kinds of actions are immoral unless they can be justified. Given this understanding, all moral agents agree that there are moral rules prohibiting such actions as killing and deceiving.

Usually it is wrong to lie, but if the only way to save an innocent person from being murdered is to lie to the assailant about that person's whereabouts, then most would agree that lying is justified. The main point is not that moral rules and principles have no exceptions; it is that taking exception to them requires having a justification, or good reason, for doing so. This contrasts with, for example, deciding whether to take a walk, go to the movies, or read a book. Breaking a promise, however, does call for a justification, as does injuring others.

3.6 GENERAL PRINCIPLES

To some it may appear that, at least as we have characterized it so far, common morality is too loosely structured. Everyone can agree that, other things being equal, we should keep our promises, be truthful, not harm others, and so on. But all too frequently, other things are not equal. Sometimes keeping a promise will harm someone, as will telling the truth. What do we do then? Are there any principles that might frame our thinking in ways that can help us resolve such conflicts?

There is a basic concept that is especially important to keep in mind in answering these questions. This is the idea of *universalizability*: Whatever is right (or wrong) in one situation is right (or wrong) in any relevantly similar situation.[9] Although this does not by itself specify what is right or wrong, it requires us to be consistent in our thinking. For example, in considering whether or not it would be morally acceptable to falsify data in a particular project, a scientist or engineer needs to think about not just this particular situation but all situations relevantly like it. Falsifying data is, essentially, a form of lying or cheating. When we broaden our focus to consider what kind of act is involved, the question of whether it is all right to falsify data is bound to appear quite different than when thinking only about the immediate situation.

In the next sections, we consider two general ways of thinking about moral issues that make use of the idea of universalizability and that attempt to provide underlying support for common morality while at the same time offering guidelines for resolving conflicts within it.[10] The first appeals to the *utilitarian* ideal of maximizing good consequences and minimizing bad consequences. The second appeals to the ideal of *respect for persons*. For some time now, philosophers have debated whether one of these ideals is so basic that it can provide a comprehensive, underlying ground for common morality. We will not enter into this debate here. It will be enough to show that both these approaches can be helpful in framing much of our moral thinking about ethical issues in engineering.

To illustrate how utilitarian and respect for persons ideals might come into play, let us consider the following situation:

David Parkinson is a member of the Madison County Solid Waste Management Planning Committee (SWMPC). State law requires that one of the committee members be a solid waste expert, David's area of specialization. SWMPC is considering recommending a specific plot of land in a sparsely populated area of Madison County for a needed public landfill. However, next to this site is a large tract of land that a group of wealthy Madison County residents wish to purchase in order to develop a private golf course surrounded by luxurious homes. Although small, this group is highly organized and it has managed to gather support from other wealthy residents in the county, including several who wield considerable political power.

Informally recognized as the Fairway Coalition, this influential group has bombarded the local media with expensive ads in its public campaign against the proposed landfill site, advocating instead a site that borders on one the least affluent areas of Madison City. The basic argument is that a landfill at the site SWMPC is considering will destroy one of Madison County's most beautiful areas. Although as many as 8000 of Madison City's 100,000 residents live within walking distance of the site proposed by the Fairway Coalition, they lack the political organization and financial means to mount significant opposition.

SWMPC is now meeting to discuss the respective merits of the two landfill sites. Members of the committee turn to David for his views on the controversy.

In this fictional case, David Parkinson is in a position of public trust, in part, because of his engineering expertise. It is evident that one of his responsibilities is to use his expertise in ways that will aid the committee in addressing matters of broad public concern—and controversy.

How might he try to take into consideration what is at stake? First, it might occur to him that locating the landfill in the more heavily populated area will benefit a relatively small number of wealthy people at the expense of risking the health and well-being of a much larger number of people. Although there may be many other factors to consider, this is a utilitarian concern to promote, or at least protect, the greatest good for the greatest number of people. Second, it might occur to David that favoring the urban site over the rural site would be basically unfair because it would fail to respect the rights of the poor to a reasonably healthy environment while providing even more privilege to a wealthy minority. This is basically an appeal to the notion of equal respect for persons.

Thus far, utilitarian and respect for persons considerations seem to lead to the same conclusion. It is important to realize that different moral principles often do converge in this way, thereby strengthening our conclusions by providing support from more than one direction. Nevertheless, even when they do reach the same conclusion, two rather distinct approaches to moral thinking are involved—one taking the greater total good as the primary concern, and the other taking protection of the equal moral standing of all members in the community as the primary concern. Also, as we shall see, sometimes these two approaches are in serious tension with one another.

3.7 UTILITARIAN THINKING

In its broadest sense, taking a utilitarian approach in addressing moral problems requires us to focus on the idea of bringing about "the greatest good for the greatest number." However, there is more than one way to attempt this. We consider three prominent ways.

The Cost–Benefit Approach

How are we to determine what counts as the greater good? One approach that has some appeal from an engineering perspective is *cost–benefit analysis*. The course of action that produces the greatest benefit relative to cost is the one that should be chosen. Sometimes this is a relatively straightforward matter. However, making this sort of determination can present several difficulties. We consider three here.

First, in order to know what we should do from the utilitarian perspective, we must know which course of action will produce the most good in both the short and the long term. Unfortunately, this knowledge is sometimes not available at the time decisions must be made. For example, we do not yet know whether permitting advertising and competitive pricing for professional services will lead to some of the problems suggested by those who oppose it. Therefore, we cannot say for sure whether these are good practices from a utilitarian perspective. Sometimes all we can do is try a certain course of action and see what happens. This may be risky in some circumstances.

Second, the utilitarian aim is to make choices that promise to bring about the greatest amount of good. We refer to the population over which the good is maximized as the *audience*. The problem is determining the scope of this audience. Ideally, it might be thought, the audience should include all human beings, or at least all human beings who might be affected by the action to be evaluated. Perhaps the audience should even include all beings capable of experiencing pleasure or pain. But then it becomes virtually impossible to calculate which actions actually produce the most good for so large an audience. If we limit the audience so that it includes only our country, our company, or our community, then we face the criticism that others have been arbitrarily excluded. Therefore, in practice, those with utilitarian sympathies need to develop acceptable ways of delimiting their range of responsibility.

A third difficulty with the utilitarian standard is that it seems sometimes to favor the greater aggregate good at the expense of a vulnerable minority. Imagine the following: A plant discharges a pollutant into the local river, where it is ingested by fish. If humans eat the fish, they experience significant health problems. Eliminating the pollutant will be so expensive that the plant will become, at best, only marginally profitable. Allowing the discharge to continue will save jobs and enhance the overall economic viability of the community. The pollutant will adversely affect only a relatively small proportion of the population—the most economically deprived members of the community who fish in the river and then eat the fish.

Under these conditions, allowing the plant to continue to discharge the pollutant might seem justifiable from a utilitarian perspective, even though it would be unjust to the poorer members of the community. Thus, there is a problem of justly distributing benefits and burdens. Many would say that the utilitarian solution should be rejected for this reason. In such cases, utilitarian reasoning seems, to some, to lead to implausible moral judgments, as measured by our understanding of common morality.

Despite these problems, cost–benefit analysis is often used in engineering. This approach attempts to apply the utilitarian standard in as quantifiable a manner as possible. An effort is made to translate negative and positive utilities into monetary terms. Cost–benefit analysis is sometimes referred to as *risk–benefit analysis* because much of the analysis requires estimating the probability of certain benefits and harms. It is possible to determine the actual cost of installing equipment to reduce the likelihood of certain health problems arising in the workplace. However, this does not guarantee that these health problems (or others) will not arise anyway, either from other sources or from the failure of the equipment to accomplish what it is designed to do. In addition, we do not know for sure what will happen if the equipment is not installed; perhaps money will be saved because the equipment will turn out not to have been necessary, or perhaps the actual consequences will turn out to be much worse than predicted. So factoring in probabilities greatly complicates cost–benefit analysis.

Cost–benefit analysis involves three steps:

1. Assess the available options.
2. Assess the costs (measured in monetary terms) and the benefits (also measured in monetary terms) of each option. The costs and benefits must be assessed for the entire audience of the action, or all those affected by the decision.

3. Make the decision that is likely to result in the greatest benefit relative to cost; that is, the course of action chosen must not be one for which the cost of implementing the option could produce greater benefit if spent on another option.

There are serious problems with using cost–benefit analysis as a sole guide for protecting the public from pollution that endangers health. One problem is that the cost–benefit analysis assumes that economic measures of cost and benefit override all other considerations. Cost–benefit analysis encourages the elimination of a pollutant only when it can be done in an economically efficient manner. However, suppose the chemical plant we have been considering is near a wilderness area that is damaged by one of the plant's emissions. It might not be economically efficient to eliminate the pollutant from the cost–benefit standpoint. Of course, the damage to the wilderness area must be included in the cost of the pollution, but the quantified cost estimate might still not justify the elimination—or even the reduction—of the pollution. Yet it is not necessarily irrational to hold that the pollutant should be eliminated, even if the elimination is not justified by the analysis. The economic value that anyone would place on saving the wilderness is not a true measure of its value.

Another problem is that it is often difficult to ascertain the costs and benefits of the many factors that should enter into a cost–benefit analysis. The most controversial issue is how to assess in cost–benefit terms the loss of human life or even serious injury. How, we may ask, can a dollar value be placed on a human life? Aside from the difficulty of determining the costs and benefits of known factors (such as immediate death or injury), it is also difficult to predict what factors will be relevant in the future. If the threat to human health posed by a substance is not known, then it is impossible to execute a definitive cost–benefit analysis. This problem becomes especially acute if we consider long-term costs and benefits, most of which are impossible to predict or measure. In addition, cost–benefit analysis often does not take into account the distribution of costs and benefits. Using our previous example, suppose a plant dumps a pollutant into a river in which many poorer members of the community fish to supplement their diets. Suppose also that after all of the known costs and benefits are calculated, it is concluded that the costs of eliminating the pollutant outweigh all of the health costs to the poor. Still, if the costs are paid by the poor and the benefits are enjoyed by the rich, then the costs and benefits are not equally shared. Even if the poor are compensated for the damage to their health, many would say that an injustice has still been done. After all, the wealthy members of the community do not have to suffer the same threat to their health.

Finally, cost–benefit analysis might seem to justify many practices in the past that we have good reason to believe were morally wrong. In the 19th century, many people opposed child labor laws, arguing that they would lead to economic inefficiencies. They pointed out, for example, that tunnels and shafts in coal mines were too small to accommodate adults. Many arguments in favor of slavery were also based on considerations of economic efficiency. When our society finally decided to eliminate child labor and slavery, it was not simply because they became economically inefficient but also because they came to be considered unjust.

Despite these problems, cost–benefit analysis can make an important contribution to moral problem solving. We can hardly imagine constructing a large engineering project, such as the Aswan High Dam in Egypt, without performing an elaborate cost–benefit analysis. Although cost–benefit analysis may not always succeed in quantifying values in ways that do justice to them, it can play an important role in utilitarian analysis. Its ability to evaluate many conflicting considerations in terms of a single measure, monetary value, makes it invaluable in certain circumstances. As with all other tools for moral analysis, however, we must keep its limitations in mind.

The Act Utilitarian Approach

Utilitarian approaches to problems do not necessarily require that values always be rendered in strictly quantitative terms. However, they do require trying to determine what will, in some sense, maximize good consequences. If we take the *act utilitarian* approach of focusing our attention on the consequences of particular actions, we can ask, "Will this course of action result in more good than any alternative course of action that is available?" To answer this question, the following procedure is useful:

1. Identify the available options in this situation.
2. Determine the appropriate audience for the options, keeping in mind the problems in determining the audience.
3. Bear in mind that whatever option is selected, it sets an example for others, and anyone else in relevantly similar circumstances would be justified in making a similar selection.
4. Decide which available option is likely to bring about the greatest good for the appropriate audience, taking into account harms as well as benefits.

This act utilitarian approach is often helpful in analyzing options in situations that call for making moral decisions. For example, assuming the economic costs are roughly equal, the choice between two safety devices in an automotive design could be decided by determining which is more likely to reduce the most injuries and fatalities. Also, road improvements might be decided on the basis of the greater number of people served. Of course, in either case, matters could be complicated by considerations of fairness to those who are not benefited by the improvements or might be put at even greater risk. Nevertheless, the utilitarian determinations seem to carry considerable moral weight even if, in some particular cases, they turn out not to be decisive. How much weight these determinations should be given cannot be decided without first making careful utilitarian calculations.

The Rule Utilitarian Approach

One of the difficulties facing the act utilitarian approach is that often there are serious problems in trying to determine all of the consequences of our actions. Not everyone is especially good at estimating the likely consequences of the options before them. This is complicated by the fact that it is also often difficult to determine what others will do. In many areas there are coordination problems that are best resolved by having commonly accepted rules that enable us to predict reliably what others will do. A clear example is rules of the road. Traffic lights, stop signs, yield signs, and other conventions of the road promote both safe and efficient travel. In general, it is better for all of us that we guide our driving by conforming to these rules and conventions rather than trying in each circumstance to determine whether, for example, it is safe to go through a red light. Furthermore, as noted for the act utilitarian approach, what one does in a particular situation can serve as an example for others to do likewise. Therefore, an important question is, "Would utility be maximized if everyone acted similarly?"

Admittedly, there are times when it would be safe for a driver to go through a red light or stop sign, but this may be only because others can be counted on to comply with the rules. If everyone, or even very many, decided for themselves whether to stop or go through the red light, the result would probably be a sharp increase in accidents, as well as less efficient travel. The *rule utilitarian* approach to this sort of problem is to propose rules that are justified by their utility. When such rules are reasonably well understood and generally accepted, there are advantages for individuals using rules as a guide to action rather than attempting directly to calculate the likely consequences of the various alternative courses of actions in each situation.

Traffic rules, in fact, pose interesting and important questions from an engineering standpoint. Useful traffic rules need to allow for exceptions that are not stated in the rules. For example, the rule that one come to a full stop at a stop sign allows for exceptional circumstances, such as when a large van is running out of control and will crash into your car if you come to a full stop and you can see that there is no crossing traffic approaching the intersection. Stating all possible exceptions in the rule would be impossible and, in any case, make for a very cumbersome rule. Still, some kinds of exceptions are understood to be disallowed. For example, treating a stop sign as if it permitted simply slowing down and proceeding without stopping if no crossing cars are observed is disallowed (otherwise it would be replaced by a yield sign)—that is, individual discretion as a general rule is ruled out when there is a stop sign (or red light). However, estimates of the overall utility of traffic rules are sometimes adjusted, thereby leading to changes. For example, years ago most states determined that using individual discretion in turning right on a red light (after coming to a full stop) is reasonably safe and efficient (except when a "No Turn on Red Light" sign is posted).

From a rule utilitarian perspective, then, in situations covered by well-understood, generally observed rules or practices that serve utilitarian ends, one should justify one's actions by appealing to the relevant rules or practices. The rules or practices, in turn, are justified by their utility when generally observed.

There are complications. If there are widespread departures from rules or practices, then it is less clear whether overall utility is still promoted by continuing to conform to the rules or practices. To preserve the beauty of a grassy campus quad, a "Please Use Sidewalks" sign may be posted. As long as most comply with this request, the grassy area may retain its beauty. But if too many cut across the grass, a worn path will begin to form. Eventually, the point of complying with the sign may seem lost from a utilitarian standpoint—the cause has been lost.

However, in situations in which the rule utilitarian mode of analysis is useful, the following procedure could be employed. Suppose engineer Karen is facing a decision regarding whether to unilaterally substitute cheaper parts for those specified in a contract. In deciding what she should do from a rule utilitarian standpoint, she must first ask whether there are well-understood, generally observed rules that serve utilitarian ends that cover such situations. In thinking this through, she might consider the following possibilities:

Rule 1: Engineers may unilaterally substitute cheaper parts for those specified in the contract.

Rule 2: Engineers may not unilaterally substitute cheaper parts for those specified in the contract.

Note that rules chosen to analyze the case must be directly relevant to the case circumstances and must not trivialize the case. For example, Karen should not use a rule such as "It is always desirable to maximize company profits" because this ignores the specific issues of the case being tested.

Next, Karen must determine the audience, which in this case includes not only the producers and purchasers but also the general public. She should then ask which of these two rules comes closest to representing the audience's common expectations and whether meeting these expectations generally serves overall utility. If she decides (as she surely will) on Rule 2, then she should follow this rule in her own action and not substitute the cheaper parts.

Notice that the rule utilitarian approach does not consider directly the utility of a particular action unless no generally observed rules or practices that serve utilitarian ends are available.[11] Unlike the act utilitarian approach, the rule utilitarian approach judges the moral acceptability of particular actions by whether they conform to rules: those whose general observance promotes utilitarian ends.

The rule utilitarian approach is often appealed to in responding to critics who say that utilitarian thinking fails to accord appropriate respect for individuals. Utilitarian thinking, critics say, can approve violating the rights of some groups of individuals in order to promote the greater good of the majority. A rule utilitarian response might argue that there is greater utility in following a rule that disallows this than one that permits it. After all, if it is understood that the rights of some groups of individuals may be violated for the sake of the greater good, this will engender fear and insecurity throughout society because we can never be certain that we will not end up in an unfortunate minority whose rights are violated. In general, it might be argued, more good overall is served by providing people with assurances that they will be treated in accordance with rules and practices that treat them justly and with respect for individual rights.

The rule utilitarian approach to problems brings to our attention an important distinction in moral thinking. Sometimes we are concerned with making decisions in particular situations: Should I accept this gift from a vendor? Should I ignore data that may raise questions about my preferred design? Should I take time to do more testing? However, sometimes we have broader concerns with the adoption or support of appropriate rules, social policies, or practices. Rule utilitarian thinking is commonly employed in this broader setting. Here, the concern is not just with the consequences of a particular action but also with the consequences of consistent, sustained patterns of action. Whether or not engineers are themselves policy makers, many have opportunities to advise those who are by providing them with the type of information they need to determine the likely long-term consequences of developing and implementing certain policies. Thus, engineers have opportunities to play a vital role at this level, even if only in consulting or advisory roles.

Whether a rule utilitarian approach to these broader concerns is fully adequate is still a matter of controversy. Critics note that the rules and practices approved by rule utilitarian thinking are not necessarily exceptionless, and they worry that some exceptions may occur at the expense of respect for the rights of individuals. People, they insist, have rights because, as individuals, they are entitled to respect, not simply because treating them as if they have rights might maximize overall utility. We explain this view more thoroughly in the next section, which discusses the moral notion of respect for persons.

3.8 RESPECT FOR PERSONS

The moral standard of the ethics of respect for persons is as follows: Those actions or rules are right that regard each person as worthy of respect as a moral agent. This equal regard for moral agents can be understood as a basic requirement of justice. A moral agent must be distinguished from inanimate objects, such as knives or airplanes, which can only fulfill goals or purposes that are imposed externally. Inanimate objects certainly cannot evaluate actions from a moral standpoint. A paradigm example of a moral agent is a normal adult human being who, in contrast to inanimate objects, can formulate and pursue goals or purposes of his or her own. Insofar as we can do this, we are said to have *autonomy*.

From the standpoint of respect for persons, the precepts of common morality protect the moral agency of individual human beings. Maximizing the welfare of the majority must take second place to this goal. People cannot be killed, deceived, denied their freedom, or otherwise violated simply to bring about a greater total amount of utility. As with our treatment of utilitarian thinking, we consider three approaches to respect for persons thinking.

The Golden Rule Approach

Like utilitarian approaches to moral thinking, respect for persons approaches employ the idea of universalizability. Universalizability is grounded in an idea that is familiar to all of us. Most of us would acknowledge that if we think we are acting in a morally acceptable fashion, then we should find it morally acceptable for others to do similar kinds of things in similar circumstances. This same insight can lead us to ask questions about fairness and equal treatment, such as "What if everyone did that?" and "Why should you make an exception of yourself?"

The idea of universalizability implies that my judgment should not change simply because the roles are reversed. When we broaden our focus to consider what kind of act is involved, the question of whether it is all right to falsify data is bound to appear quite different than when thinking only about the immediate situation. *Reversibility* is a special application of the idea of universalizability: In thinking about treating others as I would have them treat me, I need to ask what I would think if the roles were reversed. If I am tempted to tell a lie in order to escape a particular difficulty, then I need to ask what I would think if I were the one to whom the lie is told.

Universalizing our thinking by applying the idea of reversibility can help us realize that we may be endorsing treating others in ways we would object to if done to us. This is the basic idea behind the Golden Rule, variations of which appear in the religious and ethical writings of most cultures.

Suppose that I am a manager who orders a young engineer to remain silent about the discovery of an emission from the plant that might cause minor health problems for people who live near the plant. For this order to satisfy the Golden Rule, I must be willing to have my supervisor give a similar order to me if I were the young engineer. I must also be willing to place myself in the position of the people who live near the plant and would experience the health problem if the emission were not eliminated.

This example reveals a possible problem in using the Golden Rule in resolving a moral problem. On the one hand, am I the kind of manager who believes that employees should obey their supervisors without question, especially if their supervisors are also professionals who have many years of experience? Then I would not object to remaining silent in accordance with my supervisor's orders if I were in the young engineer's position. Am I a member of the public whose health might be affected by the emission? Am I also concerned with economic efficiency and skeptical of environmental regulations? Then I might even be willing to endure minor health problems in order to keep the plant from having to buy expensive new pollution-control equipment. Thus, it seems that the Golden Rule could be satisfied. On the other hand, if I do not have these beliefs, then I cannot justify my action by the Golden Rule. The results of using the Golden Rule as a test of morally permissible action seem to vary, then, depending on the values and beliefs of the actor.

One way of trying to avoid some of these problems is to interpret the Golden Rule as requiring not only that I place myself in the position of the recipient but also that I adopt the recipient's values and individual circumstances. Thus, not only would I have to put myself in the young engineer's place but also I would have to assume her values and her station in life. Because she was evidently troubled by my order to remain silent and probably is in a low position in the firm's hierarchy, I have to assume that I would find the order contrary to my own adopted wishes and values as well, and that I believe a professional has the right to question her supervisor's judgment. Thus, I would not want to be ordered to remain silent, and my action as a manager in ordering the young engineer to remain silent would fail the requirements of the Golden Rule. I also have to assume the position of the people who would experience the minor health problems. Many of them—especially those whose health would be most directly affected—would be as concerned for economic considerations as I am and would object to the emissions.

Unfortunately, this tactic does not resolve all the problems. In other situations, placing myself in the position of the other people and assuming their values creates a new set of problems. Suppose I am an engineer who supervises other engineers and I find that I must dismiss one of my supervisees because he is lazy and unproductive. The engineer whom I want to dismiss, however, believes that "the world owes me a living" and does not want to be punished for his irresponsibility. Now if I place myself in the position of the recipient of my own action—namely, the unproductive engineer—but retain my own values, then I might use the Golden Rule to justify dismissing him. This is because I might believe that irresponsible employees should be dismissed and even be willing to be dismissed myself if I am lazy and unproductive. If I place myself in my supervisee's position and assume his values, however, I must admit that I would not want to be dismissed. Thus, dismissing the young engineer fails this interpretation of the Golden Rule requirement, even though most of us probably believe that this is the right thing to do.

We have identified two kinds of problems with the Golden Rule: those that result from exclusive attention to what the agent is willing to accept and those that result from exclusive attention to what the recipient is willing to accept. However, both perspectives (agent and recipient) seem important for an appropriate interpretation of the Golden Rule.

Rather than focus simply on what a particular individual (agent or recipient) wants, prefers, or is willing to accept, we need to consider matters from a more general perspective—one in which we strive to treat others in accordance with standards that we can share.[12] We must keep in mind that whatever standards are adopted, they must respect all affected parties. Viewing oneself as, potentially, both agent and recipient is required. This process certainly requires attempting to understand the perspectives of agents and recipients, and the Golden Rule provides the useful function of reminding us of this. Understanding these perspectives does not require us to find them acceptable, but at some point these perspectives can be evaluated in terms of the standard of respect for persons. Is the manager respecting the young engineer's professional autonomy when attempting to silence her? Understanding what the manager might be willing to accept if put in the position of the engineer does not necessarily answer this question.

The Self-Defeating Approach

The Golden Rule does not by itself provide all the criteria that must be met to satisfy the standard of respect for persons. But its requirements of universalizability and reversibility are vital steps in satisfying that standard. Next, we consider additional features of universalizability as they apply to the notion of respect for persons.

Another way of applying the fundamental idea of the universalizability principle is to ask whether I would be able to perform the action in question if everyone else performed the same action in the same or similar circumstances: If everyone else did what I am doing, would this undermine my own ability to do the same thing?[13] If I must say "yes" to this question, then I cannot approve others doing the same kind of thing I have done, and thus universalizing one's action would be *self-defeating*. To proceed anyway, treating myself as an exception to the rule is to pursue my own good at the expense of others. Thus, it fails to treat them with appropriate respect.

A universalized action can be self-defeating in either of two ways. First, sometimes the action itself cannot be performed if it is universalized. Suppose John borrows money, promising to pay it back at a certain time but having no intention of doing so. For this lying promise to work, the person to whom John makes the promise must believe that he will make good on his word. But if everyone borrowed money on the promise to return it and had no intention of keeping the promise, promises would not be taken seriously. No one would loan money on the basis of a promise. The very practice of promising would lose its point and cease to exist. Promising, as we understand it, would be impossible.

Second, sometimes the purpose I have in performing the action is undermined if everyone else does what I do, even if I can perform the action itself. If I cheat on an exam and everyone else cheats too, then their cheating does not prevent me from cheating. My purpose, however, may be defeated. If my purpose is to get better grades than other students, then it will be undermined if everyone else cheats because I will no longer have an advantage over them.

Consider an engineering example. Suppose engineer John decides to substitute an inferior and cheaper part in a product he is designing for one of his firm's large customers. He assumes that the customer will not check the product closely enough to detect the inferior part or will not have enough technical knowledge to know that the part is inferior. If everyone practiced this sort of deception and expected others to practice it as well, then customers would be far more inclined to have products carefully checked by experts before they were purchased. This would make it much less likely that John's deception would be successful.

It is important to realize that using the self-defeating criterion does not depend on everyone, or even anyone, actually telling promises without intending to keep them, cheating on exams, or substituting inferior and cheaper parts. The question is, What *if* everyone did this? This is a hypothetical question—not a prediction that others actually will act this way as a result of what someone else does.

As with other approaches, the self-defeating criterion also has limitations. Some unethical actions might avoid being morally self-defeating. Engineer Bill is by nature an aggressive person who genuinely loves a highly competitive, even brutal, business climate. He enjoys an atmosphere in which everyone attempts to cheat the other person and to get away with as much deception as they can, and he conducts his business in this way. If everyone follows his example, then his ability to be ruthless in a ruthless business is not undermined. His action is not self-defeating, even though most of us would consider his practice immoral.

Engineer Alex, who has no concern for preserving the environment, could design projects that were highly destructive to the environment without his action's being self-defeating. The fact that other engineers knew what Alex was doing and even designed environmentally destructive projects themselves would not keep him from doing so or destroy the goal he had in designing such projects, namely, to maximize his profit.

However, as with the Golden Rule, we need to remember that the universalizability principle functions to help us apply the respect for persons standard. If it can be argued that Bill's ruthlessness fails to respect others as persons, then it can hardly be universalized; in fact, Bill would have to approve of being disrespected by others (because, by the same standard, others could treat him with disrespect). Still, the idea of universalizability by itself does not generate the idea of respect for persons; it says only that if some persons are to be respected, then this must be extended to all. We turn to a consideration of rights to determine if this can give further support to the idea of respect for persons.

The Rights Approach

Many theorists in the respect for persons tradition have concluded that respecting the moral agency of others requires that we accord others the rights necessary to exercise their agency and to pursuing their well-being. A right may be understood as an entitlement to act or to have another individual act in a certain way. Minimally, rights serve as a protective barrier, shielding individuals from unjustified infringements of their moral agency by others. Beyond this, rights are sometimes asserted more positively as requiring the provision of food, clothing, and education. Here, we focus on rights as requiring only noninterference with another person, not active support of that person's interests.

When we think of rights as forming a protective barrier, they can be regarded as prohibiting certain infringements of our moral agency by others. Some jurists use the expression "penumbra of rights" to refer to this protective barrier that gives individuals immunity from interference from others. Thinking of rights in this way implies that for every right we have, others have corresponding duties of noninterference. So, for example, if Kelly has a right to life, others have a duty not to kill Kelly; Kelly's right to free speech implies others have a duty not to prevent Kelly from speaking freely; and so on.

Just what rights people have, and exactly what they require from others, can be controversial. However, the general underlying principle is that an individual should not be deprived of certain things if this deprivation interferes seriously with one's moral agency. If someone takes your life, then you cannot exercise your moral agency at all. If someone harms your body or your mental capacities, then that person has interfered with your capacity to act as a moral agent. In the case of some rights, interference with them is perhaps not wholly negating your moral agency, but it is diminishing your power to exercise it effectively.

One problem any account of rights must face is how to deal with conflicting rights. Suppose a plant manager wants to save money by emitting a pollutant from his plant that is carcinogenic. The manager, acting on behalf of the firm, has a right to free action and to use the plant (the firm's property) for the economic benefit of the firm. But the pollutant threatens the right to life of the surrounding inhabitants. Note that the pollutants do not directly and in every case kill surrounding inhabitants, but they do increase the risk of the inhabitants getting cancer. Therefore, we can say that the pollutant infringes on the right to life of the inhabitants rather than violates those rights. In a rights violation, one's ability to exercise a right in a certain situation is essentially wholly denied, whereas in a rights infringement, one's ability to exercise a right is only diminished. This diminishment can occur in one of two ways. First, sometimes the infringement is a potential violation of that right, as in the case of a pollutant that increases the chance of death. Second, sometimes the infringement is a partial violation, as when some, but not all, of a person's property is taken.

The problem of conflicting rights requires that we prioritize rights, giving greater importance to some than to others. A useful way of doing this is offered by philosopher Alan Gewirth.[14] He suggests a three-tiered hierarchy of rights, ranging from more basic to less basic. The first tier includes the most basic rights, the essential preconditions of action: life, physical integrity, and mental health. The second tier includes rights to maintain the level of purpose fulfillment an individual has already achieved. This category includes such rights as the right not to be deceived or cheated, the right to informed consent in medical practice and experimentation, the right not to have possessions stolen, the right not to be defamed, and the right not to suffer broken promises. The third tier includes those rights necessary to increase one's level of purpose fulfillment, including the right to try to acquire property.

Using this hierarchy, it would be wrong for the plant manager to attempt to save money by emitting a pollutant that is highly carcinogenic because the right to life is a first-tier right and the right to acquire and use property for one's benefit is a third-tier right. Sometimes, however, the hierarchy is more difficult to apply. How shall we balance a slight infringement of a first-tier right against a much more serious infringement or outright violation of a second-tier or third-tier right?

The hierarchy of rights provides no automatic answer to such questions. Nevertheless, it provides a framework for addressing them. We suggest a set of steps that could be taken:

1. Identify the basic obligations, values, and interests at stake, noting any conflicts.
2. Analyze the action or rule to determine what options are available and what rights are at stake.
3. Determine the audience of the action or rule (those whose rights would be affected).
4. Evaluate the seriousness of the rights infringements that would occur with each option, taking into account both the tier level of rights and the number of violations or infringements involved.
5. Make a choice that seems likely to produce the least serious rights infringements.

3.9 CHAPTER SUMMARY

Most of us agree about what is right or wrong in many particular situations, as well as over many moral rules or principles. Nevertheless, we are all familiar with moral disagreement, whether it occurs with respect to general rules or principles, or with respect to what should be done in a particular situation.

It is possible to isolate several sources of moral disagreement. We can disagree over the facts relevant to an ethical problem. If two people disagree over the relevant facts, then they may disagree as to what should be done in a particular situation, even thought they have the same basic moral beliefs. There can also be conceptual issues about the basic definitions of key ideas (e.g., "What is *bribery*?"). Finally, there can be application issues regarding whether a certain concept actually fits the case at hand (e.g., "Is *this* a case of bribery?"). These issues may pivot around the particular facts of the case, as well as how a concept should be defined.

Good moral thinking requires applying relevant facts (including laws and regulations), concepts, and the criteria of common morality to the case in question. Carefully organizing one's thinking around these requirements often yields straightforward moral conclusions. However, sometimes it causes us to rethink matters, especially when we discover that there are unknown facts that might affect our conclusions.

We have seen in this chapter that utilitarian and respect for persons approaches to moral problems sometimes assist us in framing moral problems. At the same time, we have been alerted to possible shortcomings of these approaches.

NOTES

1. This case is based on a much more extensive presentation by Tom L. Beauchamp, Joanne L. Jurmu, and Anna Pinodo. See "The OSHA-Benzene Case," in Tom L. Beauchamp, *Case Studies in Business, Society, and Ethics,* 2nd ed. (Englewood Cliffs, NJ: Prentice Hall, 1989), pp. 203–211.
2. 29 U.S.C. S655(b)(5).
3. Ibid.
4. *Industrial Union Department, AFL-CIO v. American Petroleum Institute, et al.,* 100 Sup. Ct. 2884 (1980).
5. Ibid.
6. W. D. Ross, *The Right and the Good* (Oxford: Oxford University Press, 1930), pp. 20–22.
7. Bernard Gert, *Common Morality: Deciding What to Do* (New York: Oxford University Press, 2004).
8. Ibid., p. 9.
9. Universalizability is widely discussed among moral philosophers. See, for example, Kurt Baier, *The Moral Point of View* (Ithaca, NY: Cornell University Press, 1958), Ch. 8; Marcus G. Singer, *Generalization in Ethics* (New York: Knopf, 1961), Ch. 2; and any of the writings of R. M. Hare.
10. These are by no means the only important traditions in ethics that might be usefully applied to practice. For a more comprehensive treatment of these and philosophical traditions in ethics, see C. E. Harris, *Applying Moral Theories,* 4th ed. (Belmont, CA: Wadsworth, 2002); James Rachels, *Elements of Morality,* 4th ed. (New York: Random House, 2003); and Hugh LaFollette, *The Practice of Ethics* (Oxford: Blackwell, 2007).
11. What if there are such rules and practices but we can think of other rules or practices that, if generally observed, would promote even greater utility? This might provide us with a good utilitarian reason for advocating changes in existing rules or practices, but it

would not necessarily justify treating these merely ideal rules or practices as our guide to action. This is because the utility of observing them depends on others doing likewise. In general, acting unilaterally is unlikely to bring about the desired changes; in fact, it might have the opposite effect.

12. For a defense of this possibility, see Marcus G. Singer, "Defense of the Golden Rule," in Marcus G. Singer, ed., *Morals and Values* (New York: Scribners, 1977).

13. This version of the universalizability criterion is suggested by Immanuel Kant. See his *Foundations of the Metaphysics of Morals, with Critical Essays* (Robert Paul Wolff, ed.) (Indianapolis: Bobbs-Merrill, 1969). For another exposition of it, see Harris, *Applying Moral Theories*, 4th ed.

14. Alan Gewirth, *Reason and Morality* (Chicago: University of Chicago Press, 1978), especially pp. 199–271 and 338–354.

CHAPTER FOUR

Resolving Problems

Main Ideas in this Chapter

- In analyzing a case, first identify the relevant facts and relevant ethical considerations.
- Ethical problems can be compared with design problems in engineering: There are better and worse solutions, even if we cannot determine the best solution.
- Line-drawing, comparing problematic cases with clear-cut cases (paradigms), sometimes helps in resolving unclear cases.
- In cases in which there are conflicting values, sometimes a creative middle way can be found that honors all of the relevant values to at least some extent.
- Utilitarian and respect for persons approaches sometimes can be used together to resolve ethical problems in ways that yield a creative middle way.
- However, sometimes difficult choices must be made in dealing with moral conflicts.

THIRTY-FOUR-YEAR-OLD STEVEN SEVERSON was in his last semester of his graduate program in mechanical engineering. Father of three small children, he was anxious to get his degree so that he could spend more time with his family. Going to school and holding down a full-time job not only kept him from his family but also shifted more parental responsibility to his wife Sarah than he believed was fair. But the end was in sight, and he could look forward both to a better job and to being a better father and husband.

Steven was following in the footsteps of his father, who received a graduate degree in mechanical engineering just months before tragically dying in an automobile accident. Sarah understood how important getting a graduate degree was to Steven, and she never complained about the long hours he spent studying. But she, too, was anxious for this chapter in their lives to end.

As part of his requirement to complete his graduate research and obtain his advanced degree, Steven was required to develop a research report. Most of the data strongly supported Steven's conclusions as well as prior conclusions developed by others. However, a few aspects of the data were at variance and not fully consistent with the conclusions contained in his report. Convinced of the soundness of his report and concerned that inclusion of the ambiguous data would detract from and distort the essential thrust of the report, Steven wondered if it would be all right to omit references to the ambiguous data.

4.1   INTRODUCTION

This chapter focuses on the task of ethical analysis with an eye on resolving ethical issues facing engineers. We begin with the fictional case of Steven Severson. It seems clear why Steven is tempted to omit references to the ambiguous data. He is understandably anxious to graduate and move on to other challenges in his professional life. He is worried that full disclosure of his findings could slow down this process, a process that has imposed a heavy burden on his family. However, his question is whether it would be *right* to omit reference to the data.

In Chapter 3, we suggested that the ethical analysis of a situation begin with two questions: What are the relevant facts? and What are the relevant kinds of ethical considerations that should be brought to bear on the situation? We also suggested that the first question cannot be answered independently of the second. Psychologically speaking, Steven is tempted, for evident reasons. Ethically speaking, should he do it? To answer this second question, we need to try to clarify what is at stake ethically, not just psychologically.

Although this case is about Steven's academic work rather than his work as a professional engineer, he is preparing for a career in engineering. Therefore, we might look at the National Society of Professional Engineers' (NSPE) Code of Ethics for Engineers for guidance. One of its fundamental canons states that in fulfilling their professional duties, engineers shall "avoid deceptive acts." Is omitting the ambiguous data deceptive? Steven might think it is not, because it is not his intention to deceive. Apparently he is still convinced of the overall soundness of his report. He does not want readers to be misled by the discrepant data. However, here a conceptual question needs to be raised. Can the omission of data be deceptive even when there is no intention to deceive?

In answering this question, we can look at another provision in the NSPE code. Under its rules of practice, provision 3 states,

> Engineers shall issue public statements only in an objective and truthful manner.
>
> a. Engineers shall be objective and truthful in professional reports, statements, or testimony. They shall include all relevant and pertinent information in such reports, statements, or testimony, which should bear the date when it was current.

Therefore, would Steven be objective if he omits the ambiguous data? Again, this might be his intent. But just as he worries that readers might be misled by the inclusion of the data, we might worry about Steven being misled by the psychological factors that tempt him to omit it. Can he be certain that he is not simply rationalizing? One thing is clear. If he keeps his examiners from seeing the ambiguous data, he is presuming that he is capable of making these sorts of determinations on his own. But, if he is right in concluding that the data is of no consequence, why should he fear that his examiners will be misled? Wouldn't they draw the same conclusions from his data that he does?

Common morality should remind Steven Severson of the importance of honesty. From this vantage point, his examiners can be seen as having a right to expect him not to distort his data. Misrepresentation of the data would be seen by them as a breach of the trust they place in students to do honest work and not interfere with their ability to assess his qualifications for an advanced degree.

Although the primary focus of this case is on the question of what Steven should do, how this question is answered has implications for other cases as well. If Steven is justified in leaving out data when he is convinced that it doesn't really discredit his conclusion, so are others who feel the same way about their research data. This is an application of the concept of universalizability. What would be the consequences of such a general practice? Notice that Steven cannot simply assume that his case is different because he believes he is right in interpreting his data, whereas others in similar situations cannot assume that they are correct. He should realize that the strong pressure he feels to finish his work successfully could compromise his judgment. Therefore, he is really not in a good position to determine this for himself. Subjective certainty in his own case is not a defensible criterion, and he should be wary of generalizing the use of this criterion to others who might be similarly tempted. A more sound position would be for him to concede that if he actually is right, a full presentation of the data should convince others as well. By withholding the data from his examiners, Steven seems to be saying that he is more capable than they are of assessing the significance of his data. Here, he might try a thought experiment: What would he think if the roles were reversed—if he were one of the examiners and he learned that one of his students omitted data in this way? This is an application of the concept of reversibility.

There is an additional concern. If Steven thinks he is justified in leaving out the data in this case, he might also think this will be acceptable in the workplace as well. There, the stakes will be much higher, risking not only economic costs to his employer but also product quality and possibly the health, safety, or welfare of the public. After all, it is possible that Steven has overlooked something important that others will notice if given the more complete set of data.

Steven may think that his is a special case. Given his family circumstances, the pressure to graduate is unusually great, and he may think that he would not repeat this behavior in the workplace. However, this seems to be more a rationalization of his action than a realistic assessment of the challenges that will face him as a practicing engineer—challenges such as meeting the pressure of deadlines.

At this point it should be noted that a great deal of the information provided in the Steven Severson case has been treated as irrelevant to our ethical analysis. In fact, despite their human interest, the first two paragraphs have no real bearing on the ethical question. Even though they explain why Steven is doing the research, and why he is anxious to bring it to a successful close, none of this seems relevant to the question of whether it is right to omit possibly important data from his report. No doubt there is also a great deal of irrelevant, unmentioned information, such as the size and color of the paper on which he prepared the report, whether or not he wears eyeglasses, how tall he is, what he ate for breakfast on the day he completed the report, and so on.

In short, to resolve an ethical question, we should focus only on those facts that are relevant to it. Sometimes this may be an easy task, and sometimes the facts make the resolution seem obvious. But in these cases, ethical criteria guide the sorting out of relevant from irrelevant facts. These criteria may come from our common morality, professional codes, or our personal morality. Hence, we should remind ourselves of all three.

From the standpoint of engineering codes of ethics, the case of Steven Severson seems to be quite straightforward. Actually, it is simply an embellishment of a fictional case prepared and discussed by the Board of Ethical Review (BER) of the NSPE.[1] The BER case consists of basically only the last paragraph of the Steven

Severson case; that is, the BER streamlined its presentation to include only relevant facts. In any actual case, however, much other information will have to be sifted through. In the original BER case, the presentation of the scenario is followed by several relevant provisions in NSPE's code of ethics. These provisions—calling for objectivity, truthfulness, and cooperative exchange of information—seem to settle the matter decisively. Steven should not omit the data.

In regard to Steven's personal morality, we can only speculate, of course. But it is quite possible that, as he reflects on his circumstance, he will realize that his personal integrity is on the line. Still, if he really is convinced of the overall soundness of his report, in omitting the data he would not be trying to convince his examiners of something he thinks is untrue or unsupportable. Thus, he may still value truthfulness. But he would be underestimating what it requires.

The ethical analysis of the Steven Severson case seems rather unproblematic. Sorting out factual, conceptual, and ethical issues is often straightforward enough that it is not difficult to resolve questions about what, from an ethical standpoint, one should do. This is not always the case, however. Fortunately, there are some ways of thinking that can help us in these more challenging cases. To illustrate this, we offer a brief account of the development of current federal guidelines for research involving human subjects, or participants. Then, we consider two useful methods of analysis: line-drawing and searching for a creative middle way.

4.2   RESEARCH INVOLVING HUMANS

The National Commission for the Protection of Human Subjects of Biomedical and Behavioral Research was established by the U.S. Congress in 1974. Its task was to develop ethical guidelines for research that makes use of human subjects, or participants. The commission was created in response to the public revelation of a number of research projects in which the treatment of human participants seemed ethically questionable. In 1978, the commission issued what is known as *The Belmont Report,* which contains the guidelines now used by institutional review boards (IRBs) at colleges, universities, and other institutions that receive federal funding for research involving human subjects. It is the responsibility of IRBs to examine research proposals to make certain that the rights and welfare of the participants are protected.

In setting up the commission, Congress selected a broadly representative group of members:

> The eleven commissioners had varied backgrounds and interests. They included men and women; blacks and whites; Catholics and Protestants, Jews, and atheists; medical scientists and behavioral psychologists; philosophers; lawyers; theologians; and public representatives. In all, five commissioners had scientific interests and six did not.[2]

The commission began by trying to "get to the bottom of things" in morality rather than simply employing common morality. However, as much we might want to secure the ultimate foundations of morality, we may find that attempting to do so actually makes it more difficult to identify what we have in common. Not surprisingly, this was experienced by the commission. Although philosophical and religious traditions throughout the world have long sought to articulate the ultimate foundations of morality, thus far no consensus has been reached. Still, it is worth noting that morality is not unusual in this regard. Specifying the ultimate philosophical foundations of virtually any discipline (e.g., mathematics, engineering, the sciences, history, and even philosophy) is highly controversial. Yet this only rarely interferes with a discipline successfully operating at less "foundational" levels.

Initially frustrated, the commission eventually decided to talk about specific examples rather than their more foundational concerns. They discussed many of the kinds of disturbing experiments that had caused Congress to convene the commission in the first place: the Tuskegee study of untreated syphilis, the injection of cancer cells into elderly persons without their knowledge or consent, experiments on children and prisoners, and so on.

Members of the commission found that they basically agreed on what was objectionable in these experiments. Eventually, they formulated a set of guidelines that emphasize three basic kinds of concern. One is a utilitarian concern for *beneficence,* which involves trying to maximize benefits and minimize harms to the participants. Insofar as they insist on acknowledging the moral status of each participant in an experiment, the other two can be placed under the idea of respect for persons discussed in Chapter 3. The commission's notion of *respect for persons* includes respect for autonomy by requiring the informed consent of participants in an experiment. Its notion of *justice* requires avoiding the use of discrimination in the selection of research participants, with special attention given to particularly vulnerable groups such as prisoners, children, and the elderly. Commissioners might have disagreed about the ultimate foundations of these general considerations, but they agreed that they are basic in addressing areas of concern in research involving humans.

Thus, despite their differences, the commissioners discovered that they had much in common morally, and they were able to put this to good use in formulating a national policy. At the same time, they realized that they had not developed a set of guidelines that eliminate the need for good judgment or that eliminate controversy:

> Three principles, or general prescriptive judgments, that are relevant to research involving human subjects are identified in this statement. Other principles may also be relevant. These three are comprehensive, however, and are stated at a level of generalization that should assist scientists, subjects, reviewers, and interested citizens to understand the ethical issues inherent in research involving human subjects. These principles cannot always be applied so as to resolve beyond dispute particular ethical problems. The objective is to provide an analytical framework that will guide the resolution of ethical problems arising from research involving human subjects.[3]

Insofar as it counsels both confidence and modesty in addressing ethical issues in research, *The Belmont Report* provides a model for deliberation in engineering ethics. There are no algorithms that can resolve ethical problems in engineering, but there are ample resources available for making good judgments.

4.3 ETHICS AND DESIGN

In many respects, the ethical problems of engineers are like the ethical problems facing moral agents in general: They call for decisions about what we should do, not simply reflection on what we or others have already done or failed to do. Of course, evaluating what has already happened can be helpful in deciding what to do. If I can see that the situation I am in now is very much like situations that I or others have faced in the past, evaluating what was done before (and what the consequences were) can be very helpful in deciding what to do now. If a situation was handled well, this can provide positive guidance for what to do now. If it was not handled well, this can serve as a lesson about what not to do.

As important as lessons from the past can be, they are limited. The present may resemble the past in important respects but not in every respect. The future may resemble the past in many respects, too; however, there is no guarantee that it will this time. We live in a complex world filled with change and uncertainty. Although it may not be difficult to determine that particular choices would be inappropriate, determining what is best from a moral point of view can be anything but clear. In fact, often it is quite possible that there is more than one available choice that can reasonably be made—and that others might reasonably decide differently than we would.

With regard to deciding what it is morally best to do, we might wish for a sure-fire method for determining the one best choice. But what if we cannot find such a method? Here is where a comparison with problems of engineering design can be helpful. Caroline Whitbeck notes,[4]

> For interesting or substantive engineering design problems, there is rarely, if ever, a uniquely correct solution or response, or indeed, any predetermined number of correct responses.

She illustrates this with a design problem regarding a travel seat for small children. The seat must be fitted onto a suitcase with wheels that can be taken on an airplane. It must be detachable so that it can be fitted onto the airplane seat or folded up and stored. In considering such a product, it would seem that there are many design possibilities that could adequately meet these requirements, in addition to having other useful features (e.g., storage places for bottles, pacifiers, or small toys). Ease of attaching the seat to the suitcase and removing it for separate use, the seat's weight, and its overall safety are obvious additional considerations. Some possible designs will clearly fail to meet minimal requirements for a good seat; but Whitbeck's main point is that although no design is likely to be "perfect," any number of designs might be quite good. Coming up with one that is quite good, although not necessarily the best imaginable, is a reasonable objective. Furthermore, among the possible designs actually being considered, there may be no "best" design.

Next, consider the challenge of developing a good design for safety belts to be worn by those who wash the windows of high-rise buildings. Window washers go up and down the sides of buildings on scaffolding, and they need both security and freedom of movement. While interviewing employees at a small firm whose main product is such a safety belt, one of the authors of this book was told that the chief design engineer sometimes worked weekends on his own time trying to improve the design of the company's belt. He did this even though the belt was more than adequately meeting the safety standards for such belts and it was selling very well. Asked why he kept working on the design, he replied, "People are still getting hurt and even dying." How does this happen? He explained that although high-rise window washers are required by law to wear safety belts when on the job, some take them off when no one is looking. They do this, he said, in order to gain more freedom of movement. The belt constrains them from raising or lowering the scaffolding as quickly as they would like.

Asked whether he thought that, at some point, responsibility for accidents falls on the workers, especially when they choose not to use a safety belt, the engineer agreed. But, he added, "You just do the best you can, and that's usually not good enough." Although not denying that the company's current belt was a good one, he was convinced that a better one is possible. Meanwhile, neither he nor his company was inclined to withdraw the current belt from the market until the company developed the best design imaginable.

As we will discuss in Chapter 7, "absolutely safe" is not an attainable engineering goal. Furthermore, safety, affordability, efficiency, and usability are different and often competing criteria for a good product. At some point, a safer car will not be affordable for most people. An even safer car (e.g., one whose engine cannot be started) will not be usable. These extremes will easily be excluded from serious consideration. However, combining factors that deserve serious consideration into a single, acceptable design is not an easy matter, and as Whitbeck observes, there may be no "uniquely correct solution or response" to this challenge.

Similar observations can be made about ethical problems. For example, in the following case, Brad is in the second year of his first full-time job after graduating from Engineering Tech.[5] He enjoys design, but he is becoming increasingly concerned that his work is not being adequately checked by more experienced engineers. He has been assigned to assist in the design of a number of projects that involve issues of public safety, such as schools and overhead walkways between buildings. He has already spoken to his supervisor, whose engineering competence he respects, and he has been told that more experienced engineers check his work. Later he discovers that his work is often not adequately checked. Instead, his drawings are stamped and passed on to the contractor. Sometimes the smaller projects he designs are under construction within a few weeks after the designs are completed.

At this point, Brad calls one of his former professors at Engineering Tech for advice. "I'm really scared that I'm going to make a mistake that will kill someone," Brad says. "I try to over-design, but the projects I'm being assigned to are becoming increasingly difficult. What should I do?" Brad's professor tells him that he cannot ethically continue on his present course because he is engaging in engineering work that surpasses his qualifications and may endanger the public. What should Brad do?

Brad's case illustrates one of the most common conflicts faced by engineers— one in which an engineer's obligations to an employer seem to conflict with obligations to the public. These dual obligations are stated in engineering codes. Canons 1 and 4 of the NSPE code illustrate this conflict:

Engineers, in the fulfillment of their professional duties, shall:

Canon 1: Hold paramount the safety, health, and welfare of the public in the performance of their professional duties.

Canon 4: Act in professional matters for each employer or client as faithful agents or trustees.

Although the obligation to the public is paramount, Brad should also honor his obligation to his employer if possible. A range of options are open to him:

1. Brad could go to his supervisor again and suggest in the most tactful way he can that he is uncomfortable about the fact that his designs are not being properly checked, pointing out that it is not in the firm's interests to produce designs that may be flawed.

2. He might talk to others in the organization with whom he has a good working relationship and ask them to help him persuade his supervisor that he (Brad) should be given more supervision.

3. He might tell his supervisor that he does not believe that he can continue to engage in design work that is beyond his abilities and experience and that he might have to consider changing jobs.

4. He could find another job and then, after his employment is secure, reveal the information to the state registration board for engineers or others who could stop the practice.

5. He could go to the press or his professional society and blow the whistle immediately.

6. He could simply find another job and keep the information about his employer's conduct to himself, allowing the practice to continue with another young engineer.

7. He could continue in his present course without protest.

To be ethically and professionally responsible, Brad should spend a considerable amount of time thinking about his options. He should attempt to find a course of action that honors both his obligation to protect the public and his obligation to his employer. It is also completely legitimate for Brad to try to protect and promote his own career, insofar as he can while still protecting the public.

With these guidelines in mind, we can see that the first option is probably the one he should try first. The second is also a good choice if the first one is ineffective. The third option is less desirable because it places him in a position of opposition to his employer, but he may have to choose it if the first two are unsuccessful. The fourth option produces a break in the relationship with his employer, but it does protect the public and Brad's career. The fifth also causes a break with his employer and threatens his career. The sixth and seventh are clearly unjustifiable because they do not protect the public.

There are, of course, still other options Brad can consider. The important point is that Brad should exercise his imagination to its fullest extent before he takes any action. He must "brainstorm" to find a number of possible solutions to his problem. Then he should attempt to rate the solutions and select from among those he finds best. Only after this fails is he justified in turning to less satisfactory options.

There is another important connection between ethics and engineering design. The Accreditation Board for Engineering and Technology (ABET 2000) directs that engineering students be exposed to design in ways that include consideration of ethical as well as economic, environmental, social, and political factors. In other words, students are to be encouraged to see that ethical considerations, too, are integral to the design process.

This can be seen in efforts by automobile manufacturers to address the problem of young children getting locked in car trunks. In response to a rash of trunk-related deaths of young children in the summer of 1998, General Motors (GM) sought a solution.[6] In addressing the problem, GM engineers engaged the assistance of a consulting psychologist and more than 100 children and their parents. The children participated in the research by trying to escape from enclosures made to resemble locked trunks that were equipped with different escape devices. The children were volunteered by their parents, who were paid a small sum of money for their children's participation. Researchers had to make the setting realistic for the children but not so frightening that psychological harm might result. Consent to participate was sought, but the children (ages 3–6 years) were not old enough to give fully informed consent. This was acquired from their parents. However, remuneration for the family could not be so great that parents might be willing to place their children at risk in ways contrary to the best interests of the children. Thus, given the pivotal role of children in the research, the experimental setting required considerable ethical sensitivity.

GM tested nine different types of trunk releases—a variety of handles, knobs, cords, and light switches. To the researchers' surprise, many of the children did not make effective use of the mechanisms that their designers thought would be readily available. Some children avoided glowing cords and handles because they worried that they were hot or otherwise dangerous. Light switches were sometimes associated with the idea of turning lights on or off rather than with escaping. Some easily gave up when the mechanism did not respond immediately. Some simply rested passively in the trunk, making no effort to escape.

In the end, the winner was an easily graspable handle with a lighting source that made it appear green rather than a "hotter" color. Even so, only 53 percent of the children between ages 3 and 6 years demonstrated an ability to escape by using the handle. Therefore, GM added a latch to the trunk lock that kept the lock from engaging unless manually reset. Resetting the lock required the finger strength of adults. However, some young children were still strong enough to lock themselves in the trunk. To solve this problem, GM introduced an infrared system that is sensitive to the motions and temperature of human bodies and that opens the trunk automatically if someone is trapped inside. Of course, this is not "perfect" either, because the similar motions and temperature of other objects could open the trunk as well.

The GM adjustments suggest another important point about engineering design that can complicate ethical decision making. Design changes are often made during the process of implementation; that is, design itself can be seen as a work in process rather than as a final plan that precedes and guides implementation.[7] This is illustrated in the fictional case study *An Incident in Morales,* which is a video developed by the National Institute for Engineering Ethics.[8] While implementing a design for a chemical plant in Mexico, the chief design engineer learns that his budget is being cut by 20%. To fall within the new budget, some design changes are necessary. Next, the engineer learns that the effluent from the plant will likely cause health problems for local residents. The current design is consistent with local standards, but it would be in violation of standards across the border in Texas. A possible solution is to line the evaporation ponds, an additional expense. Implementing this solution provides greater protection to the public; however, as it turns out, this comes at the expense of putting some workers at the plant at greater risk because of a money-saving switch to cheaper controls within the plant—another design change. Therefore, a basic question facing the engineer is, given the tight budgetary constraints, which standards of practice take priority? The moral of the story is that from the very outset of this project, the engineer failed to take sufficiently into account signs of trouble ahead, including warnings from senior engineers at another facility that taking certain shortcuts would be unwise (if not unethical).

Our brief discussion of design problems is intended to encourage readers to take a constructive attitude toward ethical problems in engineering. Design problems have better and worse solutions but perhaps no best solution. This is also true of ethical problems, including ethical problems in engineering design and practice. In Chapter 3, we discussed considerations that we should bear in mind when trying to frame the ethical dimensions of problems facing engineers. Bringing these considerations into play in an engineering context is challenging in ways that resemble the challenges of engineering design. In neither case should we expect "perfection," but some success in sorting out the better from the worse is a reasonable aim. To assist us in this sorting process, we next discuss two special strategies: line-drawing and seeking a creative middle way.

4.4   LINE-DRAWING

An appropriate metaphor for line-drawing is a surveyor deciding where to set the boundary between two pieces of property: We know the hill to the right belongs to Jones and the hill to the left belongs to Brown, but who owns this particular tree? Where, precisely, should we draw the line?

Consider the following example. The NSPE says about disclosure of business and trade secrets, "Engineers shall not disclose confidential information concerning the business affairs or technical processes of any present or former client or employer without his consent (III.4)."

Suppose Amanda signs an agreement with Company A (with no time limit) that obligates her not to reveal its trade secrets. Amanda later moves to Company B, where she finds a use for some ideas that she conceived while at Company A. She never developed the ideas into an industrial process at Company A, and Company B is not in competition with Company A, but she still wonders whether using those ideas at Company B is a violation of the agreement she had with Company A. She has an uneasy feeling that she is in a gray area and wonders where to draw the line between the legitimate and illegitimate use of knowledge. How should she proceed?

Although definitions of concepts are open-ended, this does not mean that every application of a concept is problematic. In fact, it is usually quite easy to find clear-cut, unproblematic instances. We can refer to these as *paradigm* cases. For example, here is a paradigm case of bribery: A vendor offers an engineer a large sum of money to get the engineer to recommend the vendor's product to the engineer's company. The engineer accepts the offer and then decides in favor of the vendor. The engineer accepts the offer for personal gain rather than because of the superior quality of the vendor's product (which actually is one of the worst in industry). Furthermore, the engineer's recommendation will be accepted by the company because only this engineer makes recommendations concerning this sort of product.

In this case, we can easily identify features that contribute heavily in favor of this being a clear-cut instance of bribery. Such features include gift size (large), timing (before the recommendation is made), reason (for personal gain), responsibility for decision (sole), product quality (poor), and product cost (highest in market) (Table 4.1).

The advantage of listing major features of clear-cut applications of a concept such as bribery is that these features can help us decide less clear-cut cases as well. Consider the following case, which we will call the *test case* (the case to be compared with clear-cut cases).

TABLE 4.1 Paradigm Case of Bribery

Features of Bribery	Paradigm Instances of Features of Bribery
Gift size	Large (>$10,000)
Timing	Before recommendation
Reason	Personal gain
Responsibility for decision	Sole
Product quality	Worst in industry
Product cost	Highest in market

Victor is an engineer at a large construction firm. It is his job to specify rivets for the construction of a large apartment building. After some research and testing, he decides to use ACME rivets for the job. On the day after Victor's order was made, an ACME representative visits him and gives him a voucher for an all-expense paid trip to the ACME Forum meeting in Jamaica. Paid expenses include day trips to the beach and the rum factories. If Victor accepts, has he been bribed?

As we examine the features identified in the first case, we can see similarities and differences. The gift is substantial because this is an expensive trip. The timing is after, rather than before this decision is made. However, this may not be the last time Victor will deal with ACME vendors. Therefore, we can worry about whether ACME is trying to influence Victor's future decisions. If Victor accepts the offer, is this for reasons of personal gain? Certainly he will have fun, but he might claim that he will also learn important things about ACME's products by attending the forum. Victor seems to be solely responsible for making the decision. Because Victor made his decision before receiving the voucher, we may think that he has made a good assessment of the product's quality and cost compared with those of competitors. However, we may wonder if his future judgments on such matters will be affected by acceptance of the voucher.

Although Victor's acceptance of the voucher might not constitute a paradigm instance of a bribery, Table 4.2 suggests that it comes close enough to the paradigmatic case to raise a real worry. In looking at the various features, it is important to bear in mind just what is worrisome about bribery. Basically, bribery offers incentives to persuade someone to violate his or her responsibilities—in this case, Victor's responsibility to exercise good judgment in behalf of his company. Here, the worry

TABLE 4.2 Line-Drawing Test of Concepts

Feature	Paradigm (Bribery)	Test Case	Paradigm (*Not* bribery)
Gift size	Large	——(X)———————	Small (<$1.00)
Timing	Before decision	———(X)————————	After decision
Reason	Personal gain	————————X————————	Educational
Responsibility	Sole	——(X)—————————	None
Product quality	Worst	————————————X—	Best
Product cost	Highest	—X—————————————	Lowest

is more about future decisions he might make rather than the one he has already made, but it is a real worry nevertheless. In any case, assessing the bribe requires more than determining where on the scale the various factors fall. The importance of each factor in particular cases must be weighted. Those two or three features that are judged most important in a particular case can be identified by drawing a circle around the appropriate X's. (For example, in Table 4.2, the X's for gift size, timing, and responsibility might be circled.)

So far, line-drawing has been applied to the analysis of concepts. It can be helpful both in clarifying the basic meanings of concepts and in their applications in particular circumstances. It can also be used to try to determine the rightness or wrongness of a course of action. Again, the notion of paradigms at opposite ends of a spectrum comes into play, with the action at one end being clearly right and the action at the other end being clearly wrong. The task is to determine whether the situation in question is more like the one in which the action is clearly right or more like the one in which the action is clearly wrong. We can also list features of these cases that make these diagnoses clear. These features can then be compared with the features of cases that fall between the two ends of the spectrum.

Cases that are uncontroversially wrong we shall call *negative paradigm* cases, and cases that are uncontroversially acceptable are *positive paradigm* cases. We shall call related, but controversial, cases that are in dispute (and that are clustered near the middle of the spectrum) *problematic* cases. We shall designate as the *test* case the one on which the analysis is to focus.

To illustrate, let us return to the case of Amanda wondering whether it is morally acceptable to use ideas at Company B that she developed while working at Company A. Because she feels she is in a gray area, it may be useful for her to compare her circumstance with a negative and positive paradigm in regard to taking one's ideas to a new place of employment. In determining what these paradigms might be, she should try to construct a list of key features that can be placed on a spectrum ranging from negative to positive. For example, violating a trade secret policy would be a negative feature, counting strongly against the appropriateness of taking her ideas to Company B. Acquiring permission from Company A would be a positive feature, counting strongly in favor of the appropriateness of taking her ideas to Company B. Schematically, Table 4.3 represents this part of Amanda's strategy.

A case dominated by negative features would be a negative paradigm, a clear instance of wrongdoing. A case dominated by positive features would be a positive paradigm, a clear instance of acceptable behavior. Amanda's situation is the test case. Once Amanda identifies the key features of her negative and positive paradigms, she can begin comparing the features of her situation with those of the paradigms. For example, a negative feature of her situation is that she signed a trade secret agreement that may include her ideas and apparently she has not sought permission from Company A to use her ideas at Company B. A positive feature is that Company A and Company B are not competitors.

As Amanda engages in this comparative analysis, she may find that she has not thought thoroughly enough about certain features. For example, she may not have thought much about the extent to which others at Company A might also have helped develop her ideas. Or, although she developed her ideas on her own time, she might realize that Company A's lab and equipment played a crucial role in their development. Or, although Company A and B were not competitors when

TABLE 4.3 Features of Paradigms

Negative Paradigm (Clearly wrong)	Positive Paradigm (Clearly acceptable)
Negative feature 1 (Vs. signed agreement)	Positive feature 1 (Permission granted)
Negative feature 2 (A and B competitors)	Positive feature 2 (A and B not competitors)
Negative feature 3 (Ideas jointly developed)	Positive feature 3 (Amanda's ideas only)
Negative feature 4 (All ideas developed on job)	Positive feature 4 (All ideas developed off job)
Negative feature 5 (Heavy use of A's lab/equipment)	Positive feature 5 (A's lab/equipment not used)
Negative feature n (Etc.)	Positive feature n (Etc.)

Amanda worked at A, they might become competitors in the area in which she developed her ideas, especially if those ideas were jointly developed with others at Company A. Table 4.4 represents some of these possible complexities.

At this point, although Amanda may feel she has a clearer understanding of her situation, she may still be unsure what to conclude. Some features of her case lean in the direction of features of the negative paradigm, whereas others lean in the direction of the positive paradigm. Furthermore, in this particular case some of the negative and positive features may be more important than others and should be more heavily weighted. Therefore, Amanda still has to assess the importance of the various negative and positive features she is considering. She may think of other possible scenarios that fall somewhere between the negative and positive paradigms, and she can compare the features of her case with those of the intermediate cases.

TABLE 4.4 Paradigm and Test Case Features

Negative Paradigm (Clearly wrong)	Test Case	Positive Paradigm (Clearly acceptable)
Negative feature 1 (Vs. signed agreement)	—X———————	Positive feature 1 (Permission granted)
Negative feature 2 (A and B competitors)	——————X—	Positive feature 2 (A and B not competitors)
Negative feature 3 (Ideas jointly developed)	—————X———	Positive feature 3 (Amanda's ideas only)
Negative feature 4 (Ideas developed on job)	—————X——	Positive feature 4 (Ideas developed off job)
Negative feature 5 (Used A's lab/equipment)	———X—————	Positive feature 5 (A's equipment not used)
Negative feature n (Etc.)	?-?-?-?-?-?-?-	Positive feature n (Etc.)

Although line-drawing techniques are often useful, we do not wish to underestimate the complexities that might be involved. Several general points need to be made. First, the more ambiguous the case, the more we must know about its particular circumstances in order to determine whether it is morally acceptable or morally wrong. In everyday affairs, determining whether failing to return money borrowed for a soda is wrong may be decided only by reference to the particular lender and his or her relationship to the borrower. Similarly, whether it is acceptable to use some ideas you developed at Company A for a completely different chemical process at Company B may be decided only by knowing the nature of the ideas and the policies of Company A and Company B. Also, whether to consider a payment of money as a bribe will depend on the amount and timing of the payment, the influence it exerts on the person who accepts the payment, the appearance and taking of the action, and other factors.

Second, imposing a line of demarcation between some of the cases in a series involves an element of arbitrariness. It is erroneous to conclude from this, however, that there is no real moral difference between *any* of the cases in a series. The precise line between night and day may be arbitrary, but this does not mean there is no difference between night and day. Nevertheless, sometimes arbitrary conventions to separate acceptable from wrong actions are in order. Companies—and in some cases professional societies—should have policies that, for example, specify in some detail just what kinds of transfers of proprietary information from one job to the other are legitimate. Despite the rules, however, there will be many instances in which we cannot avoid an exercise of judgment. And, of course, judgment is called for in making rules.

Third, in using the method of line-drawing it is important to keep in mind that concentrating on only one feature will usually be insufficient to determine where on the continuum to place a given case. The method of line-drawing is based on the identification of analogies and disanalogies between various examples in a series of cases. Unfortunately, we cannot depend on any single analogy or disanalogy to carry through all the examples.

Fourth, we need to bear in mind that the method of line-drawing resembles a kind of "common-law ethics" in which, as in law, what one decides in one case serves as a precedent for similar cases.[9] Thus, although one begins with the particular case and tries to determine relevant paradigms with which to compare and contrast it, eventually one links the case in question with relevant moral rules or principles, paying special attention to the importance of consistency—treating similar cases similarly. This is an application of the concept of universalizability.

4.5 CONFLICTING VALUES: CREATIVE MIDDLE WAY SOLUTIONS

We have already pointed out that values of common morality (e.g., being honest and preventing harm) can conflict with one another. There are situations in which two or more moral rules or duties seem to apply and in which they appear to imply different and incompatible moral judgments. This occurs often in engineering ethics, as in other areas.

When we take a closer look at such a situation, we may find that one value clearly has a higher priority than the other. From a moral standpoint, we then have what we can call an *easy choice*. Suppose you are driving along a freeway on your way to a dinner engagement. You have promised to meet a friend at 6 p.m. and are almost late. You see a person waving for help and realize there has been an accident. If you stop to assist, you will not be on time for your dinner. In a situation like this, you might well stop even though you have promised to meet your friend at 6 p.m. The need to render assistance has a higher priority than keeping the date on time.

Examples occur in engineering ethics also. James is an engineer in private practice. He is approached by a client who asks him to design a project that both know clearly involves illegal activity. Engineer Susan is asked to design a product that will require the use of outmoded technology that, although less expensive and still legal, poses substantially greater risk to human life. James and Susan should simply reject such requests out of hand, even though they could dramatically increase the profits of their firms. The obligations to obey the law and to protect human life so clearly outweigh any obligation to maximize profits that James and Susan should have no difficulty in deciding what is right to do.

In such situations, it may sometimes be difficult to *do* what is right, but it is not difficult to *know* what is right. We might not even want to refer to this as a serious moral conflict at all because the obligations involved have very different weights. In many real-life situations, however, the values are more evenly matched, and no hierarchy of values can give an easy answer. For example, the value of human life normally overrides other considerations, but this is often not the choice we face. Usually, the trade-off is between a slightly increased risk to human life as opposed to some other value. We make trade-offs like this all the time. Automobile manufacturers could make their products much safer if they could sell them for $100,000, but then few people could afford automobiles.

Sometimes we may be forced to make some *difficult choices*—choices in which we are not able to honor some real and important values in a way that we consider desirable. However, before concluding this, it is best to look for a creative middle way between conflicting values, a resolution in which all the conflicting demands are at least partially met. In many situations, all of the values make legitimate claims on us so that the ideal resolution of the conflict is to find some way to honor each of them. This approach might suggest new possibilities for Amanda's situation. After employing line-drawing techniques, Amanda may still be unsure whether it would be all right for her to make use of ideas she developed while working for Company A. She could explain her concerns to Company A and consider its response. If Company A does not object, Amanda has successfully resolved her problem. If Company A objects, Amanda has a strong indication that had she gone ahead without consulting Company A and Company A discovered this, she and Company B could have run into serious problems.

One of our students provided us with an example of a creative middle way solution to a moral challenge he faced as a co-op student. His supervisor did not have adequate funds to pay the student for his work on a particular project, but he had an excess of funds for another project. So the supervisor asked the student to fill out his time sheets stating that he had worked on a project that had excessive funding—even though the student had not worked on that project at all. The student really needed the money to continue his college education, and he knew his supervisor had a short temper and would probably fire him if he did not do as requested. However, the student also abhorred lying.

The student came up with the following creative middle way solution. He told his supervisor, "I know you don't have money budgeted from the project I worked on to pay me. But my conscience will not permit me to sign a false statement on my time sheet. How about if I just don't put in a time sheet for my work last week; and if you can, in the future please assign me to projects with budgets sufficient to pay me." His supervisor was so embarrassed and moved by this response that not only did he never again put the student in this kind of situation but also he paid the student's salary for the previous week out of his own pocket.

To take another example, suppose an engineer, John, is representing his company in a foreign country where bribery is common.[10] If John does not pay a bribe, valuable business opportunities may be lost. If he makes payments, he may be doing something illegal under the Foreign Corrupt Practices Act, or he may at the very least be violating his own conscience. Instead of yielding to either of these unattractive alternatives, one writer has proposed a "donation strategy," according to which donations are given to a community rather than to individuals. A corporation might construct a hospital or dig new wells. In the 1970s, for example, Coca-Cola hired hundreds of Egyptians to plant orange trees on thousands of acres of desert, creating more goodwill than it would have generated by giving bribes to individuals. In 1983, the British gained goodwill for themselves in Tanzania by assembling thousands of dollars worth of tools and vehicle parts. They also trained the Tanzanians to service the vehicles, enabling the Tanzanians to continue patrolling their wild game preserves, which they had almost stopped doing due to the weakened economy. This gift was given in place of a cash donation, which might well have been interpreted as a bribe. We can, of course, object to this solution. Not all creative middle ways are satisfactory, or at least equally satisfactory. We might argue that such gifts are still really bribes and are morally wrong. The evidence for this is that the effect of the gift is the same as the effect of an outright bribe: The person giving the gift gets the desired business contract. Furthermore, the motivation of the gift-giver is the same as the motivation of the briber—securing the business. There are also certain disanalogies, such as the gift-giving not being done in secret and its satisfying something more than the self-interest of an individual. We shall not attempt to resolve the problems raised by this solution, which depend heavily on the details of particular circumstances. We simply point out that it is an example of an attempted creative middle way solution (and that line-drawing techniques can be useful in bringing it to a final resolution).

4.6 CONVERGENCE, DIVERGENCE, AND CREATIVE MIDDLE WAYS

As noted previously, although utilitarian and respect for persons standards are different, they often lead to the same conclusions about what should be done in particular cases. This convergence strengthens those conclusions because more than one kind of basic reason supports those conclusions. However, as we have also seen, occasionally these standards seem to lead to conflicting conclusions. This divergence may leave us in some doubt about what should be done in those cases. Sometimes a creative middle way solution can be worked out that makes it unnecessary to make a difficult choice between the two standards. We offer the following case to illustrate this possibility.

In 1993, it was publicly revealed that Germany's Heidelberg University used more than 200 corpses, including those of eight children, in automobile crash tests.[11] This revelation drew immediate protests in Germany. Rudolph

Hammerschmidt, spokesperson for the Roman Catholic German Bishops' Conference objected, "Even the dead possess human dignity. This research should be done with mannequins." ADAC, Germany's largest automobile club, issued a statement saying, "In an age when experiments on animals are being put into question, such tests must be carried out on dummies and not on children's cadavers."

In reply, the university claimed that, in every case, relatives granted permission, as required by German law. It added that although it had used children in the past, this practice had been stopped in 1989. The rationale for using corpses is that data from crash tests are "vital for constructing more than 120 types of instrumented dummies, ranging in size from infants to adults, that can simulate dozens of human reactions in a crash." These data, it claimed, have been used to save many lives, including those of children.

Similar testing has also been conducted in the United States at Wayne State University's Bioengineering Center. Robert Wartner, a Wayne State spokesperson, indicated that this has been done as a part of a study by the federal government's Centers for Disease Control. However, he added, "Cadavers are used only when alternatives could not produce useful safety research."

Clarence Ditlow, head of the Center for Auto Safety, a Washington, DC, public advocacy group, said that the center advocates three criteria for using cadavers in crash testing: (1) assurance that the data sought by the tests cannot be gained from using dummies, (2) prior consent by the deceased person, and (3) informed consent of the family.

These three criteria for using cadavers in crash testing combine utilitarian and respect for persons concerns. Criterion 1 is essentially utilitarian. It implies that benefits (saving lives and reducing injuries) can result from the use of cadavers that are not obtainable from using dummies alone. Criteria 2 and 3 acknowledge the importance of respect for persons—both the deceased person and his or her family. If we focus only on adults, assuming that enough cadavers are available, then it seems that the consent requirement incurs no utilitarian loss. Criterion 2 rules out the use of the cadavers of children too young to have given their informed consent. This may come at some utilitarian expense because data on adults may not provide a reliable enough basis for determining how children fare in crashes. (An important illustration is the concern about the special vulnerability of small children in cars equipped with air bags.) However, another utilitarian consideration is the level of public concern about the use of children's cadavers.

Does this creative middle way solution satisfactorily resolve the issues? For most, it may. For others (e.g., those who would deny the right to volunteer one's own body), perhaps nothing short of a total cessation of the practice will suffice. However, from both utilitarian and respect for persons perspectives, it is not clear how the imposition of the protesters' desire for further restrictions can be justified. Without consent restrictions, the utilitarian and respect for persons standards seem to conflict. With consent restrictions, a high degree of convergence seems obtainable.

4.7 CHAPTER SUMMARY

As noted in Chapter 3, good moral thinking requires making careful use of relevant facts, concepts, and moral rules or principles. Often, this enables us to draw moral conclusions in a rather straightforward manner. However, sometimes it does not, and further reflection is needed. Here, is it useful to compare ethical problems with problems in engineering design. In seeking the best design in addressing an engineering problem, we can make a distinction between better and worse without necessarily knowing what is best. In fact, there may not be any uniquely best design. The same is true in resolving ethical problems.

Line-drawing techniques can be used in cases in which we are unsure how to distinguish between acceptable and unacceptable actions. By comparing problematic cases with those in which it is clear what we should do, we can often decide what we should do in the problematic cases.

Often, we face two or more conflicting morally important values. Sometimes, one value seems to be so much more important than the others that we must choose to honor the more important and, at least for the moment, neglect the others. Morally speaking, this is an easy choice. At other times, however, we may be able to come up with a creative middle way, a solution to the conflicting values that enables us to honor all of the relevant values. However, sometimes we must make difficult choices between competing values.

Utilitarian and respect for persons approaches can be combined in various ways with the methods for resolving line-drawing and conflict problems. The person who is skilled in moral thinking must determine which approaches to moral problem solving are the most appropriate in a given situation.

Often, the utilitarian and respect for persons approaches lead to the same conclusions. Because both approaches have initial plausibility, this convergence should strengthen our conviction that those conclusions are defensible, even though the two approaches proceed differently. Sometimes, however, these two approaches lead to different conclusions, and this divergence can lead to particularly difficult problems.

Several suggestions may aid in resolving divergence problems. First, when unfairness to individuals is minimal, utilitarian considerations may sometimes prevail. Second, in cases of divergence, it may be useful to employ line-drawing or creative middle way techniques. Third, when unfairness to individuals is serious, respect for persons considerations take on greater weight, and utilitarian considerations are more difficult to sustain.

However, it is not our task in this book to provide algorithms for determining which approach (if either) should prevail in any given case. Those who are skilled in ethical thinking will have their own views on how best to resolve problems of divergence. For better or worse, they must bear the responsibility of deciding for themselves.

NOTES

1. This is BER Case No. 85-5 in NSPE's *Opinions of the Board of Ethical Review,* Vol. VI (Alexandria, VA: National Society of Professional Engineers, 1989). The BER discussion is on pp. 67–69.
2. Albert R. Jonsen and Stephen Toulmin, *The Abuse of Casuistry: A History of Moral Reasoning* (Berkeley: University of California Press, 1988), p. 17.
3. *The Belmont Report: Ethical Principles and Guidelines for Protection of Human Subjects of Biomedical and Behavioral Research,* pub. No. OS-78-00012 (Washington, DC: U.S. Department of Health, Education, and Welfare, 1978), pp. 1–2.
4. Caroline Whitbeck, *Ethics in Engineering Practice and Research* (New York: Cambridge University Press, 1998), p. 57.

5. This case is suggested by the experience of a former engineering student at Texas A & M University.

6. This account is based on Catherine Strong, "GM Researches Methods for Kids to Escape Trunks," *Kalamazoo Gazette,* July 12, 1999.

7. For a detailed discussion, see Louis Bucciarelli, *Designing Engineers* (Cambridge, MA: MIT Press, 1994).

8. *An Incident at Morales: An Engineering Ethics Story,* developed and distributed by the National Institute for Engineering Ethics, the Murdough Center for Engineering Professionalism, and the Texas Tech University College of Engineering (2003). Further information is available at http://www.niee.org.

9. This point is made by David Boeyink in his "Casuistry: A Case-Based Method for Journalists," *Journal of Mass Media Ethics,* Summer 1992, pp. 112–113.

10. For this example, see Jeffrey A. Fadiman, "A Traveler's Guide to Gifts and Bribes," *Harvard Business Review,* July/August 1986, pp. 122–126, 130–136.

11. This account is based on Terrence Petty, "Use of Corpses in Auto-Crash Test Outrages Germans," *Time,* Dec. 6, 1993, p. 70.

C H A P T E R F I V E

The Social and Value Dimensions of Technology

Main Ideas in this Chapter

- Technology is embedded in a social context and both influences and is influenced by the larger society.
- Engineers should take a critical attitude toward technology, appreciating and taking pride in its benefits while being aware of the problems it can create.
- Computer technology illustrates the benefits that technology can confer as well as the social policy issues that one type of technology can raise.
- Engineering design often raises social and ethical issues that engineers must address and shows how engineering is a kind of social experimentation.

TROY, LISA, AND PAUL WERE engineering students at a prominent North American university. Troy and Lisa were graduate students, and Paul was an undergraduate.[1] They were chosen for a study of the attitudes of engineering students toward the social dimension of engineering because they had shown interest in topics such as the effect of technology on workers, especially with regard to occupational health and safety. Yet even these students had difficulties integrating such concerns into their engineering studies. Commenting on a class that focused on the humanistic aspects of technology, Troy remarked, "We've got enough to worry about [as engineers] and now we've got to worry about this."[2] On the final exam of a course on technology and society, he wrote to the teaching assistant, "My life was great until I met you."[3]

Commenting on the same topic, Lisa said,

> My engineering education didn't give me really a political context, and it sort of denied a political *context* by denoting that everything was objective and quantifiable and could be sort of foreseen. And if it couldn't be foreseen, we didn't measure it and we didn't take account of it.[4]

There was a difference in how these three students perceived the introduction of the social and humanistic elements into their thinking as engineers. Paul saw the socially oriented view as an extension of his engineering education. Lisa perceived a fundamental conflict between what she learned in engineering class and the social and humanistic orientation. Troy fluctuated between the views of Paul and Lisa.

Troy and Lisa spoke of a moment when the importance of the social and humanistic dimension of engineering "just hit me" (Troy) or "sort of clicked with me" (Lisa). Before that time, they could make no sense of what their teachers and project leaders were telling them about the social context of engineering. After that moment of insight, they had a foothold on the broader perspective.[5]

5.1 THINKING ABOUT TECHNOLOGY AND SOCIETY

Becoming a Socially Conscious Engineer

Despite these moments of insight, Troy evidently could never feel completely comfortable thinking about the social and value dimensions of his own profession. Maybe this is the reason he decided to take a job with a company in his university town that develops and sells computer software to other engineers. He evidently managed to convince himself—erroneously, we suspect—that he would not face any of the larger social and value issues in this job.

Troy's attitude, however, is increasingly out of step with the leaders of the engineering profession. In 2000, the Accreditation Board for Engineering and Technology (ABET), the organization that accredits engineering schools in the United States, formulated a new set of criteria that it would use in deciding whether to accredit the academic programs in engineering schools. Criterion 6 requires engineering students to have "an understanding of professional and ethical responsibility." Criterion 8 requires the same students to have "the broad education necessary to understand the impact of engineering solutions in a global and societal context." Criterion 10 requires students to have "a knowledge of contemporary issues." Assuming that engineering students are supposed to carry some of this knowledge beyond their college years, this suggests that ABET expects engineers in their professional work to take account of the social and value implications of technology.[6]

This call for greater attention to the broader social issues raised by technology comes from other sources in the engineering profession as well. In October 2000, William A. Wulf, then president of the National Academy of Engineering (NAE), gave a talk at the NAE annual meeting. In a revised version of the talk, he referred to the NAE's selection of the 20 greatest engineering achievements of the 20th century. The criterion for selection was

> *not* technical "gee whiz," but how much an achievement improved people's quality
> of life. The result is a testament to the power and promise of engineering to improve
> the quality of human life worldwide.[7]

Looking to the 21st century, Dr. Wulf believed that the greatest challenge may be *engineering ethics.* Specifically, he argued that the issues may not be as much problems for individual engineers (what he called "microethical" issues) but, rather, issues of social policy regarding technology (what he called "macroethical" issues). Although macroethical issues, he believed, must ultimately be decided by the larger society, engineers and the engineering profession must furnish information and guidance in making these decisions. One such example of a social issue given by Dr. Wulf is how to manage the Everglades. Here, we are addressing a complex entity in the field known as earth systems engineering, and we cannot know all of the consequences of our intervention in a natural system. No doubt we all can think of many other social policy issues involving technology.

We may also not agree that all issues involving the interface of technology, society, and values are "macroethical," or large social policy issues. Sometimes individual engineers face design issues that involve technology and society, as our engineering students did. However, there is considerable high-level support for the view that engineers should think about their work in the context of the society that supports it and is affected by it. The best way to begin this process is to think about the nature of technology.

What Is Technology?

The philosophy of technology is a subdiscipline in philosophy that comprises the results of philosophical reflection on the nature, effects, and evaluation of technology. One of these themes raises the following question: How should technology be defined? When we begin to think about a definition of technology, the first idea that comes to us is probably that technology is the making and using of "tools."[8] Humans have, in fact, often been called tool-making beings. So-called primitive humans made arrowheads, instruments of war, and, later, carts and other implements that we could call tools.

It is important to understand that this definition of technology suggests that humans are firmly in control of technology: We make tools to accomplish purposes that we set. Tools do not use us; we use tools. The technology-as-tools idea may also suggest that technology is neutral from the standpoint of social and value issues. Yes, humans can create weapons, but weapons can be used not only to commit aggression and promote tyrannical regimes but also to defend against aggression and to defend democracies. Yes, computers can be used to invade privacy and engage in identity theft, but they can also facilitate communication, store vast amounts of information, and accomplish many computing tasks that would not otherwise be possible.

A problem with this first definition, however, is that some technologies do not involve tools in any ordinary sense. The behavioral technology of B. F. Skinner used primarily verbal manipulation and rewards. Lewis Mumford, a well-known student of the history of technology, believed that the organization of large numbers of human beings for such projects as building pyramids, dams, and irrigation projects should be regarded as technology, even though the organization itself was not a tool.[9]

Another definition—one that some engineers may like—is that technology is the application of science to the solution of practical problems. Although this definition gives an important insight into the nature of modern technology, it is not applicable to all technology. The medieval Chinese had a highly developed technology, but they had neither the notion of laws of nature nor a concept of controlled experiment. Historically, most technology in the West did not derive directly from science in its modern sense either. The inventors of the 17th and 18th centuries were not usually well-versed in mathematical physics. Instead, they were practical people—tinkerers who found solutions to problems by intuition and trial and error. Thomas Edison did his creative work without knowing the electromagnetic theory of James Clark Maxwell. In fact, Edison thought physicists had little to contribute to technology.

Still another definition, favored by many contemporary scholars, is that technology is best understood as a "system" composed of physical objects and tools, knowledge, inventors, operators, repair people, managers, government regulators, and others.[10] Notice that this definition views technology as firmly embedded in a social network. It implies that we cannot understand a technology without under-standing the society of which it is a part, and it carries at least the suggestion that technology both influences and is influenced by the larger society. This is the defini-tion that we shall favor and that will be expanded in the rest of this chapter.

A second theme in the philosophy of technology that is relevant to our discus-sion is the controversy regarding what is usually called technological determinism. *Technological determinism* holds that technological development has a life of its own, an internal logic that cannot be controlled by individual humans or even the whole society. The steamship was developed from prior wind-driven vessels, and diesel-powered ships could not have been built without the steamship. Furthermore, according to technological determinism, a technology that can be developed usually will be developed, and a technology that can be put to some use almost always will be put to that use.

Technological determinism, if it were true, would have important implications for the argument over engineering responsibility for the social and value aspects of technology. If there is little individuals or even the society can do about the course of technological development, if technology is going to go on its merry way regardless of what we might do, why assume responsibility for it? Why take re-sponsibility for what we cannot control? Technological determinism, then, raises im-portant issues that we must consider.

A third theme in the philosophy of technology is the controversy between tech-nological optimism and technological pessimism. *Technological optimism* is the view that the effects of technology on human well-being are almost altogether good. Technology enables us to provide for our basic needs and even some luxuries, and it does so without our having to spend all of our waking hours merely trying to sur-vive. Even if technology does have some negative effects, such as pollution and harm to the environment, the overall effects of technology are overwhelmingly on the pos-itive side. *Technological pessimism,* on the other hand, takes a more negative view of the effects of technology on human life. Even though many technological pessimists say they do not want to be considered as simply against technology, they are much more likely to point out the undesirable aspects of technological development. In the next two sections, we discuss this controversy in more detail.

5.2 TECHNOLOGICAL OPTIMISM: THE PROMISE OF TECHNOLOGY

As previously noted, in February 2000, the NAE identified the 20 greatest engineer-ing achievements of the 20th century: electrification, the automobile, the airplane, water supply and distribution, electronics (vacuum tubes, transistors, etc.), radio and television, agricultural mechanization, computers, the telephone, air condition-ing and refrigeration, highways, spacecraft, the Internet, imaging (especially in med-icine), household appliances, health technologies, petroleum and petroleum technologies, laser and fiber optics, nuclear technologies, and high-performance materials.[11] These technologies have vastly improved the quality of our lives. Surely, technology is a gift to the millions of human beings whom it has liberated from lives of bone-grinding toil, poverty, starvation, and early death.

We could illustrate the case for technological optimism by examining the history of America and Europe since the industrial revolution, but it is more exciting to consider the example of a country in which technological development is taking place before our eyes. India is a good example.[12] India seems finally poised to experience the kind of explosive economic development necessary to lift its millions out of poverty. Rapid growth is essential if the country is to find jobs for the 70 million young people who will enter the labor force in the next 5 years and if the 260 million people who live on less than $1 a day are to escape their current condition.

Signs of development are everywhere. Approximately 5 million new mobile phone connections are added each month. Approximately 450 shopping malls are under construction. Perhaps one-third of India's prime office space has been built in the past 15 months. In India's technology capital, Bangalore, a hotel room can rent for $299 a day. Three of India's largest companies by stock market valuation are in an industry that barely existed in 1991—information technology. This is an industry that, according to some experts, can do for India what automobiles did for Japan and oil did for Saudi Arabia. A young software engineer who has graduated from one of India's elite institutes of technology is in high demand.

We are all familiar with India's telemarketing and call centers, but the country is also moving into areas that involve more judgment and higher levels of expertise. For example, there is a growing business in "litigation support" for American multinationals, such as DuPont, in which thousands of documents and e-mails are examined with regard to their relevance to a particular case. With their background in the English language and the common law tradition, Indians are particularly suited to legal outsourcing. India has also become a favored place for outsourcing clinical trials, a type of service that is projected to reach $1 billion by 2010.

Manufacturing is probably an essential ingredient in the rise from poverty, and India has some shining success stories. Tata Steel is the world's lowest cost steel producer. India also has outstanding capacity in concrete, pharmaceuticals, and automotive parts. Its strength is generally as a high-value rather than a low-cost producer.

India's story has been repeated in many other areas of the world. China is somewhat further along the road to development than India, and South Korea and Japan have already achieved industrialized status. No doubt technological growth will be accompanied by harm to the environment and social disruption, but insofar as we are considering the role of technological development in the liberation of millions of people from disease and poverty, most of us would probably consider the effects of technology to be overwhelmingly on the positive side.

5.3   TECHNOLOGICAL PESSIMISM: THE PERILS OF TECHNOLOGY

Technology and Human Experience

Technological pessimists want to enter a cautionary note to this unbridled optimism. The effects of technology, they argue, are not altogether good. Many ancient myths, such as the stories of the Tower of Babel, Prometheus, and Icarus, warn humans that technology carries risks. These risks include an excessive pride in human power, a failure to acknowledge dependence on providence, a tendency to forsake the pursuit of personal excellence in favor of enjoying the luxuries that technology brings, and a tendency to confuse the "unreal" world of material things with ultimate reality. In his poem *Milton*, William Blake (1757–1827) referred to the "dark and satanic mills" produced by the industrial revolution. Earlier, in *Paradise Lost,* John Milton (1608–1674) identified Satan with the technologies of mining and smelting, which are used to construct the city of Pandemonium, and he associated Satan's legions with engineering.[13]

We briefly examine two themes often sounded by technological pessimists. The first theme is that technology is associated with a dominating, controlling frame of mind, obsessed with achieving ever greater efficiency, that harms the environment and obscures certain aspects of human experience, especially the aesthetic and spiritual aspects. Nature is seen as a storehouse of resources to be exploited for human well-being. In the words of the German philosopher Martin Heidegger, an important critic of technology, technology leads us to view nature as a "standing reserve," as significant only insofar as it can be used for our purposes. This is in contrast to an attitude toward nature as deserving caring and reverence, much like the attitude we try to have toward other human beings, especially to those we love.

A major mechanism for efficient exploitation of nature is the quantification and standardization of the natural world and human experiences. The standardization of time provides an interesting example. Biological rhythms speed up and slow down. Our most intimate experience of time is similar. Sometimes (as in pleasant experiences) time "flies," whereas at other times (as in painful experiences) time drags on exceedingly slowly. Uniformity must be imposed upon, not read out of, our experience of time. Clocks—those paradigmatic technical artifacts—produce an artificial experience of time. Minutes and hours are of uniform duration. Units of clock time are not affected by our experience of the world, although they do affect our experience of the world. They can lead us to be insensitive to the way our life is actually experienced. Uniform measurement of time did not exist in the same way before the invention of the clock, although uniformity has always been loosely tied to astronomical events. Thinking of time in quantitative terms, as a collection of hours and minutes rather than as a sequence of experiences, is a modern invention. Prior to the 14th century, the day was divided into hours of light and darkness, generally of 12 hours each. This meant that in London the hour varied from 38 to 82 minutes.

Standardization of other units of measure shows a similar history. Witold Kula gives an example from late 18th-century France, where a *seteree* of land was larger or smaller depending on the quality of the soil.[14] Until the 19th century in Europe, many measures of commodities and objects varied, depending both on the local traditions and on the qualities of what was measured.[15] Measurements of space in terms of a "stone's throw" or a "day's walk" were common in earlier cultures, but they have generally been replaced by standardized and "objective" measures.

As technology evolves, nature changes from that which surrounds and encloses us to that which we surround and enclose. Consider the difference in our relationship to the natural world when we are driving on a highway in an air-conditioned car and when we are struggling to climb a mountain, backpacking in the wilderness, or camping by ourselves in a lonely place in the middle of the night. In the first type of experience, we feel almost literally "on top" of nature, fully in control. In the second type of experience, we feel surrounded by, and even overwhelmed by, the natural world. Yet many find this second type of experience of value as well as the first and consider it a vehicle for experiencing transcendence in other forms.

A second theme of technological pessimism is that technology tends to fragment human experience and thus destroy the meaningfulness of much that we do.[16] In former times, at least according to a commonly accepted idealized picture, the family gathered around the evening meal as a unit. A blessing was commonly offered, thereby recognizing a form of transcendence, problems and issues of the day were commonly discussed, and family bonds were renewed. Today, due to the microwave and pre-prepared foods, family members often eat by themselves and do so "on the run." This mode of living tends to fracture family bonds or, at the very least, not renew them in a meaningful way. Although it is true that nutrition can be supplied by a microwaved meal consumed in solitude, the family meal is much more than a simple means to nutrition. It gathers the family in meaningful activities that have value for their own sake, apart from supplying nutrition.

This same kind of fragmentation takes place in many other areas of modern life. Because of the automobile, individuals can pursue their own goals and purposes without taking account of the needs of family members or, for that matter, any others. Computers facilitate communication that not only does not involve an actual encounter of two people but also does not involve even the auditory encounter of the telephone. It seems that even communication is becoming increasingly abstract and impersonal. Given the brevity of e-mail encounters, with the emphasis on efficiency, modern communication does not even convey emotion and feeling as well as letter writing. We are all familiar with the difficulties we often have in conveying the attitude with which something is said so that we often put "smile" at the end of a sentence. In place of actual encounters with people who serve to be the focal point of our attention, we have "relationships" by way of e-mail, for example, that can be turned on and off at will. It is easy to pass from one encounter to another by e-mail, and in fact we can have several "conversations" at the same time.

In a well-known example among critics of technology, Albert Borgmann contrasts a fireplace with a modern furnace.[17] The fireplace, like the family meal, was a focal point for the family. The family gathered there for conversation and story-telling. Often, the mother built the fire, the children brought in the wood, and the father cut the wood. Contrast the hearth with the modern furnace, in which heat appears without effort and without any involvement of family members.

We can accuse Borgmann and other technology critics of romanticizing the past and praising forms of life that involved a great deal of drudgery, which often kept people from more creative and fulfilling activities that they might well have found more meaningful. Nevertheless, the critics of technology may have something important to say. Let us suppose that an important part of what we call "meaningfulness" in human life consists of a network of activities in which we relate to one another in terms of a common goal or set of activities. This need is at least a partial explanation of the existence of churches, clubs, political parties, and other social institutions. Insofar as technology eliminates some traditional focal things and focal activities, perhaps it should replace them with others.

Taking a Critical Attitude toward Technology

The controversy between technological pessimists and technological optimists is not a controversy in which we have to take sides. Most of us would probably want to say that both positions have some merit. Technology is an enormously liberating force for human beings, but it may obscure some aspects of reality, and it may have other negative consequences as well. The previous discussion did not raise the issue of environmental pollution and depletion of natural resources. The larger lesson, however, is that we should adopt a critical attitude toward technology. We should be conscious of its important and enormously valuable place in human society while being aware that its effects may not be always and altogether good. It is this critical attitude toward technology that we shall attempt to promote in the rest of this chapter.

5.4 COMPUTER TECHNOLOGY: PRIVACY AND SOCIAL POLICY

We can begin by examining an area in which technology, although undoubtedly bestowing some great benefits on society, raises some problems for social policy. The technology is computing, and the social policy issues have to do with privacy and the ownership of computing software.

Privacy and Boundary-Crossing

A good way to start thinking about the concept of privacy as it relates to computers is to think of a "privacy fence" that my neighbor and I might construct between our backyards. The fence keeps him from looking into my yard and me from looking into his. We might refer to this kind of privacy as *informational privacy* because it keeps me from gaining the information about my neighbor that I might obtain by looking into his backyard and him from gaining similar information about me. Informational privacy can be violated when others obtain information about us that we would prefer that they did not have. Computing technology can violate our informational privacy by serving as the means for the construction of databases about our income, purchasing habits, and perhaps even more intimate characteristics, such as political and religious affiliations and sexual orientation.

A privacy fence also serves another purpose. It keeps my dog from running into my neighbor's yard and his dog from running into mine, and it also helps to keep the sounds produced in his yard (e.g., the sounds of his children shouting or his dog barking) from coming into mine and the sounds produced in my yard from entering his. Let us call this *physical privacy* because it involves the movement of something physical (dogs or sound waves) across the line between my property and his. Physical privacy can be violated by an invasion of my (or his) physical space. The most obvious examples of violations of physical privacy are the invasions of bodily integrity such as unwanted touching or rape. Unwanted telephone solicitations are also a form of invasion of physical privacy. For some, the sending of unwanted cookies or spam is also an invasion of physical privacy.

Two other conceptions of privacy have less importance for computing technology but are sometimes relevant. *Decisional privacy* is freedom from governmental or other outside interference to make decisions about such issues as one's political and religious beliefs and practices or general lifestyle. The use of computers to monitor the activities of individuals can sometimes be used to intimidate them, by the threat of being exposed, into repressing their true beliefs. *Proprietary privacy* is the ability to control the use of one's name, likeness, or other aspects of one's identity. Computers assist the violation of proprietary privacy when they are used in "identity theft."[18]

One way in which computers are involved in the invasion of privacy is in computer "matching." Apparently unrelated information from several different sources is put in a single data bank. For example, a person's credit record can be combined with his or her employment record and criminal or traffic violation record, and perhaps with various other records, to form a composite picture. Defenders of computer matching argue that the merged files do not contain any "new" information but, rather, information that is already available, some of it of public record. Computer matching, however, may be an example of the whole being greater than the sum of its parts. For example, there may be a record of extensive purchases of liquor, combined with a record of arrests for driving while intoxicated, and dismissal from several jobs for excessive absences. Such a record provides strong evidence that the person has a problem with alcohol.

Privacy versus Social Utility

An understanding of the theories of utilitarianism and respect for persons can help us anticipate and in fact construct most of the major arguments regarding social policy that have been put forth with regard to computers and privacy. For example, some of the strongest arguments for limiting the ability of others to cross the boundaries protecting our privacy come from the tradition of respect for persons. One writer has said that "our interest in privacy ... is related to our concern over our accessibility to others: the extent to which others have physical access to us, and the extent to which we are the subject of others' attention."[19] Here, informational privacy and physical privacy are central. James Rachels has argued for informational privacy by maintaining that people need to control information about themselves in order to control the intimacy of their relationships with others.[20] We do not ordinarily have the same degree of intimacy with a store clerk as we have with our accountant or physician, and we do not ordinarily have the same degree of intimacy with our accountant or physician that we have with our spouse.

Computers can take control of information about us and give it to others, or they can simply make information about us available to almost everyone indiscriminately. Although this information can rarely, if ever, reproduce the intimacy we have with those with whom we are closest, such as our spouses, it could give relative strangers the kind of information we would want only our best friends to have. At the age of 40 years, we might not want our employer to know that we got a traffic ticket for driving while intoxicated when we were 20 years old. We may or may not want our employer or an insurance company to know our sexual orientation, propensity to engage in dangerous pastimes, views about the environment, or religious affiliation.

Respecting the moral agency of individuals also requires that their political rights be respected, including the right to be presumed innocent until proven guilty. Some people believe that computer matching violates this right. One of the purposes of matching is to identify individuals who might be guilty of wrongdoing before there is any evidence against them, even though the constitutional system of the United States forbids investigating citizens who are not otherwise suspected of crimes. Furthermore, people suspected of wrongdoing because of a computer match do not have an adequate opportunity to contest the results of the match. In one example, matching federal employees with welfare rolls turned up many "hits." Those who were identified were supposedly receiving welfare payments while still being employed by the federal government and thus were presumed guilty of a serious crime. Those identified, however, were not immediately informed of the match because the entire procedure was designed as an attempt to detect criminal violations, and prosecutors wanted to wait until a solid case against the supposed offenders was built.[21]

There are, however, several utilitarian arguments that point to the values involved in collecting information. We can use credit cards because there are credit records that distinguish good from bad credit risks. We can prevent the sale of handguns to convicted felons because computerized criminal records are easily accessible. We can cash checks because computerized records enable retailers to have information about checking accounts. Databases allow targeted marketing, which not only is more efficient for businesses but also keeps people from being subject to irrelevant advertising. In addition to commercial usages, governmental agencies also find computer databases useful. The proposed National Information System would be useful in eliminating welfare fraud and in identifying physicians who double-bill for Medicare and Medicaid services. The FBI has proposed a national computerized criminal history system that would combine in a single database the 195 million criminal history records in the United States. The system would no doubt be of great benefit to the criminal justice system.

Finding a Creative Middle Way

The issue of computers and privacy presents a conflict of values. The abilities of computers to collect and match data can provide significant benefits to the public. Few of us, for example, would want to do without credit cards, and yet it would be unreasonable to issue credit cards without being able to perform credit checks. Most of us would admit many other legitimate uses of databases. Furthermore, advocates of the creation of computerized databases can point out that the public may not be as concerned about the invasion of privacy as some critics suppose. Many people enjoy television programs in which the daily lives of ordinary people are followed in detail, and "confessional" programs and magazines have great popularity. Many people are willing to hold intimate conversations on cell phones within earshot of others.

All of these examples, however, involve the consent of those whose privacy is invaded. Even those who use cell phones in public presumably could seek more privacy if it was important to them. People most typically become disturbed when their privacy is invaded without their consent. So the uses of computerized databases has the clear potential for violating important values, especially those values associated with respecting the moral agency of individuals. The essence of the concept of privacy is that moral agents are—or should be—protected by a "privacy fence" that should be reasonably impervious to penetration without the consent of the moral agent.

Is it possible to find a creative middle way in public policy that will do justice to the conflicting values on this issue? How do we combine the utility of having computer databases with an adequate protection of privacy? One of the earliest and still most important attempts to limit the creation of large computerized databases was the Privacy Act passed by the U.S. Congress in 1974, which prohibited the executive branch from using information gathered in one program area in an entirely different and unrelated area. Operating with the assumption that one way to limit power is to divide it, the act prohibited the creation of integrated national information systems. Enforcement was left to the agencies, however, and the agencies have interpreted the act to suit their own purposes. As a result, the act has lost much of its force.

An approach that mirrors some of the considerations in the Privacy Act and that appears to be an attempt to find a creative middle way solution is a set of guidelines for "fair information practices." These guidelines include the following provisions:[22]

1. The existence of data systems containing personal information should be public knowledge.
2. Personal information should be collected for narrow, specific purposes and used only in ways that are similar to and consistent with the primary purposes for its collection.
3. Personal information should be collected only with the informed consent of the persons about whom the information is collected or their legal representatives.
4. Personal information should not be shared with third parties without notice or consent of those about whom the information is collected.
5. To ensure accuracy, the time information can be stored should be limited, and individuals should be permitted to review the information and correct errors.
6. Those who collect personal date should ensure the security and integrity of personal data systems.

As with any creative middle way solution, competing values should be honored as fully as possible. In this case, that means that the solution should both protect personal privacy and promote the social goods to which computer databases can contribute. Critics might argue that some of the provisions of this solution concede too much to the demands of individual privacy. It would be difficult, for example, to secure the informed consent of everyone from whom the information is collected. Because consent must be obtained before information may be shared with third parties (including, presumably, other databases), repeated requests to share the same information could become a nuisance. Furthermore, the requirement that information about an individual should be limited in duration means that some of the same information would have to be collected repeatedly, presumably with the consent of the person from whom the information was derived. Critics could also point out that the provision that gives individuals the right to "correct errors" in the information about them is subject to abuse. Some individuals would probably want to change information about them that does not reflect well on them, even if it is correct. Safeguards against this kind of abuse would have to be established. Nevertheless, these guidelines suggest the possible shape of a creative middle way solution to the problem of computers and privacy.

This issue shows how technology raises issues of social policy that the larger society and its legal system must resolve. Computer engineers cannot presume that their work is without social implications. It does not follow, of course, that engineers should not develop computer technology, but it does follow that they should at the very least be aware that what they do affects the larger society. They may also be called upon to contribute to political and legal decisions in these areas.

5.5 COMPUTER TECHNOLOGY: OWNERSHIP OF COMPUTER SOFTWARE AND PUBLIC POLICY

On June 28, 1990, an important decision was rendered with regard to a lawsuit between Lotus Development Corporation, the creator of the Lotus 1-2-3 spreadsheet, and Paperback International, the creator of the VP-Planner spreadsheet. Lotus had sued Paperback International for infringement of its copyright on Lotus 1-2-3. Paperback had copied the entire menu structure of the Lotus 1-2-3 program. The manual of the VP-Planner even contained the following statement:

> VP-Planner is designed to work like Lotus, 1-2-3, keystroke for keystroke .... VP-Planner's worksheet is a feature-for-feature workalike for the 1-2-3. It does micros. It has the same command tree. It allows the same kind of calculations, the same kind of numerical information. Everything 1-2-3 does, VP-Planner does.[23]

Paperback, in turn, alleged that only the part of a computer program written in some computer language, such as C, is copyrightable. It argued that the more graphic parts of a program, such as the overall organization of the program, the structure of the program's command system, the menu, and the general presentation of information on the screen, are not copyrightable. Lotus countered by arguing that copyright protection extends to all elements of a computer program that embody original expression.

The judge ruled that even though the idea of an electronic spreadsheet is not copyrightable, the Lotus spreadsheet was original and nonobvious enough to be copyrightable, and that VP-planner had infringed on the copyright. Accordingly, District Judge Keaton ruled that VP-Planner had infringed on the Lotus copyright.

Should Software Be Protected?

Computer programs are often worth a lot of money in the marketplace. Should they receive legal protection? Let us begin by considering some conceptual issues related to ownership and intellectual property.

What does it mean to own something—in particular, a piece of software? A central idea is that if I own something, I may control its use and (as a consequence) exclude others from its use if I wish. Because I own my car, I may control its use and exclude you from using it if I wish. If I want to make money from your use of my car, then I have the right to charge you a fee for its use.

What is the justification for granting similar discretionary powers to an owner of software? One justification—a utilitarian one—is that it promotes the progress of technology. The U.S. Constitution gives a utilitarian argument of this type when it authorizes Congress "to promote the progress of science and the useful arts, by securing for limited times to authors and inventors the exclusive right to their respective writings and discoveries." Some people believe that, in fact, rates of technological innovation have been greatest in those social systems that recognize intellectual property rights.

It is worth pointing out that there is also a utilitarian argument for *not* granting legal protection to the creators of software. Some people have maintained that there was more innovation and experimentation in the early days of software development when software was free.[24] Between this utilitarian argument and the one given previously, there is a factual disagreement on what policy most effectively promotes the growth of technology. There is also a utilitarian argument of a different kind against granting legal protection to software. Granting legal protection tends to increase the price and may reduce the quality of software because competition is limited or reduced.

Another type of justification for giving legal protection to software owners is based on the ethics of respect for persons. We might call it the *labor theory of ownership*. This view starts with the idea that the first and most important thing we own is our own body. We can, however, "mix" the labor of our bodies with other things in nature that are unowned or purchased from the former owners and thereby establish our ownership of them as well. For example, if a pioneer clears unowned land and plants crops on it, he has mixed his labor with the land and can claim ownership of it. Similarly, if a person begins with basic ideas of science and logic that are common property and thus in a sense unowned, and then adds to them her intellectual labor so that she produces a new computer program, then she may lay claim to ownership of that program. She has mixed her labor with unowned property to produce the program and therefore may legitimately claim to "own" the program. In a slight modification of this story, if she mixes her intellectual labor with the intellectual property of others that she has the right to use (perhaps by paying a fee to the owner), she may claim ownership of the new product.

Both the utilitarian arguments and the respect for persons arguments have considerable moral force. Because of its basis in the Constitution, the utilitarian argument is prominent in legal controversies. Because of its great intuitive appeal, however, the argument based on the labor theory of ownership, which is based on the perspective of respect for persons, has an important place in our thinking about ownership. Probably most of us would combine the two arguments. Given the relative ease of free riding on the work of others and the considerable expense and effort involved in creative innovation, most of us would probably conclude that the creators of software deserve protection for several reasons. First, technological innovation has considerable value to society, and innovation is probably stimulated when innovators know that they can claim ownership of their work and require compensation for its use. Second, if we believe that owners generally have the right to exclude others from the use of their own creations, except on terms that they specify, then it would seem that this right should extend to software. The question now arises as to how this protection should be implemented.

How Should Software Be Protected?

Two principal options have been proposed for protecting intellectual property: copyrights and patents. The peculiar nature of software, however, makes both of these options problematic. Software does not fit the paradigm or typical case of either something that should be copyrighted or something that should be patented. In some ways, software is like a "work of authorship" and should be appropriate for copyright. A program is, after all, written in a "language" and has a logical sequence like a story or a play. In other ways, software is more analogous to an invention because it is a list of ways to react to certain conditions, much like a machine might react to certain conditions. Because of these problems of classification, some people have suggested that software should be classified as a "legal hybrid" and that special laws should be made for the protection of software that are different from the laws applicable to either copyright or patent.[25] Just as there should be special laws to protect some products of biotechnology, such as plant hybrids, so should there be special laws to protect software.

However, there are disadvantages to creating special laws just to protect software, one of which is that such laws in the United States might not be recognized throughout the rest of the world. So the legal hybrid approach has not gained wide acceptance, and we must look at what might well be called the copyright–patent controversy. Because software partakes of characteristics that are appropriate to copyrightable material and characteristics that are appropriate to patentable material, there is a line-drawing problem that involves an application issue. Does software fit more closely the paradigm for a patentable creation or for a copyrightable creation? We might begin by examining copyrights more closely, which have been the most popular form of protection for software.

Software is appropriate for copyright only if we view programs as literary works. However, a central tenet of U.S. law is that copyright can protect only the expression of an idea, not the idea itself. Copyright law holds that a basic idea is not copyrightable, but a particular expression of an idea might be copyrightable. One cannot copyright an idea for a novel in which boy meets girl, they fall in love, and they live happily thereafter. An author can only copyright a particular story embodying this idea, which is written or "expressed" in considerable detail. The author must describe the background of the boy and girl, the circumstances of their meeting, the events that led to their engagement and marriage, and the reasons why their lives were filled with bliss. To refer to an actual case, the idea of a jewel-encrusted lifelike bee pin cannot be copyrighted because this idea is inseparable from its expression. The expression may not be copyrighted because protecting the expression would be conferring a monopoly on the use of the idea.[26]

In determining whether an expression is, in fact, copyrightable, the courts use several tests. First, the expression must be original—that is, it originated with the author. Second, the expression must be functional in the sense of having some usefulness. Third, the expression must be nonobvious: "When an expression goes no further than the obvious, it is inseparable from the idea itself."[27] Fourth, there must be several or many different ways of expressing the idea. If there are no other ways—or few other ways—of expressing it, then the software creator cannot lay a significant claim to originality in the sense of uniqueness.

Let us assume that Lotus 1-2-3 is distinct enough from the basic idea of a spreadsheet to be classified as an expression of an idea. Is it a copyrightable expression? Using the method of line-drawing, Table 5.1 gives an analysis of the issue.

TABLE 5.1 Is Lotus 1-2-3 Copyrightable?

Feature	Copyrightable	Lotus 1-2-3	Noncopyrightable
Originated with author	Yes	X——————	No
Functional	Yes	X——————	No
Nonobvious	Yes	X——————	No
Alternate expressions	Yes	X——————	No

According to this analysis, because the X's are all on the left side of the spectrum, Lotus 1-2-3 is copyrightable, as the judge in the case concluded. To establish the additional claim that VP-Planner infringed on Lotus's copyright, we must show that VP-Planner has simply appropriated the original expression of the idea of a spreadsheet that Lotus developed. As we have seen, the judge concluded that it did.

How far can software depart from the paradigm of copyrightable material and still hold a valid copyright? In paradigmatic copyrightable material such as a literary work, the material is expressed in a language that can be read. In an important legal case, however, the action of the court implied that a program is copyrightable even though it cannot be read. *Apple v. Franklin* involved a company (Franklin) that sold computers that included copies of a program used to run Apple computers. It justified its practice by claiming that Apple's copyrights were invalid because the code was unreadable and therefore not a proper subject for copyright. The court was so hostile to Franklin that Franklin gave up, conveying the message to software producers that one can obtain a copyright on an unreadable, executable computer program. The court also determined that copyright could cover both source code (the higher level language) and the object code (machine language).[28]

As the previous case indicates, copyright protection can be expanded beyond its usual boundary to cover software, but there are limits. In particular, copyright does not cover algorithms, which are often the most creative parts of computer software. If we assume that the products of creativity should be protected, then there is reason to look at patents that might cover the algorithm itself. To obtain a patent on a program, the author must show that the program is (1) useful, (2) novel, (3) nonobvious, and (4) the type of thing that is generally accorded patents. This last category contains the following types of items: "processes" (such as a way to cure rubber), "machines" (such as a hearing aid), and "manufactures" or "compositions of matter" (such as a new composite for aircraft). As is often the case, computer programs do not fit neatly into any of these categories. Programs have usually been considered a process, but a process must change or transform something. This raises the question of what is changed by a program. Is it the data or the internal structure of a computer?

Another problem is specifying the subject matter of the patent. The U.S. Supreme Court has held that mathematical algorithms may not be patented.[29] Because mathematical algorithms may not be patented, the patent applicant must show (1) that no mathematical algorithm is involved or (2) that the algorithm does not form the basis of the patent claim. Applications of mathematical algorithms, however, may be patented. Unfortunately, it is not always easy to draw the line between these two situations. For example, in one important case, *In re Iwahashi* (1989), a patent was granted over a computerized method for estimating solutions to a computerized mathematical formula.[30]

As we have seen, there are strong moral arguments in favor of protecting software, and both copyright and patent protection are necessary in order to do this. Both types of protection, therefore, are morally justified. Which is more effective and feasible depends on the nature of the software, as well as other circumstances. Here again, technology raises moral issues involving public policy that our society must debate and resolve.

5.6 ENGINEERING RESPONSIBILITY IN DEMOCRATIC DELIBERATION ON TECHNOLOGY POLICY

The issues of privacy and ownership of software raised by computer technology are only a few of the questions for social policy that technology poses for the public. There are many more. Is the space program worthwhile? What about the supercollider, as opposed to a very expensive optical or radio telescope? How do these compare with nanotechnology, with its promise of providing new products, replacing dependence on foreign imports, and minimizing the effect of technology on the environment?

There are also more abstract and general questions. What criteria should govern science and technology policy? Research that violates fundamental moral norms of course should not be allowed. The infamous Tuskegee experiments, in which black men were allowed to go untreated for syphilis in order to determine the long-term effects of the disease, were morally unacceptable. But a technological project that is morally acceptable might still not be significant enough to warrant large public expenditures. Also, there are questions of justice. What should be done about the fact that science and technology-based economic growth often seems to be accompanied by increasing inequality in the distribution of economic benefits?

In a democracy, debates about public policy regarding science and technology encounter a dilemma. Let us call this the *democratic dilemma*. On the one hand, the public has the prerogative of making the final decisions about science and technology. On the other hand, this same public has difficulty in understanding something as complex and forbidding as science and technology, and the simplifications necessary to get understandable information across to the public may involve serious distortions. This conflict between democracy and scientific and technological elitism is present in Vannevar Bush's *Science—The Endless Frontier,* which is commonly acclaimed as the most important single document in science policy.[31] The document focuses on science policy rather than technology policy, but the issues are fundamentally the same. In this document, Bush discusses a variation on essentially the same dilemma. For Bush, the problem is a conflict between the desire of scientists to appeal for public support by showing the benefits that science can confer on society and the equally strong desire of scientists to protect science from interference by citizens whose understanding of science is minimal.

What are the responsibilities of engineers with regard to the democratic dilemma? We believe the responsibilities can be summarized in three words: alert, inform, and advise. First, as the primary creators of technology, engineers have a special responsibility to alert the public to issues raised by technology. In particular, engineers have a special responsibility to alert to potential dangers from technology. In the case of the issues of privacy and ownership raised in the previous section, the public and especially the business community are already aware of the problems without any alarm having to be sounded by engineers. For many other problems, however, this is not the case. Short of being alerted to the problem by experts, the public may not know the dangers of a new automobile design or the hazards to the environment imposed by a new chemical process. This responsibility may sometimes involve whistleblowing, but an engineer should always first try to work through organizational means to alert the public. We consider this issue further in a later chapter. Second, engineers also have a responsibility to inform the public of the issues on both sides of a debate. A new technology may pose dangers, but it may also have great potential benefits. Apart from tutoring by experts, the public has little chance of gaining even a minimal insight into such issues. Finally, engineers should in some instances offer advice and guidance on an issue, especially when there is some degree of consensus in the engineering community.

We can hear an obvious objection to these proposals: "It is unfair to impose these heavy responsibilities on individual engineers." For the most part, we agree. Alerting the public to dangers from technology may of necessity be a responsibility that often falls on individual engineers. Often, it is only individual engineers who are close to a technology itself and are aware of the problems with a new product or a new process. We believe, however, that sometimes the responsibility of alerting the public to dangers and informing and advising the public should fall on professional engineering societies. Unfortunately, with a few exceptions, engineering societies have not adequately taken up these responsibilities. One notable exception is the involvement of the American Society of Mechanical Engineers in establishing the specifications for pressure vessels in the United States. After a series of tragic boiler explosions, the U.S. Congress decided it was time to write specifications for safe boiler construction into federal law. The professional expertise of mechanical engineers was essential in establishing these criteria.

One explanation for the reluctance of engineering societies to involve themselves in public debates regarding technology is the fact that the membership may be divided on the correct course of action. Some members may favor one policy and some members may favor another. Also, members may be reluctant to support an organization that advocates policies with which they disagree. This is true with regard to environmental issues, but it is probably true of other issues as well. However, this objection does not apply to the obligation to inform the public about the issues on both sides of a public debate. The public may be woefully ignorant of, or confused about, the issues, and engineering societies are one of the best sources of information.

One of the advantages of taking a more active role in informing and, in some cases, advising the public is that it would raise the visibility of the engineering profession. Engineering is in some ways the most invisible of the major professions, primarily because ordinary citizens do not encounter engineers in their daily lives in the same way they encounter physicians, lawyers, pharmacists, and other professionals. A more active and visible participation in public policy debates on technology policy would be one way of overcoming this relative invisibility.

5.7   THE SOCIAL EMBEDDEDNESS OF TECHNOLOGY
The Social Interaction of Technology and Society

So far, we have considered primarily the influence of technology on society and the social policy issues that such influence creates. However, in contrast to technological determinism, most scholars believe that social forces also influence the direction of technological development so that the truth seems to be that there is two-way causal interaction between technology and society: Technology influences society, but society also influences the development of technology. From the common-sense standpoint, this may be the most plausible position. There are many examples of social, political, and economic influences on technology. The abolition of child labor stimulated the development of technologies that could do the jobs that small bodies once did. The abolition of slavery created a demand for labor-saving agricultural machinery that was not needed when there was an abundant supply of cheap human labor. In our own time, moral and political considerations are influencing the funding of stem cell research. Partially for economic reasons, construction of the superconducting supercollider was abandoned after approximately $2 billion had been spent. Research and development on human space flights has been slow or fast, depending on changes in funding, which has in turn been influenced by political considerations.

This two-way or interactionist view gives us a picture of the social embeddedness of technology that we believe is the true account of the relationship of technology to society. This is in some way a more subtle understanding of the relationship of technology to society and issues of value than we have considered so far. In order to explore the nature of this interaction and to understand how social and value issues enter into the development of technology, we need the assistance of a field of study called science and technology studies (STS). STS is based on empirical studies of technology by historians, sociologists, and anthropologists. Let us take a quick look at this important field of study.

Science and Technology Studies: Opening the Black Box of Technology

Many traditional engineering ethics cases give short descriptions and then identify a decision point at the end of the case where an engineer has to make a crucial decision. Even many of the more extended descriptions, such as the description of the *Challenger* case, often focus primarily on what we might call an "endpoint decision." In the *Challenger* case, the decision point was the teleconference the night before the launch, in which Roger Boisjoly and other engineers attempted to get Morton Thiokol to hold to their no-launch recommendation. STS researchers contrast his approach, which might also be called the "externalist" account of a case, to what we might call a "process" account, in which various points throughout the case can be identified in which ethical issues arise. Since this approach focuses on points throughout the narrative, it is also sometimes called an internalist account. STS researchers criticize traditional engineering ethics for too often giving an endpoint or externalist account of a case and argue that, in doing so, less is learned about how to avoid the crisis situations depicted in endpoint accounts.[32]

The internalist account of a situation requires what STS researchers call "thick descriptions"—that is, extended descriptions of cases that show the entire process of development up to the endpoint decision. Thus, STS researchers follow the development of products from design to production, observe in detail what goes on in laboratories and how scientists and engineers describe their activities in the laboratories, study the rhetoric of journal articles and public statements in science and technology, and in many other ways subject every aspect of the scientific and technological enterprise to detailed investigation.

One theoretical orientation in STS that emphasizes the social embeddedness of technology is actor–network theory (ANT), developed by Bruno Latour and others. According to this theory, scientists and engineers are "actors" who pursue their work in a social network. The network consists not only of such elements as other scientists and engineers but also of members of the public, funding agencies, business institutions, and government.

Here is an example of the application of ANT to a specific issue. Michael Callon, an ANT theorist, describes the attempt of a group of engineers to introduce the electric car in France.[33] The engineers articulated not only the technical details of the fuel cells but also a vision of French society into which the cars would fit. Engineers at Renault, who wanted to preserve internal combustion engines as the primary source of power for automobiles, criticized both the technical details and the social feasibility of the alternative engines. In this example, social considerations and public perceptions were important causal factors in the development of the technologies, as well as considerations of the feasibility of the new technologies.

In another example of social embeddedness, Latour describes the efforts of the engineer Rudolf Diesel to build a new type of engine that would use any fuel.[34] Diesel found that only kerosene ignited, and so he had to modify the elements in his social network by shifting alliances to include providers of kerosene and pumps, other scientists and engineers, financiers, and entrepreneurs. Success in science and technology is the result of managing the interaction of these factors in the network in a successful way toward a goal set by the scientist or engineer who is the primary actor in the network. Although technological restraints played a part in the development of the diesel, social factors were also important.

Detailed internalist accounts of technology have led to another common STS theme, namely that there are usually several workable solutions to a technical problem, and social and value factors usually make the final selection of the solution that is adopted. Sociologists of technology Trevor Pinch and Wiebe Bijker illustrate this theme with the early history of the bicycle.[35] The early evolution of the bicycle had two "branches": a sportsman's bike with a high front wheel that was relatively unstable and a more utilitarian version with a smaller front wheel that was more stable. The sportsman's version was designed for speed and was especially attractive to young athletic males. The utilitarian version was more appropriate for pleasure riding and ordinary transportation. Eventually, the utilitarian design came to be more widely accepted and the high-wheeled bike disappeared. Later, the more stable version was designed for greater speed. Taking the retrospective view, technological determinists might argue that the high-wheelers were clumsy and less efficient and simply dropped out of the evolutionary path of development, and that the entire process of development was governed by an "internal logic" that took only technological considerations into account. In fact, the high-wheeled and utilitarian bikes both existed for years and neither was a stage in the evolution of the other. High-wheelers simply represented an alternative path of development that was eventually abandoned. Most people evidently decided that producing a sportsman's toy was not as important as producing a useful means of transportation.[36]

On a still more subtle level of analysis, STS researchers have found that even concepts that are usually thought to have a purely technical definition often have a social and value dimension. For example, what constitutes "effective functioning" or "efficiency" in a technological device is not determined wholly by technical considerations but also in part by social considerations:

> In engineering, the efficiency of a device is taken to be a purely quantitative ratio of the energy input and energy output. However, in practice, whether a device is considered to "work well" is a product of the character and interests of a user group.[37]

Child labor was in some ways more "efficient" than the use of adults, but when it was decided that the use of child labor was immoral, children were no longer taken into account as a possible source of more efficient labor. Child labor was not even considered an option to improve efficiency. Instead, children were redefined as learners and consumers, not laborers. This is an example of how "efficiency" is a socially relative concept.[38]

The following is another example of how social considerations define the concept of efficiency. Boiler explosions took the lives of many people, especially on steamboats, in the early 19th century. In 1837, at the request of the U.S. Congress, the Franklin Institute undertook a rigorous examination of boiler construction. Boilermakers and steamboat owners resisted higher standards, and Congress did not impose the higher standards until 1852, after many more people had been killed in steamboat accidents. The accident rate decreased dramatically after thicker walls and safety valves were mandated. What *constituted* or defined a proper boiler was redefined by the new standards, which were issued by the American Society of Mechanical Engineers. Although it might be more "efficient" to build boilers by the old standards, this is no longer an option.

A similar process seems to be occurring in many areas related to the environment. Standards for consumption of gasoline are changing. Even if it may be more "efficient" to build automobiles by older standards or to use less environmentally friendly standards in other areas, society will almost certainly continue to change the standards in favor of the environment. Many design standards were controversial at one time but no longer. Design standards already incorporate many safety and environmental considerations that probably cannot be justified economically or even by a consideration of "trade-offs." Society has simply made certain decisions that are no longer in dispute. They become part of the definition of what it means to design a product, such as an automobile.

5.8 HOW SHALL WE DESIGN?

Ethical Issues in Design

In pursuing the internalist program, STS researchers have often focused on design as a crucial stage of development in which value issues can arise where either decisions are made that affect society or social forces have affected the course of design. A study by two researchers from The Netherlands illustrates how issues for ethical reflection arise in the design process.[39] Following another scholar, W. G. Vincinti, they identify two types of design challenges.[40] Without going into the elaborate classification they use, we present some of their examples.

In one project, the challenge was to design a sustainable car, where sustainability was closely associated with light weight. Light weight tended to make the car less safe, however. Thus, the nature of the values of safety and sustainability had to be discussed, as well as the relative moral weight of these two values. Even with respect to sustainability, there were moral considerations. Is the value of sustainability derived from an obligation to respect nature for its own sake or from an obligation to allow future generations to fulfill their needs? In another example, the challenge was to develop a coolant that did not contribute to ozone depletion like the traditional chlorofluorocarbons, such as CFC 12. It turns out, however, that there is a trade-off between flammability and environmental harm. The more environmentally friendly coolants tended to be more flammable. Although one might think that flammability poses a safety issue, some challenged even that assumption. Thus, the nature of these two values and their relative importance had to be discussed. In a third example, the challenge was to create a housing system for laying hens. Important ethical issues were the environmental emissions of the chickens, the labor circumstances for farmers, and the health and welfare of the laying hens.

These three examples of design problems raise many significant moral issues regarding safety, sustainability, the environment, animal welfare and health, labor conditions, and so forth. As we have seen in our previous discussions of design, there is not usually a single correct way to resolve the ethical problems raised by these design projects, although there are probably some solutions that are not necessarily acceptable.

Designing for the Environment and for Human Community

Previously, we considered the technological pessimist's criticism of technology as being oriented toward efficient domination of nature and leading to a fracturing of human experience. Some critics of technology have noted that these features may not be essential to technology itself. Technologies can be designed that stress sustainability, the use of renewable resources, and minimal pollution. Technologies can also serve to promote human community rather than fracture it. If modern technology destroys or makes irrelevant such traditional focal things as the fireplace, it can replace them by others. Computerized networks for those with AIDS or with a thousand other interests can provide a focus for activities. Urban and housing design can provide humane living spaces that allow greater access to others and to the natural world. Running shoes and backpacking equipment can enable us to again establish access to the natural world.

The advance of technology does not necessarily destroy values that we consider of great importance, such as a relationship to the natural world and focused human activities. It does, however, change the forms and contexts in which these activities take place. It is not too much to ask engineers to think about these issues, as many engineers are certainly doing. Such thought can lead to creative designs and a more humanly satisfying life.

5.9 CONCLUSION: ENGINEERING AS SOCIAL EXPERIMENTATION

In their textbook *Ethics in Engineering,* philosopher Mike W. Martin and engineer Roland Schinzinger develop the idea of *engineering as social experimentation.*[41] There are several analogies between engineering and experimentation. First, engineering works—whether consumer products, bridges, or buildings—have experimental subjects, like scientific experiments. In engineering, however, the subjects are the public who utilize the products of engineering. Second, as in any experiment, there is always an element of uncertainty about the outcome. Engineers never know for sure how well a new automobile will perform on the road, or whether a new building will withstand a hurricane. Yet, there is a necessity of gaining new knowledge, which can only come by experimentation. Only by innovation can technology advance. Third, like experimenters, engineers must assume responsibility for their experiments. They must think about the possible consequences, both good and bad, and attempt to eliminate as many bad consequences as possible.

With engineering works, however, this obligation to take responsibility is especially grave because of a unique feature of engineering. Unlike ordinary scientific experimentation, the requirement of free and informed consent of the experimental subjects cannot be fully honored. The public is usually not able to sufficiently comprehend the risks and social implications inherent in an engineering work to give a fully free and informed consent. Few airplane passengers understand aerospace engineering well enough to know the risks they are assuming when they walk into the passenger compartment. An automobile driver does not usually understand mechanical engineering well enough to be able to fully assess the risk of getting behind the wheel of a new car, nor is she enough of an expert in civil engineering to understand the risks she is taking in driving over a new bridge. Similar considerations, of course, apply to the use of virtually any product of engineering design. Of course, few patients understand medicine well enough to give fully informed consent to medical procedures used on them, but a physician is obligated to explain the procedures in as much detail as the patient requires. In engineering, partially because engineers do not directly relate to the general public, even this degree of explanation is usually impossible.

The concept of engineering as social experimentation, then, is a useful way of summarizing most of the responsibilities of engineers with regard to technology. First, engineers should recognize that technology is not socially neutral but, rather, embedded in a social network as both cause and effect. Second, as the debate between technological optimism and pessimism shows, technology has both conferred great goods on humankind and raised problems and issues that demand solutions, many of them from engineers. Therefore, engineers must adopt a critical attitude toward technology. Third, engineers and especially engineering societies have a responsibility to alert the public to the dangers and risks imposed by technology and to inform and sometimes advise the public on policy issues regarding technology. Fourth, as the primary creators of technology, engineers have a responsibility to design with consideration of the social and value implications of their designs. In these ways, engineers can more adequately fulfill their responsibilities with regard to the social and value dimensions of technology.

5.10   CHAPTER SUMMARY

Technology has been defined as the making of tools. Another definition is that it is the application of science to the solution of practical problems. Still another definition, favored by many contemporary scholars, is that technology is best understood as a "system" composed of physical objects and tools, knowledge, investors, operators, repair people, managers, government regulators, and others. This definition highlights the social interaction of technology with the larger society.

Technological determinism denies the effect of society on technology by holding that technological development is controlled by its own internal "logic" and is relatively insulated from social forces. STS offers empirical evidence to deny this claim. Technological optimism emphasizes the benefits that technology confers on humankind, and technological pessimism emphasizes the dangers and problems that it creates. The fact that there are valid insights in both positions suggests that engineers should take a critical attitude toward technology—appreciating and taking pride in its benefits while being aware of the problems it can create.

Computer technology illustrates the benefits that technology can confer while also illustrating the issues and problems it can raise. Some of the issues have to do with privacy and the ownership of computer software. Many of these issues raised by computer technology are probably best resolved by creative middle way solutions. The issues raised by computer technology are social policy issues that must, in a democracy, ultimately be addressed by democratic deliberation. Engineers have a responsibility to participate in democratic deliberation regarding technology by alerting the public to dangers and issues, giving the public relevant information, and in some cases, especially when there is consensus in the engineering community, advising the public on a course of action.

STS scholars have shown that contrary to technological determinism, there is both a causal effect of society on technology and a causal effect of technology on society. ANT is one theoretical perspective that illustrates the social embeddedness of technology. STS researchers have shown that there are often several workable solutions to a technological problem, and that social and value factors often tip the scales. On the conceptual level, even concepts such as "efficiency" often have a value dimension.

Engineers, in their design work, have a responsibility to design in light of the social embeddedness of technology. Many design projects, as some simple examples show, raise ethical issues for designers that cannot be avoided. The concept of engineering as social experimentation is an apt way of summarizing the interaction of society and technology and of highlighting the responsibilities of engineers in light of this interaction.

NOTES

1. Sarah Kuhn, "When Worlds Collide: Engineering Students Encounter Social Aspects of Production," *Science and Engineering Ethics*, 1998, pp. 457–472.
2. Ibid., p. 461.
3. Ibid., p. 465.
4. Ibid., p. 466.
5. Ibid., p. 467.
6. See http://www.abet.org.
7. William A. Wulf, "Great Achievements and Grand Challenges," revised version of a lecture given on October 22 at the 2000 annual meeting of the National Academy of Engineering, p. 1. Used with permission.
8. For some ideas on the discussion in this section, see Val Dusek, *Philosophy of Technology* (Malden, MA: Blackwell, 2006), especially pp. 31–36.
9. Lewis Mumford, *Technics and Civilization* (New York: Harcourt Brace Jovanovich, 1963).
10. Val Dusek, *Philosophy of Technology: An Introduction* (Malden, MA: Blackwell, 2006), pp. 32–36. Dusek discusses the other themes mentioned here as well.
11. Paul Wulf, op. cit., pp. 2–3.
12. This account is taken mostly from "Now for the Hard Part: A Survey of Business in India," *The Economist*, June 3–9, 2006, special insert, pp. 3–18. For a somewhat more negative account, see "The Myth of the New India," *New York Times*, July 6, 2006, p. A23.
13. For the first reference, see *Paradise Lost*, bk. 1, line 670. For the reference to engineering, see bk. 1, line 750 and bk. 6, line 553.

14. Witold Kula, *Measure and Men,* R. Szreter, trans. (Princeton, NJ: Princeton University Press, 1986), pp. 30–31.

15. Sergio Sismondo, *An Introduction to Science and Technology Studies* (Malden, MA: Blackwell, 2004), pp. 32–36.

16. For this and the following paragraphs, we have found Andrew Feenberg's "Technology and Meaning" in his *Questioning Technology* (New York: Routledge, 1999) to be useful.

17. "Focal Things and Practices," in David M. Kaplan, ed., *Readings in the Philosophy of Technology* (New York: Rowman & Littlefield, 2004), pp. 115–136. Taken from Albert Borgmann, *Technology and the Character of Contemporary Life: A Philosophical Inquiry* (Chicago: University of Chicago Press, 1984).

18. For the distinction between these four types of privacy, see Anita L. Allen, "Genetic Privacy: Emerging Concepts and Values," in Mark Rothstein, ed., *Genetic Secrets* (New Haven, CT: Yale University Press, 1997), p. 33.

19. Ruth Gavison, "Privacy Is Important," in Deborah G. Johnson and Helen Nissenbaum, eds., *Computers, Ethics & Social Policy* (Upper Saddle River, NJ: Prentice Hall, 1995), p. 333.

20. James Rachels, "Why Privacy Is Important," in Johnson and Nissenbaum, pp. 351–357.

21. John Shattuck, "Computer Matching Is a Serious Threat to Individual Rights," in Johnson and Nissenbaum, p. 307.

22. For the list of standards from which this list is taken, see Anita L. Allen, "Privacy," in Hugh La Follette, ed., *The Oxford Handbook of Practical Ethics* (Oxford: Oxford University Press, 2003), p. 500.

23. VP-Planner manual at xi, 1.11.

24. Richard Stallman, "Why Software Should Be Free," in Johnson and Nissenbaum, pp. 190–200.

25. Office of Technology Assessment, "Evolution of Case Law on Copyrights and Computer Software," in Johnson and Nissenbaum, p. 165.

26. *Herbert Rosenthan Jewelry Corp. v. Kalpakian,* 446 F.2d 738, 742 (9th Cir. 1971). Cited in *Lotus Development Corporation v. Paperback Software International and Stephenson Software, Limited,* Civ, A, No. 87-76-K. United States District Court. D. Massachusetts. June 28, 1990. Reprinted in Johnson and Nissenbaum, pp. 236–252.

27. *Lotus Development Corporation v. Paperback Software International and Stephenson Software, Limited,* p. 242.

28. 714 F.2d 1240 (3rd Cir.983).

29. Brian Kahin, "The Software Patent Crisis," in Johnson and Nissenbaum, pp. 190–200.

30. John W. Snapper, "Intellectual Property Protections for Computer Software," in Johnson and Nissenbaum, p. 189.

31. Vannevar Bush, *Science—The Endless Frontier: A Report to the President on a Program for Postwar Scientific Research* (Washington, DC: National Science Foundation, 1980).

32. W. T. Lynch and R. Kline, "Engineering Practice and Engineering Ethics," *Science, Technology, and Human Values* 25, 2000, pp. 195–225. For a criticism of this article and similar STS positions, see Michael Davis, "Engineering Ethics, Individuals, and Organizations," *Science and Engineering Ethics,* 12, 2006, pp. 223–231. Another criticism of STS claims is Langdon Winner, "Social Constructivism: Opening the Black Box and Finding it Empty," *Science as Culture,* 16, 1993, pp. 427–452. Reprinted in Robert C. Scharff and Val Dusek, eds., *Philosophy of Technology* (Malden, MA: Blackwell, 2003), pp. 233–243.

33. Michael Callon and John Law, "Agency and the Hybrid Collectif," *South Atlantic Quarterly* 94, 1994, pp. 481–607.

34. Bruno Latour, *Science in Action: How to Follow Scientists and Engineers through Society* (Cambridge, MA: Harvard University Press, 1987), p. 123.

35. Trevor J. Pinch and Wiebe E. Bijker, "The Social Construction of Facts and Artifacts: Or How the Sociology of Science and the Sociology of Technology Might Benefit Each Other," in W. E. Bijker, T. P. Hughes, and T. Pinch, eds., *The Social Construction of Technological Systems: New Directions in the Sociology and History of Technology* (Cambridge, MA: MIT Press, 1987), pp. 17–50.

36. Andrew Feenberg, "Democratic Rationalization: Technology, Power, and Freedom," in Scharff and Dusek, pp. 654–655.

37. Val Dusek, *Philosophy of Technology: An Introduction,* p. 205.

38. See Feenberg, op. cit., pp. 659–660.

39. Ibo Van De Powl and A. C. van Gorp, "The Need for Ethical Reflection in Engineering Design," *Science, Technology, and Human Values,* 31, no. 3, 2006, pp. 333–360.

40. W. G. Vincinti, "Engineering Knowledge, Type of Design, and Level of Hierarchy: Further Thoughts about What Engineers Know," in P. Kroes and M. Bakker, eds., *Technological Development and Science in the Industrial Age* (Dordrecht, The Netherlands: Kluwer, 1992), pp. 17–34.

41. Mike W. Martin and Roland Schinzinger, *Ethics in Engineering,* 4th ed. (Boston: McGraw-Hill, 2005), pp. 88–100.

CHAPTER SIX

Trust and Reliability

Main Ideas in this Chapter

- This chapter focuses on issues regarding the importance of trustworthiness in engineers: honesty, confidentiality, intellectual property, expert witnessing, public communication, and conflicts of interest.
- Forms of dishonesty include lying, deliberate deception, withholding information, and failure to seek out the truth.
- Dishonesty in engineering research and testing includes plagiarism and the falsification and fabrication of data.
- Engineers are expected to respect professional confidentiality in their work.
- Integrity in expert testimony requires not only truthfulness but also adequate background and preparation in the areas requiring expertise.
- Conflicts of interest are especially problematic because they threaten to compromise professional judgment.

JOHN IS A CO-OP STUDENT who has a summer job with Oil Exploration, Inc., a company that does exploratory contract work for large oil firms.[1] The company drills, tests, and writes advisory reports to clients based on the test results. As an upper-level undergraduate student in petroleum engineering, John is placed in charge of a field team of roustabouts and technicians who test drill at various sites specified by the customer. John has the responsibility of transforming rough field data into succinct reports for the customer. Paul, an old high school friend of John's, is the foreperson of John's team. In fact, Paul was instrumental in getting this high-paying summer job for John.

While reviewing the field data for the last drilling report, John notices that a crucial step was omitted, one that would be impossible to correct without returning to the site and repeating the entire test at great expense to the company. The omitted step involves the foreperson's adding a certain test chemical to the lubricant being pumped into the test drill site. The test is important because it provides the data for deciding whether the drill site is worth developing for natural gas protection. Unfortunately, Paul forgot to add the test chemical at the last drill site.

John believes that Paul is likely to lose his job if his mistake comes to light. Paul cannot afford to lose his job at a time when the oil business is slow and his wife is expecting a child. John learns from past company data files that the chemical additive indicates the presence of natural gas in approximately 1 percent of the tests.

Should John withhold the information that the test for natural gas was not performed from his superiors? Should the information be withheld from the customer?

6.1   INTRODUCTION

In Chapter 2, we noted William F. May's observation that as society has become increasingly professionalized, it has also become more dependent on the services of professionals whose knowledge and expertise are not widely shared or understood. What this means is that, in its ignorance, the public must place its trust in the reliable performance of engineers, both as individuals and as members of teams of engineers who work together. This chapter focuses on areas of moral concern that are especially relevant to the trustworthiness of engineers: honesty and dishonesty, confidentiality, intellectual property rights, expert witnessing, communicating with the public, and conflicts of interest.

6.2   HONESTY

The concern with truth telling extends far beyond the boundaries of the engineering profession. Religious and secular literature contain many injunctions to tell the truth. For example, one of the Ten Commandments forbids bearing false witness against one's neighbor. In Shakespeare's *Hamlet,* Polonius gives some advice regarding honesty to his son, Laertes, just before the son's first trip abroad from Denmark: "This above all: to thine own self be true. And it must follow, as the night the day, thou canst not then be false to any man." John Bartlett's *Familiar Quotations* lists in the index two columns of entries on the word *true,* another four on *truth,* and a half column on *honesty.* Miguel de Cervantes is the author of the famous aphorism, "Honesty's the best policy," which was used by George Washington in his 1796 farewell address. In 1381, John Wycliffe told the Duke of Lancaster, "I believe that in the end the truth will conquer."

In light of the long emphasis on honesty in our moral tradition, it is not surprising that engineering codes contain many references to honesty. The third canon of the code of ethics of the Institute of Electrical and Electronics Engineers (IEEE) encourages all members "to be honest and realistic in stating claims or estimates based on available data." Canon 7 requires engineers "to seek, accept, and offer honest criticism of technical work." The American Society of Mechanical Engineers (ASME) code of ethics is equally straightforward. Fundamental Principle II states that engineers must practice the profession by "being honest and impartial." The seventh Fundamental Canon states, "Engineers shall issue public statements only in an objective and truthful manner." A subsection enjoins engineers not to "participate in the dissemination of untrue, unfair, or exaggerated statements regarding engineering."

The importance of honesty in engineering practice is a major focus of this chapter. However, in addition to issues of honesty, we also explore other important aspects of professional judgment and communication. For example, the second cannon of the IEEE code requires members to avoid conflicts of interest because they can distort professional judgment. A subsection of Canon 3 of the ASCE code requires members not to issue statements on engineering matters "which are inspired or paid for by interested parties, unless they indicate on whose behalf the statements are made." Here again, the emphasis is on full disclosure. A subsection of Canon 4 of the same code speaks to the matter of confidentiality, an area in which withholding information is justified. It enjoins engineers to avoid conflicts of interest and forbids them from using "confidential information coming to them in the course of their assignments as a means of making personal profit if such action is adverse to the interests of their clients, employers, or the public."

The more detailed National Society for Professional Engineers (NSPE) code admonishes engineers "to participate in none but honest enterprise." The preamble states that "the services provided by engineers require honesty, impartiality, fairness, and equity." The third Fundamental Canon (1.3) requires engineers to "avoid deceptive acts in the solicitation of professional employment." In the Rules of Practice, there are several references to honesty. In item II.1.d, the code states the following: "Engineers shall not permit the use of their name or firm name nor associate in business ventures with any person or firm which they have reasons to believe is engaging in fraudulent or dishonest business or professional practices." Items II.2.a–II.2.c and II.3.a–II.3.c in the Rules of Practice give more detailed direction for the practice of the profession. Item II.3 states that "engineers shall issue public statements only in an objective and truthful manner." Item II.5 states that "engineers shall avoid deceptive ads in the solicitation of professional employment." Items II.5a and II.5.b give more detailed explanations regarding how to implement this statement. In Section III, "Professional Obligations," the code refers to the obligation for engineers to be honest and truthful and not to misrepresent facts—and does so in six different locations (III.1.a, III.1.d, III.2.c, III.3.a, III.7, and III.8). In a statement that speaks directly to John's situation, part (a) of the third Rule of Practice states, "Engineers shall be objective and truthful in professional reports, statements, or testimony. They shall include all relevant and pertinent information in such reports, statements, or testimony."

6.3 FORMS OF DISHONESTY

Lying

When we think of dishonesty, we usually think of lying. Ethicists have long struggled over the definition of lying. One reason for the difficulty is that not every falsehood is a lie. If an engineer mistakenly conveys incorrect test results on soil samples, she is not lying even though she may not be telling the truth. To lie, a person must intentionally or at least knowingly convey false or misleading information. But even here complications arise. A person may give information that she believes to be false, even though it is actually true. In this case, we may be perplexed as to whether we should describe her action as lying. Her intention is to lie, but what she says is actually true.

To make matters more complicated, a person may give others false information by means other than making false statements. Gestures and nods, as well as indirect statements, can give a false impression in a conversation, even though the person has not told an outright lie. Despite these complications, most people believe that lies— or at least paradigm cases of lies—have three elements: First, a lie ordinarily involves something that is believed to be false or seriously misleading. Second, a lie is ordinarily stated in words. Third, a lie is made with the intention to deceive. So perhaps we can offer the following working definition: "A lie is a statement believed to be false or seriously misleading, made with the intention to deceive." Of course, this definition leaves the phrase "seriously misleading" open for interpretation, but the open-ended nature of this working definition is deliberate. We call some misleading statements lies and others not.

Deliberate Deception

If Andrew discusses technical matters in a manner that implies knowledge that he does not have to impress an employer or potential customer, then he is certainly engaging in deliberate deception, even if he is not lying. In addition to misrepresenting one's own expertise, one can misrepresent the value of certain products or designs by praising their advantages inordinately. Such deception can sometimes have more disastrous consequences than outright lying.

Withholding Information

Omitting or withholding information is another type of deceptive behavior. If Jane deliberately fails to discuss some of the negative aspects of a project she is promoting to her superior, she engages in serious deception even though she is not lying. Failing to report that you own stock in a company whose product you are recommending is a form of dishonesty. Perhaps we can say in more general terms that one is practicing a form of dishonesty by omission (1) if one fails to convey information that the audience would reasonably expect would not be omitted and (2) if the intent of the omission is to deceive.

Failure to Seek Out the Truth

The honest engineer is one who is committed to finding the truth, not simply avoiding dishonesty. Suppose engineer Mary suspects that some of the data she has received from the test lab are inaccurate. In using the results as they are, she is neither lying nor concealing the truth. But she may be irresponsible in using the results without inquiring further into their accuracy. Honesty in this positive sense is part of what is involved in being a responsible engineer.

It would not be correct to assume that lying is always more serious than deliberate deception, withholding information, failing to adequately promote the dissemination of information, or failing to seek out the truth. Sometimes the consequences of lying may not be as serious as the consequences of some of these other actions. The order of these first four types of misusing the truth reflects primarily the degree to which one is actively distorting the truth rather than the seriousness of the consequences of the actions.

6.4 WHY IS DISHONESTY WRONG?

The term *honest* has such a positive connotation and the term *dishonest* such a negative one that we forget that telling the full truth may sometimes be wrong and concealing the truth may sometimes be the right thing to do. A society in which people are totally candid with each other would be difficult to tolerate. The requirement of total candor would mean that people would be brutally frank about their opinions of each other and unable to exercise the sort of tact and reticence that we associate with polite and civilized society. With regard to professionals, the requirement never to conceal truth would mean that engineers, physicians, lawyers, and other professionals could not exercise confidentiality or protect proprietary information. Doctors could never misrepresent the truth to their patients, even when there is strong evidence that this is what the patients prefer and that the truth could be devastating.

Despite possible exceptions, however, dishonesty and the various other ways of misusing the truth are generally wrong. A helpful way to see this is to consider dishonesty from the standpoints of the ethics of respect for persons and utilitarianism; each can provide valuable suggestions for thinking about moral issues related to honesty.

Let us review some of the major components of the respect for persons perspective. As discussed in Chapter 3, actions are wrong if they violate the moral agency of individuals. Moral agents are human beings capable of formulating and pursuing goals and purposes of their own—they are autonomous. The word *autonomy* comes from two Greek terms: *autos,* meaning "self," and *nomos,* meaning "rule" or "law." A moral agent is autonomous in the sense of being self-governing.

Thus, to respect the moral agency of patients, physicians have three responsibilities. First, they must ensure that their patients make decisions about their medical treatment with informed consent. They must see to it that their patients understand the consequences of their decisions and rationally make decisions that have some relationship to their life plans. Second, they have some responsibility to ensure that patients make decisions without undue coercive influences such as stress, illness, and family pressures. Finally, physicians must ensure that patients are sufficiently informed about options for treatment and the consequences of the options.

Engineers have some degree of responsibility to ensure that employers, clients, and the general public make autonomous decisions, but their responsibilities are more limited than those of physicians. Their responsibilities probably extend only to the third of these three conditions of autonomy, ensuring that employers, clients, and the general public make decisions regarding technology with understanding, particularly understanding of their consequences. We have seen, for example, that the IEEE code requires members to "disclose promptly factors that might endanger the public or the environment" and that when the safety, health, and welfare of the public are endangered ASCE members must "inform their clients or employers of the possible consequences." In engineering, this applies to such issues as product safety and the provision of professional advice and information. If customers do not know that a car has an unusual safety problem, then they cannot make an informed decision regarding whether to purchase it. If a customer is paying for professional engineering advice and is given misinformation, then he again cannot make a free and informed decision.

The astronauts on the *Challenger* were informed on the morning of the flight about the ice buildup on the launching pad and were given the option of postponing the launch. They chose not to exercise that option. However, no one presented them with the information about O-ring behavior at low temperatures. Therefore, they did not give their fully informed consent to launch despite the O-ring risk because they were unaware of the risk. The *Challenger* incident is a tragic example of the violation of the engineer's obligation to protect informed consent. The fault, however, was not primarily with the engineers but with the managers who supported the launch and did not inform the astronauts of the danger.

Many situations are more complex. To be informed, decision makers must not only have the relevant information but also understand it. Furthermore, nobody has all of the relevant information or has complete understanding of it so that being informed in both of these senses is a matter of degree. Therefore, the extent of the engineer's obligation regarding informed consent will sometimes be controversial, and whether or not the obligation has been fulfilled will also sometimes be controversial. We return to these considerations later, but what we have said here is enough to show that even withholding information or failing to adequately disseminate it can be serious violations of professional responsibilities.

Now let us turn to the utilitarian perspective on honesty. Utilitarianism requires that our actions promote human happiness and well-being. The profession of engineering contributes to this utilitarian goal by providing designs for the creation of buildings, bridges, chemicals, electronic devices, automobiles, and many other things on which our society depends. It also provides information about technology that is important in decision making at the individual, corporate, and public policy levels.

Dishonesty in engineering research can undermine these functions. If engineers report data falsely or omit crucial data, then other researchers cannot depend on their results. This can undermine the relations of trust on which a scientific community is founded. Just as a designer who is untruthful about the strength of materials she specifies for a building threatens the collapse of the building, a researcher who falsifies the data reported in a professional journal threatens the collapse of the infrastructure of engineering.

Dishonesty can also undermine informed decision making. Managers in both business and government, as well as legislators, depend on the knowledge and judgments provided by engineers to make decisions. If these are unreliable, then the ability of those who depend on engineers to make good decisions regarding technology is undermined. To the extent that this happens, engineers have failed in their obligation to promote the public welfare.

From both the respect for persons and utilitarian perspectives, then, outright dishonesty as well as other forms of misusing the truth with regard to technical information and judgment are usually wrong. These actions undermine the moral agency of individuals by preventing them from making decisions with free and informed consent. They also prevent engineers from promoting the public welfare.

6.5   DISHONESTY ON CAMPUS

Three students were working on a senior capstone engineering design project. The project was to design, build, and test an inexpensive meter that would be mounted on the dashboard of automobiles and would measure the distance a car could travel on a gallon of gasoline. Even though personal computers, microchip calculators, and "smart instruments" were not available at the time, the students came up with a clever approach that had a good chance of success. They devised a scheme to instantaneously measure voltage equivalents of both gasoline flow to the engine and speedometer readings on the odometer while keeping a cumulative record of the quotient of the two. In other words, miles per hour divided by gallons per hour would give the figure for the miles the automobile is traveling per gallon of gasoline. The students even devised a way to filter and smooth out instantaneous fluctuations in either signal to ensure time-averaged data. Finally, they devised a bench-top experiment to prove the feasibility of their concept. The only thing missing was a flow meter that would measure the flow of gasoline to the engine in gallons per hour and produce a proportional voltage signal.

Nowadays, customers can order this feature as an option on some automobiles, but at the time the design was remarkably innovative. The professor directing the project was so impressed that he found a source of funds to buy the flow meter. He also encouraged the three students to draft an article describing their design for a technical journal.

Several weeks later, the professor was surprised to receive a letter from the editor of a prominent journal, accepting for publication the "excellent article" that, according to the letter, he had "coauthored" with his three senior design students. The professor knew that the flow meter had not yet arrived, nor had he seen any draft version of the paper, so he asked the three students for an explanation. They explained that they had followed the professor's advice and prepared an article about their design. They had put the professor's name on the paper as senior author because, after all, it was his idea to write the paper and he was the faculty advisor. They did not want to bother the professor with the early draft. Furthermore, they really could not wait for the flow-measuring instrument to arrive because they were all graduating in a few weeks and planned to begin new jobs.

Finally, because they were sure the data would give the predicted results, they simulated some time-varying voltages on a power supply unit to replicate what they thought the flow-measuring voltages would be. They had every intention, they said, of checking the flow voltage and the overall system behavior after the flow meter arrived and, if necessary, making minor modifications in the paper.

As a matter of fact, the students incorrectly assumed that the flow and voltages would be related linearly. They also made false assumptions about the response of the professor to their actions. The result was that the paper was withdrawn from the journal, and the students sent letters of apology to the journal. Copies of the letter were placed in their files, the students received an "F" in the senior design course, and their graduation was delayed 6 months. Despite this, one of them requested that the professor write a letter of recommendation for a summer job he was seeking!

A student's experience in engineering school is a training period for his or her professional career. If dishonesty is as detrimental to engineering professionalism as we have suggested, then part of this training should be in professional honesty. Furthermore, the pressures that students experience in the academic setting are not that different from (and perhaps less than) those they will experience in their jobs. If it is morally permissible to cheat on exams and misrepresent data on laboratory reports and design projects, then why isn't it permissible to misrepresent data to please the boss, get a promotion, or keep a job?

As we shall see in the next section, there are exact counterparts in the scientific and engineering communities to the types of dishonesty exhibited by students. Smoothing data points on the graph of a freshman physics laboratory report to get an "A" on the report, selecting the research data that support the desired conclusion, entirely inventing the data, and plagiarism of the words and ideas of others all have obvious parallels in nonacademic settings.

6.6   DISHONESTY IN ENGINEERING RESEARCH AND TESTING

Dishonesty in science and engineering takes several forms: falsification of data, fabrication of data, and plagiarism. *Falsification* involves distorting data by smoothing out irregularities or presenting only those data which fit one's favored theory and discarding the rest. *Fabrication* involves inventing data and even reporting results of experiments that were never conducted. *Plagiarism* is the use of the intellectual property of others without proper permission or credit. It takes many different forms. Plagiarism is really a type of theft. Drawing the line between legitimate and illegitimate use of the intellectual property of others is often difficult, and the method of line-drawing is useful in helping us discriminate between the two. Some cases are undeniable examples of plagiarism, such as when the extended passages involving the exact words or the data of another are used without proper permission or attribution. On the other side of the spectrum, the quotation of short statements by others with proper attribution is clearly permissible. Between these two extremes are many cases in which drawing the line is more difficult.

Multiple authorship of papers can often raise particularly vexing issues with regard to honesty in scientific and technological work. Sometimes, as many as 40–50 researchers are listed as the authors of a scientific paper. One can think of several justifications for this practice. First, often a large number of scientists participate in some forms of research, and they all make genuine contributions. For example, large numbers of people are sometimes involved in medical research or research with a particle accelerator. Second, the distinction between whether someone is the author of a paper or merely deserves to be cited may indeed be tenuous in some circumstances. The fairest or at least the most generous thing to do in such circumstances is to cite such people as authors.

However, there are less honest motives for the practice, the most obvious one being the desire of most scientists for as many publications as possible. This is true of both academic and nonacademic scientists. In addition, many graduate and postdoctoral students need to be published to secure jobs. Sometimes, more senior scientists are tempted to list graduate students as authors, even though their contribution to the publication was minimal, to make the student's research record appear as impressive as possible.

From a moral standpoint, there are at least two potential problems with multiple authorship. First, it is fraudulent to claim significant credit for scientific research when, in fact, a contribution is relatively insignificant. If claims to authorship are indeed fraudulent, then those who are evaluating the scientist or engineer are not able to make informed decisions in their evaluations. Second, fraudulent claims to authorship give one an unfair advantage in the competition for jobs, promotions, and recognition in the scientific community. From the standpoint of fairness alone, unsubstantiated claims to authorship should be avoided.

6.7   CONFIDENTIALITY

One can misuse the truth not only by lying or otherwise distorting or withholding it but also by disclosing it in inappropriate circumstances. Engineers in private practice might be tempted to disclose confidential information without the consent of the client. Information may be confidential if it is either given to the engineer by the client or discovered by the engineer in the process of work done for the client.

Given that most engineers are employees, a more common problem involving the improper use of information is the violation of proprietary information of a former employer. Using designs and other proprietary information of a former employer can be dishonest and may even result in litigation. Even using ideas one developed while working for a former employer can be questionable, particularly if those ideas involve trade secrets, patents, or licensing arrangements.

Most engineers are employees of large corporations, but some, especially civil engineers, subcontract for design firms that have clients. For these engineers, there is an obligation to protect the confidentiality of the client–professional relationship, just as with lawyers and physicians. Confidentiality would ordinarily cover both sensitive information given by the client and information gained by the professional in work paid for by the client.

An engineer can abuse client–professional confidentiality in two ways. First, she may break confidentiality when it is not warranted. Second, she may refuse to break confidentiality when the higher obligation to the public requires it.

The following is an example of the first type of abuse.[2] Jane, a civil engineer, is contracted to do a preliminary study for a new shopping mall for Greenville, California. The town already has a mall that is 20 years old. The owner of the existing mall is trying to decide whether to renovate or close the old mall. He has done a lot of business with Jane and asks her detailed questions about the new mall. Jane answers the questions.

The following is another example in the first category. Suppose Engineer A inspects a residence for a homeowner for a fee. He finds the residence in generally good condition, although it is in need of several minor repairs. Engineer A sends a copy of his one-page report to the homeowner, showing that a carbon copy was sent to the real estate firm handling the sale of the residence.

This case was considered by the NSPE Board of Ethical Review, which ruled that "Engineer A acted unethically in submitting a copy of the home inspection to the real estate firm representing the owners." It cites section II.1.c of the NSPE code, which states, "Engineers shall not reveal facts, data, or information obtained in a professional capacity without the prior consent of the client or employer except as authorized by law or this Code."[3]

This opinion seems correct. The clients paid for the information and therefore could lay claim to its exclusive possession. The residence was fundamentally sound, and there was no reason to believe that the welfare of the public was at stake. The case would have been more difficult if there had been a fundamental structural flaw. Even here, however, we can argue that there was no fundamental threat to life. Prospective buyers are always free to pay for an inspection themselves.

The following hypothetical case raises more serious difficulties. Suppose engineer James inspects a building for a client before the client puts the building up for sale. James discovers fundamental structural defects that could pose a threat to public safety. James informs the client of these defects in the building and recommends its evacuation and repair before it is put up for sale. The client replies,

> James, I am not going to evacuate the building, and I am certainly not going to spend a lot of money on the building before I put it up for sale. Furthermore, if

you reveal the information to the authorities or to any potential buyer, I am going to take whatever legal action I can against you. Not only that, but I have a lot of friends. If I pass the word around, you will lose a lot of business. The information is mine. I paid for it, and you have no right to reveal it to anyone else without my permission.

James's obligation to his client is clearly at odds with his obligation to the public. Although he may have an obligation to potential buyers, his more immediate and pressing one is to protect the safety of the current occupants of the building. Note that the section of the NSPE code quoted previously requires engineers to keep the confidentiality of their clients in all cases, except where exceptions are authorized "by law or this Code." This is probably a case in which part of the code (specifically, the part emphasizing the higher obligation to the safety of the public) should override the requirement of confidentiality.

Even here, however, James should probably try to find a creative middle way that allows him to honor his obligations to his client, the occupants of the building, and potential buyers. He might attempt to persuade the client that his intention to refuse to correct the structural defects is morally wrong and probably not even in his long-term self-interest. He might argue that the client may find himself entangled in lawsuits and that surely he would find it difficult to live with himself if a catastrophe occurred.

Unfortunately, such an approach might not work. James's client might refuse to change his mind. Then James must rank his competing obligations. Most engineering codes, including the NSPE code, are clear that the engineer's first obligation is to the safety of the public, so James must make public the information about the structural defects of the building, at least according to the NSPE code as we interpret it.

The limits of client–professional confidentiality are controversial in most professions. In many states, physicians must reveal cases of child abuse, even if it violates patient–physician confidentiality. The Model Rules of Professional Conduct of the American Bar Association states that lawyers "may" reveal confidential information when there is a threat of "imminent death or substantial bodily harm" (Rule 1.6b).

One of the most famous legal cases regarding professional confidentiality involves a psychologist whose client, Prosenjit Poddar, killed his girlfriend, Tatiana Tarasoff, after informing his psychologist of his intentions. Neither Ms. Tarasoff nor her parents were warned of the danger, and after her death, the parents sued the University of California, where the psychologist was employed. A California court ruled in favor of the parents. Excerpts from the court's opinion are directly relevant to the situation sometimes faced by engineers:

> When a therapist determines, or pursuant to the standards of his profession should determine, that his patient presents a serious danger of violence to another, he incurs an obligation to use reasonable care to protect the intended victim.... We conclude that the public policy favoring protection of the confidential character of patient–psychotherapist communications must yield to the extent to which disclosure is essential to avert danger to others. The protective privilege ends where the public peril begins.[4]

The California court agrees with engineering codes in placing the interests of the public above those of clients or employers. Still, not all cases involving confidentiality will be as clear-cut as the one James faced. In fact, his situation might serve as one extreme on a spectrum of cases. The other extreme might be a case in which an engineer breaks confidentiality to promote his own financial interests. Between these two extremes are many other possible situations in which the decision might be difficult. Again, in such cases, it is appropriate to use the line-drawing method.

6.8   INTELLECTUAL PROPERTY

Intellectual property is property that results from mental labor. It can be protected in several ways, including as trade secrets, patents, trademarks, and copyrights.

Trade secrets are formulas, patterns, devices, or compilations of information that are used in business to gain an advantage over competitors who do not possess the trade secrets. The formula for Coca-Cola is an example of a trade secret. Trade secrets must not be in the public domain and the secrecy must be protected by the firm because trade secrets are not protected by patents.

Patents are documents issued by the government that allow the owner of the patent to exclude others from making use of the patented information for 20 years from the date of filing. To obtain a patent, the invention must be new, useful, and nonobvious. As an example, the puncture-proof tire is patented.

Trademarks are words, phrases, designs, sounds, or symbols associated with goods or services. "Coca-Cola" is a registered trademark.

Copyrights are rights to creative products such as books, pictures, graphics, sculptures, music, movies, and computer programs. The author's estate or heirs retain the copyright for 50 years after his or her death. Copyrights protect the expression of the ideas but not the ideas themselves. The script of *Star Wars,* for example, is copyrighted.

Many companies require their employees to sign a patent assignment whereby all patents and inventions of the employee become the property of the company, often in exchange for a token fee of $1. Sometimes, employees find themselves caught between two employers with respect to such issues.

Consider the case of Bill, a senior engineering production manager of a tire manufacturing company, Roadrubber, Inc. Bill has been so successful in decreasing production costs for his company by developing innovative manufacturing techniques that he has captured the attention of the competition. One competing firm, Slippery Tire, Inc., offers Bill a senior management position at a greatly increased salary. Bill warns Slippery Tire that he has signed a standard agreement with Roadrubber not to use or divulge any of the ideas he developed or learned at Roadrubber for 2 years following any change of employment.

Slippery Tire's managers assure Bill that they understand and will not try to get him to reveal any secrets and also that they want him as an employee because of his demonstrated managerial skills. After a few months on the job at Slippery Tire, someone who was not a part of the earlier negotiations with Bill asks him to reveal some of the secret processes that he developed while at Roadrubber. When Bill refuses, he is told, "Come on, Bill, you know this is the reason you were hired at the inflated salary. If you don't tell us what we want to know, you're out of here." This is a clear case of an attempt to steal information. If the managers who attracted Bill to Slippery Tire were engineers, then they also violated the NSPE code.

"Professional Obligations," item III.1.d of the NSPE code, says, "Engineers shall not attempt to attract an engineer from another employer by false or misleading pretenses." Some cases are not as clear. Sometimes an employee develops ideas at

Company A and later finds that those same ideas can be useful—although perhaps in an entirely different application—to her new employer, Company B.

Suppose Betty's new employer is not a competing tire company but one that manufactures rubber boats. A few months after being hired by Rubberboat, Betty comes up with a new process for Rubberboat. It is only later that she realizes that she probably thought of the idea because of her earlier work with Roadrubber. The processes are different in many ways, and Rubberboat is not a competitor of Roadrubber, but she still wonders whether it is right to offer her idea to Rubberboat.

Let's examine what the NSPE code of ethics has to say about such situations. As already noted, under Rules of Practice, item II.1.c states, "Engineers shall not reveal facts, data, or information obtained in a professional capacity without the prior consent of the client or employer except as authorized or required by law or this Code." Item III.4 states,

> Engineers shall not disclose confidential information concerning the business affairs or technical processes of any present or former client or employer without his consent. (a) Engineers in the employ of others shall not without the consent of all interested parties enter promotional efforts or negotiations for work or make arrangements for other employment as a principal or to practice in connection with a specific project for which the engineer has gained particular and specialized knowledge. (b) Engineers shall not, without the consent of all interested parties, participate in or represent an adversary interest in connection with a specific project or proceedings in which the engineer has gained particular specialized knowledge on behalf of a former client or employer.

Similarly, the Model Rules of Professional Conduct for the National Council of Examiners for Engineering and Surveying (NCEES) require engineers to "not reveal facts, data, or information obtained in a professional capacity without the prior consent of the client or employer as authorized by law" (I.1.d).

These code statements strongly suggest that even in the second case Betty should tell the management at Rubberboat that it must enter into licensing negotiations with Roadrubber. In other words, she must be honest in fulfilling all of her still existing obligations to Roadrubber.

Other cases can be even less clear, however. Suppose the ideas Betty developed while at Roadrubber were never used by Roadrubber. She realized they would be of no use and never even mentioned them to management at Roadrubber. Thus, they might not be considered a part of any agreement between her and Roadrubber. Still, the ideas were developed using Roadrubber's computers and laboratory facilities. Or suppose Betty's ideas occurred to her at home while she was still an employee of Roadrubber, although the ideas probably would never have occurred to her if she had not been working on somewhat related problems at Roadrubber.

We can best deal with these problems by employing the line-drawing method. As we have seen, the method involves pointing out similarities and dissimilarities between the cases whose moral status is clear and the cases whose moral status is less clear.

Here is a simple illustration of how such a line-drawing analysis might work. In the following tables, the Positive column refers to features that, if present, count in favor of the action's being morally acceptable. The Negative column refers to features that, if present, count against the action's being morally acceptable. The "test case" follows.

Case 1. Tom is a young engineering graduate who designs automobile brakes for Ford. While working for Ford, he learns a lot about heat transfer and materials. After 5 years, Tom leaves Ford to take a job at General Motors. While at General Motors, Tom applies his knowledge of heat transfer and materials to design engines. Is Tom stealing Ford's intellectual property? (See Table 6.1.)

TABLE 6.1 (Case 1)

Feature	Positive	Test Case	Negative
Generic Information	Yes	X———————	No
Different Application	Yes	X———————	No
Information Protected as a Trade Secret	No	X———————	Yes

Case 2. Tom is a young engineering graduate who designs automobile brakes for Ford. While working for the company, he learns a lot about heat transfer and materials. After 5 years, Tom leaves Ford to take a job at General Motors. While at General Motors, Tom applies his knowledge of heat transfer and materials to design brakes. Is Tom stealing Ford's intellectual property? (See Table 6.2.)

TABLE 6.2 (Case 2)

Feature	Positive	Test Case	Negative
Generic Information	Yes	X———————	No
Different Application	Yes	———————X	No
Information Protected as a Trade Secret	No	X———————	Yes

Case 3. Tom is a young engineering graduate who designs automobile brakes for Ford. While working for Ford, Tom helps develop a new brake lining that lasts twice as long as conventional brake linings. Ford decides to keep the formula for this brake lining as a trade secret. After 5 years, Tom leaves Ford to take a job at General Motors. While at General Motors, Tom tells the company the formula for the new brake lining. Is Tom stealing Ford's intellectual property? (See Table 6.3.)

TABLE 6.3 (Case 3)

Feature	Positive	Test Case	Negative
Generic Information	Yes	———————X	No
Different Application	Yes	———————X	No
Information Protected as a Trade Secret	No	———————X	Yes

In Case 1, Tom has not stolen Ford's intellectual property. Although it is true that he used generic scientific knowledge acquired while he was at Ford, the information is available to anyone. The application of the generic scientific knowledge is markedly different at General Motors. But because General Motors and Ford both compete in the same market sector and brakes and motors are both parts of automobiles, the "X" does appear at the extreme left of the spectrum. In Case 2, Tom applies his knowledge to the same area, brake design, but the knowledge is still generic scientific knowledge over which Ford has no claim, even if Tom acquired this knowledge while at Ford. Assume the two brake designs are different.

In Case 3, Tom applies his knowledge to the same area, brake design, and the knowledge is specific knowledge of brake design over which Ford has a rightful claim. Tom's action in Case 3 is wrong.

Additional features may come to light in analyzing a particular case. There can also be other intermediate cases between the ones presented here. The particular case of interest must be compared with the spectrum of cases to determine where the line between permissible and impermissible action should be drawn.

6.9   EXPERT WITNESSING

Engineers are sometimes hired as expert witnesses in cases that involve accidents, defective products, structural defects, and patent infringements, as well as in other areas where competent technical knowledge is required. Calling upon an expert witness is one of the most important moves a lawyer can make in such cases, and engineers are usually well compensated for their testimony. However, being an expert witness is time-consuming and often stressful.

Speaking at the winter annual meeting of the American Society of Mechanical Engineers in November 1992, Thomas A. Hunter, an engineering consultant and frequent expert witness, remarked, "Engineers must be credible in court. This credibility depends on the engineer's knowledge of engineering, the particular case, and especially the court process."[5] With regard to cases involving defective products, Hunter warned,

> To make a credible presentation to the jury, it is simply not enough to merely point out that there is a design defect. At a minimum, the expert must show three things. First, that the defect was recognizable by the designer; second, that there were means available to correct the defect when the product was designed; and third, that the costs of corrective features would not price the product out of the market or interfere with the product's effectiveness.[6]

When confronted with these demands, the expert witness faces certain ethical pitfalls. The most obvious is perjury on the witness stand. A more likely temptation is to withhold information that would be unfavorable to the client's case. In addition to being ethically questionable, such withholding can be an embarrassment to the engineer because cross-examination often exposes it. To avoid problems of this sort, an expert should follow several rules.[7]

First, she should not take a case if she does not have adequate time for a thorough investigation. Rushed preparation can be disastrous for the reputation of the expert witness as well as for her client. Being prepared requires not only general technical knowledge but also detailed knowledge of the particular case and the process of the court before which the witness will testify.

Second, she should not accept a case if she cannot do so with good conscience. This means that she should be able to testify honestly and not feel the need to withhold information to make an adequate case for her client.

Third, the engineer should consult extensively with the lawyer so that the lawyer is as familiar as possible with the technical details of the case and can prepare the expert witness for cross-examination.

Fourth, the witness should maintain an objective and unbiased demeanor on the witness stand. This includes sticking to the questions asked and keeping an even temper, especially under cross-examination.

Fifth, the witness should always be open to new information, even during the course of the trial. The following example does not involve an expert witness, but it does show how important new information gained during a trial can be. During a trial of an accident case in Kansas, the defendant discovered in his basement an old document that conclusively showed that his company was culpable in the accident. He introduced this new evidence in court proceedings, even though it cost his company millions of dollars and resulted in the largest accident court judgment in the history of Kansas.[8]

One position a potential expert witness can take with respect to a client is to say something like the following:

> I will have only one opinion, not a "real" opinion and a story I will tell for you on the witness stand. My opinion will be as unbiased and objective as I can possibly make it. I will form my opinion after looking at the case, and you should pay me to investigate the facts of the case. I will tell the truth and the whole truth as I see it on the witness stand, and I will tell you what I will say beforehand. If you can use my testimony, I will serve as an expert witness for you. If not, you can dismiss me.

This approach may not solve all the problems. If an expert witness is dismissed by a lawyer because he has damaging evidence, then is it ethically permissible to simply walk away, without revealing the evidence, even when public safety is involved? Should the witness testify for the other side if asked?

6.10   INFORMING THE PUBLIC

Some types of professional irresponsibility in handling technical information may be best described as a failure to inform those whose decisions are impaired by the absence of the information. From the standpoint of the ethics of respect for persons, this is a serious impairment of moral agency. The failure of engineers to ensure that technical information is available to those who need it is especially wrong where disasters can be avoided.

Dan Applegate was Convair's senior engineer directing a subcontract with McDonnell Douglas in 1972.[9] The contract was for designing and building a cargo hatch door for the DC-10. The design for the cargo door's latch was known to be faulty. When the first DC-10 was pressure tested on the assembly line, the cargo hatch door blew out and the passenger cabin floor buckled, resulting in the destruction of several hydraulic and electrical power lines. Modifications in the design did not solve the problem. Later, a DC-10 flight over Windsor, Ontario, had to make an emergency landing in Detroit after the cargo hatch door flew open and the cabin floor again buckled. Fortunately, no one was injured.

In light of these problems, Applegate wrote a memo to the vice president of Convair, itemizing the dangers of the design. However, Convair managers decided not to pass this information on to McDonnell Douglas because of the possibility of financial penalties and litigation if accidents occurred. Applegate's memorandum was prophetic. Two years later, in 1974, a fully loaded DC-10 crashed just outside Orly Field in Paris, killing all 346 passengers. The crash happened for the reasons that Applegate had outlined in his memorandum. There were genuine legal impediments to disclosing the dangers in the DC-10 design to the federal government or to the general public, but this story emphasizes the fact that failure to disclose information can have catastrophic consequences.

In this case, most of us would probably say that Dan Applegate's professional responsibility to protect the safety of the public required that he do something to make his professional concerns about the DC-10 known. In requiring engineers to notify employers "or such other authority as may be appropriate" if their "professional judgment is overruled under circumstances where the safety, health, property, or welfare of the public are endangered," the NSPE code seems to imply this (II.1.a). Using almost identical language, the NCEES Model Rules of Professional Conduct require registrants to "notify their employer or client and such other authority as may be appropriate when their professional judgment is overruled under circumstances where the life, health, property, and welfare of the public is endangered" (I.c). Applegate's memo was a step in the right direction. Unfortunately, his superiors did not pass his concerns on to the client (McDonnell Douglas). Who bears responsibility for the client never receiving this information is another matter. However, the failure to alert others to the danger resulted in massive expense and loss of life and denied passengers the ability to make an informed decision in accepting an unusual risk in flying in the aircraft.

Similar issues are raised in another well-known case involving the Ford Pinto gas tank in the early 1970s. At the time the Pinto was introduced, Ford was making every effort to compete with the new compact Japanese imports by producing a car in less than 2 years that weighed less than 2000 pounds and cost less than $2000.[10] The project engineer, Lee Iacocca, and his management team believed that the American public wanted the product they were designing. They also believed that the American public would not be willing to pay the extra $11 to eliminate the risk of a rupturing gas tank. The engineers who were responsible for the rear-end crash tests of early prototype models of the Pinto knew that the Pinto met the current regulations for safety requirements in rear-end collisions; however, they also knew that the car failed the new higher standards that were to go into effect in just 2 years. In fact, the car failed 11 of 12 rear-end collisions at the newly prescribed 20-miles-per-hour crash tests. In the new crashes, the gas tanks ruptured and the vehicles caught fire. Thus, many engineers at Ford knew that the drivers of the Pinto were subject to unusual risks of which they were unaware. They also knew that management was not sympathetic to their safety concerns. One of the engineers working on the Pinto test program found that the ignorance of potential drivers about the car's dangers was unacceptable and decided to resign and make the information public. The engineer thus gave car buyers the knowledge they needed to purchase the Pinto with informed consent.

There is evidence that Ford management did not necessarily have a callous disregard for safety. Only a few years earlier, Ford management voluntarily reported that some line employees, in a misguided show of company loyalty, had falsified EPA emissions data on new engines to bring Ford into compliance with EPA regulations on a new model. As a result of this honest disclosure, Ford was required to pay a stiff fine and had to substitute an older model engine on the new car at even greater expense.

The obligation of engineers to protect the health and safety of the public requires more than refraining from telling lies or simply refusing to withhold information. It sometimes requires that engineers aggressively do what they can to ensure that the consumers of technology are not forced to make uninformed decisions regarding the use of that technology. This is especially true when the use of technology involves unusual and unperceived risks. This obligation may require engineers to do what is necessary to either eliminate the unusual risks or, at the very least, inform those who use the technology of its dangers. Otherwise, their moral agency is seriously eroded. Placing yourself in the position of the seven *Challenger* astronauts, you probably would have wanted to hear all of the relevant engineering facts about the risky effects of low temperatures on the rocket booster O-ring seals before giving permission for liftoff. Similar considerations apply to those who flew the DC-10 or drove Pintos.

6.11 CONFLICTS OF INTEREST

John is employed as a design engineer at a small company that uses valves. In recommending product designs for his company's clients, he usually specifies valves made by a relative, even when valves made by other companies might be more appropriate. Should his company's clients discover this, they might well complain that John is involved in a conflict of interest. What does this mean?

Michael Davis has provided one of the most useful discussions of conflicts of interest. Using a modified version of Davis's definition, we shall say that a conflict of interest exists for a professional when, acting in a professional role, he or she has interests that tend to make a professional's judgment less likely to benefit the customer or client than the customer or client is justified in expecting.[11] In the preceding example, John has allowed his interest in maintaining a good relationship with his relative to unduly influence his professional judgment. He has betrayed the trust that his clients have placed in his professional judgment by serving his personal interest in his relative rather than the interests of his clients as he is paid to do.

Conflicts of interest can strike at the heart of professionalism. This is because professionals are paid for their expertise and unbiased professional judgment in pursuing their professional duties, and conflicts of interest threaten to undermine the trust that clients, employers, and the public place in that expertise or judgment. When a conflict of interest is present, there is an inherent conflict between a professional actively pursuing certain interests and carrying out his or her professional duties as one should.

Engineering codes of ethics usually have something to say about conflicts of interest. Cases involving conflicts of interest are the most common kinds of cases brought before the NSPE's Board of Ethical Review. Fundamental Canon 4 of the NSPE code addresses the idea that engineers should act as "faithful agents or trustees" in performing their professional duties. The first entry under the heading is that engineers should disclose all "known" or "potential" conflicts of interest to their employers or clients. Section III on professional obligations specifies some specific prohibitions:

> 5. Engineers shall not be influenced in their professional duties by conflicting interests.
>
>> a. Engineers shall not accept financial or other considerations, including free engineering designs, from material suppliers for specifying their product.
>> b. Engineers shall not accept commissions or allowances, directly or indirectly, from contractors or other parties dealing with clients or employers for the Engineer in connection with work for which the Engineer is responsible.

In considering these prohibitions and conflicts of interest more generally, however, several important points must be kept in mind.

First, a conflict of interest is not just any set of conflicting interests. An engineer may like tennis and swimming and cannot decide which interest is more important to her. This is not a conflict of interest in the special sense in which this term is used in professional ethics because it does not involve a conflict that is likely to influence professional judgment.

Second, simply having more commitments than one can satisfy in a given period of time is not a conflict of interest. Overcommitment can best be characterized as a conflict of commitment. This, too, should be avoided. However, a conflict of interest involves an inherent conflict between a particular duty and a particular interest, regardless of how much time one has on one's hands. For example, serving on a review panel for awarding research grants and at the same time submitting a grant proposal to that review panel creates an inherent conflict between one's interest in being awarded a grant and one's responsibility to exercise impartial judgment of proposal submissions.

Third, the interests of the client, employer, or public that the engineer must protect are restricted to those that are morally legitimate. An employer or client might have an interest that can be served or protected only through illegal activity (e.g., fraud, theft, embezzlement, and murder). An engineer has no professional duty to serve or protect such interests. On the contrary, the engineer may have a duty to expose such interests to external authorities.

Fourth, a distinction is sometimes made between actual and potential conflicts of interest. The following are examples:

> *Actual.* John has to recommend parts for one of his company's products. One of the vendors is Ajax Suppliers, a company in which John has heavily invested.
>
> *Potential.* Roger will have a conflict of interest if he agrees to serve on a committee to review proposals if he has already submitted his own proposal to be reviewed.

The first hypothetical case illustrates something very important about conflicts of interest. Having a conflict of interest need not, in itself, be unethical. John has a conflict of interest, but he has not necessarily done anything wrong—yet. What he does about his conflict of interest is what matters. If he tries to conceal from others that he has the conflict of interest and then recommends Ajax, he will have engaged in ethically questionable behavior. But he could acknowledge the conflict of interest and refrain from recommending in this case. Thus, his conflict of interest would not result in his judgment being compromised.

Fifth, even though it is best to avoid conflicts of interest, sometimes this cannot reasonably be done. Even then, the professional should reveal the existence of the conflict rather than wait for the customer or the public to find out about it on their own. In line with this, Fundamental Canon 4 of the NSPE code states,

> "a. Engineers shall disclose all known or potential conflicts of interest to their employers or clients by promptly informing them of any business association, interest, or other circumstances which could influence or appear to influence their judgment or the quality.

After disclosure, clients and employers can decide whether they are willing to risk the possible corruption of the professional's judgment that such a conflict of interest might cause. Thus, the free and informed consent of clients and employers is preserved.

What if an engineer is convinced that he or she does not have a conflict of interest even though others may think otherwise? Two comments should be stated regarding this issue. First, self-deception is always possible. In a case in which there actually is a conflict of interest, one may have some motivation not to acknowledge this to oneself. Second, it is important to realize that even the appearance of a conflict of interest decreases the confidence of the public in the objectivity and trustworthiness of professional services and thus harms both the profession and the public. Therefore, it is best for engineers to use caution regarding even the appearance of a conflict of interest.

An important part of any professional service is professional judgment. Allowing this to be corrupted or unduly influenced by conflicts of interest or other extraneous considerations can lead to another type of misusing the truth. Suppose engineer Joe is designing a chemical plant and specifies several large pieces of equipment manufactured by a company whose salesperson he has known for many years. The equipment is of good quality, but newer and more innovative lines may actually be better. In specifying his friend's equipment, Joe is not giving his employer or client the benefit of his best and most unbiased professional judgment. In some cases, this may be a form of dishonesty, but in any case Joe's judgment is unreliable.

6.12 CHAPTER SUMMARY

Recognizing the importance of trust and reliability in engineering practice, codes of ethics require engineers to be honest and impartial in their professional judgments. Forms of dishonesty include not only lying and deliberate deception but also withholding the truth and failing to seek out the truth.

From the standpoint of the ethics of respect for persons, dishonesty is wrong because it violates the moral agency of individuals by causing them to make decisions without informed consent. From the utilitarian perspective, dishonesty is wrong because it can undermine the relations of trust on which a scientific community is founded, as well as informed decision making, thus impeding the development of technology.

Dishonesty on campus accustoms a student to dishonesty, which can carry over into his or her professional life. There are, in fact, exact counterparts in the scientific research and engineering communities to the types of dishonesty exhibited by students on campus.

An engineer should respect professional confidentiality. The limits of confidentiality are controversial and often difficult to determine in engineering as in most professions. Decisions to the proper use of intellectual property with regard to trade secrets, patents, and copyrighted material are often difficult to make because they may involve varying degrees of use of intellectual property. The line-drawing method is useful in resolving these problems.

Integrity in expert testimony requires engineers to take cases only when they have adequate time for preparation, to refuse to take cases when they cannot testify in good conscience on behalf of their client, to consult extensively with the lawyer regarding the technical and legal details of the case, to maintain an objective and unbiased demeanor, and always to be open to new information. Engineers also misuse the truth when they fail to seek out or inform employers, clients, or the public of relevant information, especially when this information concerns the health, safety, and welfare of the public.

A conflict of interest exists for professionals when, acting in their professional roles, they have other interests that, if actively pursued, threaten to compromise their professional judgment and interfere with satisfactorily fulfilling their professional duties.

NOTES

1. We are indebted to our student, Ray Flumerfelt, Jr., for this case. Names have been changed to protect those involved.
2. We are indebted to Mark Holtzapple for this example.
3. *Opinions of the Board of Ethical Review,* Vol. VI (Alexandria, VA: National Society of Professional Engineers, 1989), p. 15.
4. California Supreme Court, July 1, 1976. 1331 *California Reporter,* pp. 14–33 (West Publishing). Cited in Joan C. Callahan, *Ethical Issues in Professional Life* (New York: Oxford University Press, 1988), pp. 239–244.
5. "Engineers Face Risks as Expert Witnesses," *Rochester Engineer,* December 1992, p. 27.
6. Ibid.
7. Ibid. For several of these suggestions, see pp. 27, 29.
8. See "Plaintiffs to Get $15.4 Million," *Miami County Republic* [Paola, Kansas], April 27, 1992, p. 1.
9. Paul Eddy, *Destination Disaster: From the Tri-Motor to the DC-10* (New York: Quadrangle/ New York Times Book Company, 1976), pp. 175–188. Reprinted in Robert J. Baum, *Ethical Problems in Engineering,* Vol. 2 (Troy, NY: Center for the Study of the Human Dimensions of Science and Technology, 1980), pp. 175–185.
10. *Grimshaw v. Ford Motor Co.,* App., 174 Cal. Rptr. 348, p. 360.
11. See Michael Davis, "Conflict of Interest," in Deborah Johnson, ed., *Ethical Issues in Engineering* (Englewood Cliffs, NJ: Prentice Hall, 1991), p. 234. For further discussion of conflicts of interest, see Michael Davis and Andrew Stark, eds., *Conflicts of Interest in the Professions* (New York: Oxford University Press, 2001); and Michael S. Pritchard, *Professional Integrity: Thinking Ethically* (Lawrence: University Press of Kansas, 2006), pp. 60–66.

CHAPTER SEVEN

Risk and Liability in Engineering

Main Ideas in this Chapter

- For engineers and risk experts, risk is the product of the likelihood and magnitude of harm.
- Engineers and risk experts have traditionally identified harms and benefits with factors that are relatively easily quantified, such as economic losses and loss of life.
- In a new version of the way engineers and risk experts deal with risk, the "capabilities" approach focuses on the effects of risks and disasters on the capabilities of people to live the kinds of lives they value.
- The public conceptualizes risk in a different way from engineers and risk experts, taking account of such factors as free and informed consent to risk and whether risk is justly distributed.
- Government regulators have a still different approach to risk because they place more weight on avoiding harm to the public than producing good.
- Engineers have techniques for estimating the causes and likelihood of harm, but their effectiveness is limited.
- Engineers must protect themselves from unjust liability for harm to risk while also protecting the public from risk.

ON THE FOGGY SATURDAY MORNING of July 28, 1945, a twin-engine U.S. Army Air Corps B-25 bomber lost in the fog crashed into the Empire State Building 914 feet above street level. It tore an 18-by-2-foot hole in the north face of the building and scattered flaming fuel into the building. New York firemen put out the blaze in 40 minutes. The crew members and 10 persons at work perished.[1] The building was repaired and still stands.

Just 10 years later, in 1955, the leaders of the New York City banking and real estate industries got together to initiate plans for the New York City World Trade Center, which would later become known as the Twin Towers, the world's tallest buildings.[2] However, as the plans emerged, it became clear that the buildings required new construction techniques and the easing of old building code requirements.[3]

On September 11, 2001, terrorists attacked the Twin Towers by flying two hijacked 727 passenger jets into them, each jet smashing approximately two-thirds of the way up its respective tower. The first consequence of the attack was the fire that started over several floors by high-octane aviation fuel. The fires isolated more than 2,000 workers in the floors above them. Only 18 of the more than 2,000 were able to descend the flaming stairwells to safety. Most of the 2,000 perished in the later collapse of the buildings. By comparison, almost all of the workers in the floors below the fire were able to make it down to safety before the towers collapsed. As reported in the *New York Times,* the present plans for the 9/11 ground-zero memorial building call for high-rise stairwell designs that would diminish the possibility of this kind of tragedy.

In the hour following the plane crashes, the intense heat of the flames (more than 1,000 degrees Fahrenheit) caused the steel floor beams in each tower to sag. As a result, the floor structures broke away from the external vertical load-carrying beams. As the floors fell, they created loads on the lower floors that the external vertical beams could not support, and both buildings fell.

For an engineer, 9/11 raises questions of how this structural failure could have happened, why the building codes did not better protect the public, and how such a disaster can be prevented in the future. There are even larger questions about acceptable risk and the proper approach to risk as an issue of public policy.

7.1 INTRODUCTION

The concern for safety is a common one for engineers. How should engineers deal with issues of safety and risk, especially when they involve possible liability for harm? In the Twin Towers case, the risk was increased by the earlier weakening of building codes and the use of new structural designs that were untested, even though the building codes required such testing.[4] This illustrates an important fact: Engineering necessarily involves risk, and innovation usually increases the risks. One cannot avoid risk simply by remaining with tried and true designs, but innovation creates technologies in which the risks are not fully understood, thereby increasing the chance of failure. Without innovation, there is no progress. A bridge or building is constructed with new materials or with a new design. New machines are created and new compounds synthesized, always without full knowledge of their long-term effects on humans or the environment. Even new hazards can be found in products, processes, and chemicals that were once thought to be safe. Thus, risk is inherent in engineering.

The relationship of safety to risk is an inverse one. Because of the laws of engineering science and statistics, the more we accept risk in an engineering project, the less safe it will become. If there were absolutely no risk in a project, then that project would be absolutely safe. So safety and risk are intimately connected. Concern for safety pervades engineering practice. One of the most common concepts in engineering practice is the notion of "factors of safety." If the largest load a walkway will have to carry at any one time is 1,000 pounds, for example, then a prudent engineer might design the walkway geometry to carry 3,000 pounds. The walkway dimensions for normal usage would then be designed with a factor of safety of three on geometry.

Accepted engineering practice goes still further. In choosing materials to build the walkway, an engineer might begin with a material that has an advertised yield stress of a given number of pounds per square inch and then treat this material as if it had only half of that capability in determining how much material to include in the walkway construction. This introduces an additional factor of safety of two on materials. The final overall factor of safety at the walkway would be the product of the two separate factors, or six in this example. Thus, a prudent engineer would design the walkway to be six times as strong as required for normal everyday use to account for unpredictably high loads or unaccountably weak construction material. This approach is taught to all engineers early in their accredited training, and factors of safety of six or higher are the norm rather than the exception.

Virtually all engineering codes give a prominent place to safety, stating that engineers must hold paramount the safety, health, and welfare of the public. The first Fundamental Canon of the National Society of Professional Engineers Code of Ethics requires members to "hold paramount the safety, health, and welfare of the public." In Section III.2.b, it instructs engineers not to "complete, sign, or seal plans and/or specifications that are not in conformity with applicable engineering standards." Section II.1.a instructs engineers that if their professional judgment is overruled in circumstances that endanger life or property, they shall notify their employer or client and such other authority as may be appropriate. Although "such other authority as may be appropriate" is left undefined, it probably includes those who enforce local building codes and regulatory agencies.

We begin this chapter by considering three different approaches to risk and safety, all of which are important in determining public policy regarding risk. Then we examine more directly the issues of risk communication and public policy concerning risk, including one example of public policy regarding risk—building codes. Next, we consider the difficulties in both estimating and preventing risk from the engineering perspective, including the problem of self-deception. Finally, we discuss some of the legal issues surrounding risk, including protecting engineers from undue liability and the differing approaches of tort law and criminal law to risk.

7.2   THE ENGINEER'S APPROACH TO RISK
Risk as the Product of the Probability and Magnitude of Harm

To assess a risk, an engineer must first identify it. To identify a risk, an engineer must first know what a risk is. The usual engineering definition of risk is "a compound measure of the probability and magnitude of adverse effect."[5] That is, risk is composed of two elements: the likelihood of an adverse effect or harm and the magnitude of that adverse effect or harm. By *compound* is meant the *product*. Risk, therefore, is the product of the likelihood and the magnitude of harm. A relatively slight harm that is highly likely might constitute a greater risk to more people than a relatively large harm that is far less likely.

We can define a *harm* as an invasion or limitation of a person's freedom or well-being. Engineers have traditionally thought of harms in terms of things that can be relatively easily quantified, namely as impairments of our physical and economic well-being. Faulty design of a building can cause it to collapse, resulting in economic loss to the owner and perhaps death for the inhabitants. Faulty design of a chemical plant can cause accidents and economic disaster. These harms are then measured in terms of the numbers of lives lost, the cost of rebuilding or repairing buildings and highways, and so forth.

Engineers and other experts on risk often believe that the public is confused about risk, sometimes because the public does not have the correct factual information about the likelihood of certain harms. A 1992 National Public Radio story on the

Environmental Protection Agency (EPA) began with a quote from EPA official Linda Fisher that illustrated the risk expert's criticism of public understanding of risk:

> A lot of our priorities are set by public opinion, and the public quite often is more worried about things that they perceive to cause greater risks than things that really cause risks. Our priorities often times are set through Congress … and those [decisions] may or may not reflect real risk. They may reflect people's opinions of risk or the Congressmen's opinions of risk.[6]

Every time Fisher refers to "risk" or "real risk," we can substitute "probability of death or injury." Fisher believes that whereas both members of the U.S. Congress and ordinary laypeople may be confused about risk, the experts know what it is. Risk is something that can be objectively measured—namely, the product of the likelihood and the magnitude of harm.

Utilitarianism and Acceptable Risk

The engineering conception of risk focuses on the factual issues of the probability and magnitude of harm and contains no implicit evaluation of whether a risk is morally acceptable. In order to determine whether a risk is morally acceptable, engineers and risk experts usually look to utilitarianism. This position holds, it will be remembered, that the answer to any moral question is to be found by determining the course of action that maximizes well-being. Given the earlier definition of risk as the product of the probability and the consequences of harm, we can state the risk expert's criterion of acceptable risk in the following way:

> An acceptable risk is one in which the product of the probability and magnitude of the harm is equaled or exceeded by the product of the probability and magnitude of the benefit, and there is no other option where the product of the probability and magnitude of the benefit is substantially greater.

One way of implementing this account of acceptable risk is by means of an adaptation of cost–benefit analysis. As we have seen, utilitarians sometimes find cost–benefit analysis to be a useful tool in assessing risk. In applying this method to risk, the technique is often called *risk–benefit analysis* because the "cost" is measured in terms of the risk of deaths, injuries, or other harms associated with a given course of action. For simplicity, however, we shall continue to use the term *cost–benefit analysis*.

Consider a case in which a manufacturing process produces bad-smelling fumes that might be a threat to health. From the cost–benefit standpoint, is the risk to the workers from the fumes acceptable? To determine whether this is an acceptable risk of death from the cost–benefit perspective, one would have to compare the cost associated with the risk to the cost of preventing or drastically reducing it. To calculate the cost of preventing the harms, we would have to include the costs of modifying the process that produces the fumes, the cost of providing protective masks, the cost of providing better ventilation systems, and the cost of any other safety measures necessary to prevent the deaths. Then we must calculate the cost of not preventing the deaths caused by the fumes. Here, we must include such factors as the cost of additional health care, the cost of possible lawsuits because of the deaths, the cost of bad publicity, the loss of income to the families of the workers, and costs associated with the loss of life. If the total cost of preventing the loss of life is greater than the total cost of not preventing the deaths, then the current level of risk is acceptable.

If the total cost of not preventing the loss of life is greater than the total cost of preventing the loss, then the current level of risk is unacceptable.

The utilitarian approach to risk embodied in risk–benefit analysis has undoubted advantage in terms of clarity, elegance, and susceptibility to numerical interpretation. Nevertheless, there are some limitations that must be kept in mind.

First, it may not be possible to anticipate all of the effects associated with each option. Insofar as this cannot be done, the cost–benefit method will yield an unreliable result.

Second, it is not always easy to translate all of the risks and benefits into monetary terms. How do we assess the risks associated with a new technology, with eliminating a wetland, or with eliminating a species of bird in a Brazilian rain forest? Apart from doing this, however, a cost–benefit analysis is incomplete.

The most controversial issue in this regard is, of course, the monetary value that should be placed on human life. One way of doing this is to estimate the value of future earnings, but this implies that the lives of retired people and others who do not work, such as housewives, are worthless. So a more reasonable approach is to attempt to place the same value on people's lives that they themselves place on their lives. For example, people often demand a compensating wage to take a job that involves more risk. By calculating the increased risk and the increased pay that people demand for more risky jobs, some economists say, we can derive an estimate of the monetary value people place on their own lives. Alternatively, we can calculate how much more people would pay for safety in an automobile or other things they use by observing how much more they are willing to pay for a safer car. Unfortunately, there are various problems with this approach. In a country in which there are few jobs, a person might be willing to take a risky job he or she would not be willing to take if more jobs were available. Furthermore, wealthy people are probably willing to pay more for a safer car than poorer citizens.

Third, cost–benefit analysis in its usual applications makes no allowance for the distribution of costs and benefits. Suppose more overall utility could be produced by exposing workers in a plant to serious risk of sickness and death. As long as the good of the majority outweighs the costs associated with the suffering and death of the workers, the risk is justified. Yet most of us would probably find that an unacceptable account of acceptable risk.

Fourth, the cost–benefit analysis gives no place for informed consent to the risks imposed by technology. We shall see in our discussion of the lay approach to risk that most people think informed consent is one of the most important features of justified risk.

Despite these limitations, cost–benefit analysis has a legitimate place in risk evaluation. When no serious threats to individual rights are involved, cost–benefit analysis may be decisive. In addition, cost–benefit analysis is systematic, offers a degree of objectivity, and provides a way of comparing risks, benefits, and cost by the use of a common measure—namely, monetary cost.

Expanding the Engineering Account of Risk: The Capabilities Approach to Identifying Harm and Benefit

As we have pointed out, engineers, in identifying risks and assessing acceptable risk, have traditionally identified harm with factors that are relatively easily quantified, such as economic losses and the number of lives lost.[7] There are, however, four main limitations with this rather narrow way of identifying harm. First, often only the immediately apparent or *focal consequences* of a hazard are included, such as the number of fatalities or the number of homes without electricity. However, hazards can have *auxiliary consequences,* or broader and more indirect harms to society. Second, both natural and engineering hazards might create opportunities, which should be accounted for in the aftermath of a disaster. Focusing solely on the negative impacts and not including these benefits may lead to overestimating the negative societal consequences of a hazard. Third, there remains a need for an accurate, uniform, and consistent metric to quantify the consequences (harms or benefits) from a hazard. For example, there is no satisfactory method for quantifying the non-fatal physical or psychological harms to individuals or the indirect impact of hazards on society. The challenge of quantification is difficult and complex, especially when auxiliary consequences and opportunities are included in the assessment. Fourth, current techniques do not demonstrate the connection between specific harms or losses, such as the loss of one's home and the diminishment of individual or societal well-being, and quality of life. Yet it is surely the larger question of effect on quality of life that is ultimately at issue when considering risk.

In their work on economic development, economist Amartya Sen and philosopher Martha Nussbaum have derived a notion of "capabilities" that the two scholars believe may be the basis of a more adequate way of measuring the harms (and sometimes the benefits) of disasters, including engineering disasters. Philosopher Colleen Murphy and engineer Paolo Gardoni have developed a capabilities-based approach to risk analysis, which focuses on the effect of disasters on overall human well-being. Well-being is defined in terms of individual capabilities, or "the ability of people to lead the kind of life they have reason to value."[8] Specific capabilities are defined in terms of *functionings,* or what an individual can do or become in his or her life that is of value. Examples of functionings are being alive, being healthy, and being sheltered. A *capability* is the real freedom of individuals to achieve a functioning, and it refers to the real options he or she has available. Capabilities are constituent elements of individual well-being.

Capabilities are distinct from *utilities,* which refers to the mental satisfaction, pleasure, or happiness of a particular individual. Often, people's preferences or choices are used to measure satisfaction. Utilities are assigned to represent a preference function. In other words, if an individual chooses A over B, then A has more utility than B. Using utilities to measure the well-being of individuals, however, is problematic because happiness or preference-satisfaction is not a sufficient indicator of an individual's well-being. For example, a person with limited resources might learn to take pleasure in small things, which are only minimally satisfying to a person with more ample means. The individual in a poverty-stricken situation might have all of his or her severely limited desires satisfied. From the utilitarian standpoint, the person would be described as happy and be said to enjoy a high standard of living. Yet this individual might still be objectively deprived. The problem here is that utilitarianism does not take into account the number and quality of options that are available to individuals, which is precisely what capabilities capture.

From the capabilities standpoint, a risk is the probability that individuals' capabilities might be reduced due to some hazard. In determining a risk, the first step is to identify the important capabilities that might be damaged by a disaster. Then, in order to quantify the ways in which the capabilities might be damaged, we must find some "indicators" that are correlated with the capabilities. For example, an indicator of the impairment of the capability for play might be the loss of parks or gym facilities. Next, the indicators must be scaled onto a common metric so that the normalized values of the indicators can be compared. Then, a summary index is constructed by combining the information provided by each normalized indicator, creating a *hazard index* (HI). Finally, in order to put the HI into the relevant context, its value is divided by the population affected by the hazard, creating the *hazard impact index*, which measures the hazard impact per person.

According to its advocates, there are four primary benefits of using the capabilities-based approach in identifying the societal impact of a hazard. First, capabilities capture the adverse effects and opportunities of hazards beyond the consequences traditionally considered. Second, since capabilities are constitutive aspects of individual well-being, this approach focuses our attention on what should be our primary concern in assessing the societal impact of a hazard. Third, the capabilities-based approach offers a more accurate way to measure the actual impact of a hazard on individuals' well-being. Fourth, rather than considering diverse consequences, which increases the difficulty of quantification, the capabilities-based approach requires considering a few properly selected capabilities.[9]

In addition to identifying more accurately and completely the impact of a hazard, its advocates believe the capabilities-based approach provides a principled foundation for judging the acceptability and tolerability of risks.[10] Judgments of the acceptability of risks are made in terms of the impact of potential hazards on the capabilities of individuals. Thus, according to the capabilities approach,

> A risk is acceptable if the probability is sufficiently small that the adverse effect of a hazard will fall below a threshold of the minimum level of capabilities attainment that is acceptable in principle.

The "in principle" qualification captures the idea that, ideally, we do not want individuals to fall below a certain level. We might not be able to ensure this, however, especially immediately after a devastating disaster. In practice, then, it can be tolerable for individuals to temporarily fall below the acceptable threshold after a disaster, as long as this situation is reversible and temporary and the probability that capabilities will fall below a tolerability threshold is sufficiently small. Capabilities can be a little lower, temporarily, as long as no permanent damage is caused and people do not fall below an absolute minimum.

7.3 THE PUBLIC'S APPROACH TO RISK

Expert and Layperson: Differences in Factual Beliefs

The capabilities approach may give a more adequate account of the harms and benefits that should be measured. However, when one encounters the lay public's approach to risk, one still seems to be entering a different universe. The profound differences between the engineering and public approach to risk have been the sources of miscommunication and even acrimony. What are the grounds for these profound differences in outlook on risk?

The first difference is that engineers and risk experts believe that the public is sometimes mistaken in estimating the probability of death and injury from various activities or technologies. Recall EPA official Linda Fisher's reference to "real risk," by which she meant the actual calculations of probability of harm. Risk expert Chauncey Starr has a similarly low opinion of the public's knowledge of probabilities of harm. He notes that people tend to overestimate the likelihood of low-probability risks associated with causes of death and to underestimate the likelihood of high-probability risks associated with causes of death. The latter tendency can lead to overconfident biasing, or *anchoring*. In anchoring, an original estimate of risk is made—an estimate that may be substantially erroneous. Even though the estimate is corrected, it is not sufficiently modified from the original estimate. The original estimate anchors all future estimates and precludes sufficient adjustment in the face of new evidence.[11]

Other scholars have reported similar findings. A study by Slovic, Fischhoff, and Lichtenstein shows that although even experts can be mistaken in their estimations of various risks, they are not as seriously mistaken as laypeople.[12] The study contrasts actual versus perceived deaths per year.[13] Experts and laypeople were asked their perception of the number of deaths per year for such activities as smoking, driving a car, driving a motorcycle, riding in a train, skiing, and so on. On a graph that plots perceived deaths (on the vertical axis) against actual deaths (on the horizontal axis) for each of several different risks, if the perception (by either laypeople or experts) of deaths were accurate, then the result would be a 45-degree line. In other words, actual and perceived deaths would be the same for the plots of the perceptions of either laypersons or experts. Instead, the experts were consistently approximately one order of magnitude (i.e., approximately 10 times) low in their perceptions of the perceived risk, and the lay public was still another order of magnitude (i.e., approximately 100 times) too low, resulting in lines of less than 45 degrees for experts and even less for laypersons.

"Risky" Situations and Acceptable Risk

It does appear to be true that the engineer and risk expert, on the one hand, and the public, on the other hand, differ regarding the probabilities of certain events. The major difference, however, is in the conception of risk itself and in beliefs about acceptable risk. One of the differences here is that the public often combines the concepts of risk and acceptable risk—concepts that engineers and risk experts separate sharply. Furthermore, public discussion is probably more likely to use the adjective "risky" than the noun "risk."

We can begin with the concepts of "risk" and "risky." In public discussion, the use of the term *risky,* rather than referring to the probability of certain events, more often than not has the function of a warning sign, a signal that special care should be taken in a certain area.[14] One reason for classifying something as risky is that it is new and unfamiliar. For example, the public may think of the risk of food poisoning from microbes as being relatively low, whereas eating irradiated food is "risky." In fact, in terms of probability of harm, there may be more danger from microbes than radiation, but the dangers posed by microbes are familiar and commonplace, whereas the dangers from irradiated foods are unfamiliar and new. Another reason for classifying something as risky is that the information about it might come from a questionable source. We might say that buying a car from a trusted friend who testifies that the car is in good shape is not risky, whereas buying a car from a used car salesman whom we do not know is risky.

Laypeople do not evaluate risk strictly in terms of expected deaths or injury. They consider other factors as well. For example, they are generally willing to take voluntary risks that are 1,000 times (three orders of magnitude) as uncertain as involuntary risks. Thus, voluntarily assumed risks are more acceptable than risks not voluntarily assumed. The amount of risk people are willing to accept in the workplace is generally proportional to the cube of the increase in the wages offered in compensation for the additional risk. For example, doubling wages would tend to convince a worker to take eight times the risk. But laypeople may also separate by three orders of magnitude the risk perceived to be involved in involuntary exposure to danger (e.g., when a corporation places a toxic waste dump next door to one's house) and the risk involved in voluntary exposure (e.g., smoking). Here, voluntarily assumed risks are viewed as inherently less risky, not simply more acceptable. Laypeople also seem to be content with spending different amounts of money in different areas to save a life. In his study of 57 risk-abatement programs at five different government agencies in Washington, DC, including the EPA and the Occupational Safety and Health Administration (OSHA), Starr shows that such programs vary greatly in the amount of money they spend to save a life. Some programs spend $170,000 per life, whereas others spend $3 million per life.[15]

Another researcher, D. Litai, has separated risk into 26 risk factors, each having a dichotomous scale associated with it.[16] For example, a risk may have a natural or a human origin. If the risk has a human origin, Litai concludes from an analysis of statistical data from insurance companies that the perceived risk is 20 times greater than a risk with a natural origin. An involuntarily assumed risk (whether of natural or human origin) is perceived as being 100 times greater than a voluntarily assumed one. An immediate risk is perceived as being 30 times greater than an ordinary one. By contrast, a regular risk is perceived as being just as great as an occasional one, and necessary risk is just as great as a luxury-induced one. Here again, there is evidence of the amalgamation of the concepts of risk and acceptable risk.

Two issues in the public's conception of risk and acceptable risk have special moral importance: free and informed consent and equity or justice. These two concepts follow more closely the ethics of respect for persons than utilitarianism. According to this ethical perspective, as we have seen, it is wrong to deny the moral agency of individuals. Moral agents are beings capable of formulating and pursuing purposes of their own. We deny the moral agency of individuals when we deny their ability to formulate and pursue their own goals or when we treat them in an inequitable manner with respect to other moral agents. Let us examine each of these concepts in more detail.

Free and Informed Consent

To give free and informed consent to the risks imposed by technology, three things are necessary. First, a person must not be coerced. Second, a person must have the relevant information. Third, a person must be rational and competent enough to evaluate the information. Unfortunately, determining when meaningful and informed consent has been given is not always easy, for several reasons.

First, it is difficult to know when consent is free. Have workers given their free consent when they continue to work at a plant with known safety hazards? Perhaps they have no alternative form of employment.

Second, people are often not adequately informed of dangers or do not evaluate them correctly. As we have seen, sometimes laypeople err in estimating risk. They underestimate the probability of events that have not occurred before or that do not get their attention, whereas they overestimate the probability of events that are dramatic or catastrophic.

Third, it is often not possible to obtain meaningful informed consent from individuals who are subject to risks from technology. How would a plant manager obtain consent from local residents for his plant to emit a substance into the atmosphere that causes mild respiratory problems in a small percentage of the population? Is the fact that the residents do not protest sufficient evidence that they have consented? What if they do not know about the substance, do not know what it does, do not understand its effects correctly, or are simply too distracted by other things?

In light of the problems in getting free and informed consent, we could compensate individuals after the fact for actual harms done to them through technology. For example, people could be compensated for harms resulting from a defective design in an automobile or a release of a poisonous gas from a chemical plant. This approach has the advantage that consent does not have to be obtained, but it also has several disadvantages. First, it does not tell us how to determine adequate compensation. Second, it limits the freedom of individuals because some people would never have consented. Third, sometimes there is no adequate compensation for a harm, as in the case of serious injury or death.

There are problems with both informed consent and compensation as ways of dealing with the ethical requirement to respect the moral agency of those exposed to risk because of technology. Nevertheless, some effort must be made to honor this requirement. Now let us return to the second requirement of the respect for persons morality with regard to risk.

Equity or Justice

The ethics of respect for persons places great emphasis on respecting the moral agency of individuals, regardless of the cost to the larger society. Philosopher John Rawls expresses this concern:[17] "[E]ach member of society is thought to have an inviolability founded upon justice ... which even the welfare of everyone else cannot override." As an example of the requirement for justice derived from the ethics of respect for persons, consider the following statement by Mrs. Talbert, whose husband's health was severely damaged by byssinosis caused by cotton dust:

> My husband worked in the cotton mill since 1937 to 1973. His breath was so short he couldn't walk from the parking lot to the gate the last two weeks he worked.
>
> He was a big man, liked fishing, hunting, swimming, playing ball, and loved to camp. We liked to go to the mountains and watch the bears. He got so he could not breathe and walk any distance, so we had to stop going anywhere. So we sold our camper, boat, and his truck as his doctor, hospital, and medicine bills were so high. We don't go anywhere now. The doctor said his lungs were as bad as they could get to still be alive. At first he used tank oxygen about two or three times a week, then it got so bad he used more and more. So now he has an oxygen concentrator, he has to stay on it 24 hours a day. When he goes to the doctor or hospital he has a little portable tank.
>
> He is bedridden now. It's a shame the mill company doesn't want to pay compensation for brown lung. If they would just come and see him as he is now, and only 61 years old.[18]

A utilitarian might be willing to trade off the great harm to Mr. Talbert that resulted from a failure to force cotton mills to protect their workers from the risk of byssinosis for the smaller advantages to an enormous number of people. After all, such protection is often highly expensive, and these expenses must eventually be passed on to consumers in the form of higher prices for cotton products. Higher prices also make U.S. cotton products more expensive and thus less competitive in world markets, thereby depriving U.S. workers of jobs. Regulations that protect workers might even force many (perhaps all) U.S. cotton mills to close. Such disutilities might well outweigh the disutilities to the Mr. Talberts of the world.

From the standpoint of the ethics of respect for persons, however, such considerations must not be allowed to obscure the fact that Mr. Talbert has been treated unjustly. Although many people enjoy the benefits of the plant, only Mr. Talbert and a few others suffer the consequences of unhealthy working conditions. The benefits and harms have been inequitably distributed. His rights to bodily integrity and life were unjustly violated. From the standpoint of the Golden Rule, probably few, if any, observers would want to be in Mr. Talbert's position.

Of course, it is not possible to distribute all risks and benefits equally. Sometimes those who endure the risks imposed by technology may not share the benefits to the same degree. For example, several years ago a proposal was made to build a port for unloading liquefied natural gas in the Gulf of Mexico off the coast of Texas. The natural gas would be shipped to many parts of the United States, so most citizens of the country would benefit from this project. Only those residents close to the port, however, would share the risks of the ships or storage tanks exploding.[19] Because there is no way to equalize the risk, informed consent and compensation should be important considerations in planning the project. Thus, informed consent, compensation, and equity are closely related considerations in moral evaluation.

Even though laypeople often combine the concept of risk with the concept of acceptable risk, we shall formulate a lay criterion of acceptable risk in the following way:

> An acceptable risk is one in which (1) risk is assumed by free and informed consent, or properly compensated, and in which (2) risk is justly distributed, or properly compensated.

We have seen that there are often great difficulties in implementing the requirements of free and informed consent, compensation, and justice. Nevertheless, they are crucial considerations from the layperson's perspective—and from the moral perspective.

7.4 THE GOVERNMENT REGULATOR'S APPROACH TO RISK

According to William Ruckelshaus, former EPA administrator, regulators face a dilemma regarding risk management.[20] On the one hand, regulators could decide to regulate only when there is a provable connection between a substance and some undesirable effect such as a risk of cancer. Because of the difficulties in establishing the acceptable levels of exposure to toxic substances at which there is no danger, this option would expose the public to unacceptable risks. On the other hand, regulators could eliminate any possible risk insofar as this is technologically possible. Choosing this option would result in the expenditure of large sums of money to eliminate minute amounts of any substance that might possibly pose risks to human beings.

This would not be cost-effective. Funds might better be spent elsewhere to eliminate much greater threats to public health.

We can illustrate this conflict with the following example. Suppose Sue, a young engineer, is worried about a compound (call it Compound X) that her company is discharging into the air. Compound X is not regulated by the EPA, and she wonders whether its emission is a health hazard for the public. Her boss says he has looked at the epidemiological literature on Compound X, and it does not show any connection between Compound X and health problems.[21] Sue, however, is more sophisticated in her knowledge of the way such connections are established.[22] Assume that a scientist wants to investigate a causal link between Compound X and cancer. In performing these studies (called *cohort studies*), the scientist would be especially concerned to avoid claiming that there is a link between Compound X and cancer when there is none. In fact, as a scientist, he is going to be more concerned about avoiding a claim that there is a link between Compound X and cancer when there is none than in claiming that there is not a link between Compound X and cancer when there is one. The reason for this is that to make a claim about a causal relationship that is false is more damaging to one's reputation as a scientist than to fail to make a claim about a causal relationship that is true.

Unfortunately, as Sue is well aware, public policy interests are not in agreement with scientific scruples at this point. From the standpoint of protecting the public from carcinogens, we are more interested in discovering a causal connection between Compound X and cancer if one actually exists than in avoiding making a claim about a causal connection that does not exist. Only by adopting this policy can the public be adequately protected from carcinogens. Thus, whereas scientists have a bias against false positives (making a claim for a causal connection when there is not one), those whose highest priority is protecting the public have a bias against false negatives (claiming there is not a causal connection when there is one). Sue knows that there is another reason why scientists place primary emphasis on eliminating false positives. From a statistical standpoint, eliminating false negatives requires a larger sample than eliminating false positives, thus making the cohort studies more expensive. Therefore, scientists avoid false positives for reasons based on economics as well as to preserve their scientific reputations.

Sue is also aware of a third reason why some scientists might favor eliminating false positives. Some scientific studies are privately funded, and many scientists have vested interests in conclusions that give compounds a clean bill of health with regard to harm to the public. Many compounds have considerable value in the marketplace, and industrial firms are not anxious to have them declared a threat to public health. Favoring the elimination of false positives tends to support the industry position.

Given these facts, Sue knows that scientific studies may not offer the public as much protection against carcinogens and other harmful substances as one might suppose. She is aware that there are value judgments involved in epidemiological estimates of risk, and these value judgments favor the discovery of scientific truth, economic efficiency, and perhaps even the interests of those who sponsor the research, rather than protecting the public. She wonders why this should be true, especially if public funding is supporting the scientific investigations. Perhaps, as one writer suggests, there should be two kinds of studies: those devoted to pure science and those that will form the basis of public policy decisions.[23]

Let us propose the following criterion of acceptable risk from the standpoint of the government regulator:

> An acceptable risk is one in which protecting the public from harm has been weighted more heavily than benefiting the public.[24]

7.5   COMMUNICATING RISK AND PUBLIC POLICY

Communicating Risk to the Public

The preceding sections show that different groups have somewhat different agendas regarding risk. Engineers are most likely to adopt the risk expert's approach to risk. They define risk as the product of the magnitude and likelihood of harm and are sympathetic with the utilitarian way of assessing acceptable risk. The professional codes require engineers to hold paramount the safety, health, and welfare of the public, so engineers have an obligation to minimize risk. However, in determining an acceptable level of risk for engineering works, they are likely to use, or at least be sympathetic with, the cost–benefit approach.

The lay public comes to issues of risk from a very different approach. Although citizens sometimes have inaccurate views about the probabilities of harms from certain types of technological risks, their different approach cannot be discounted in terms of simple factual inaccuracies. Part of the difference in approach results from the tendency to combine judgments of the likelihood and acceptability of risk. (The term *risky* seems to include both concepts.) For example, use of a technology is more risky if the technology is relatively new, and if information about it comes from a source (either expert or nonexpert) that the public has come to regard as unreliable. More important, the lay public considers free and informed consent and equitable distribution of risk (or appropriate compensation) to be important in the determination of acceptable risk.

Finally, the government regulator, with her special obligation to protect the public from undue technological risks, is more concerned with preventing harm to the public than with avoiding claims for harm that turn out to be false. This bias contrasts to some extent with the agendas of both the engineer and the layperson. Although, as a government regulator, she may often use cost–benefit analysis as a part of her method of determining acceptable risk, she has a special obligation to prevent harm to the public, and this may go beyond what cost–benefit considerations require. On the other hand, considerations of free and informed consent and equity, while important, may be balanced by cost–benefit considerations.

In light of these three different agendas, it is clear that social policy regarding risk must take into consideration wider perspectives than the risk expert approach would indicate. There are at least two reasons for this claim. First, the public and government regulators will probably continue to insist on introducing their own agendas into the public debate about technological risk. In a democracy, this probably means that these considerations will be a part of public policy regarding technological risk, whether or not engineers and risk experts approve. This is simply a fact to which engineers and risk experts must adjust. Second, we believe the two alternative approaches to risk have a genuine moral foundation. Free and informed consent, equity, protecting the public from harm—these are morally legitimate considerations. Therefore, public policy regarding risk should probably be a mix of the considerations we have put forth here as well as, no doubt, many others we have not discussed.

What, then, is the professional obligation of engineers regarding risk? One answer is that engineers should continue to follow the risk expert's approach to risk and let public debate take care of the wider considerations. We believe there is some validity to this claim, and in the next section we return to a consideration of issues in typical engineering approaches to risk. However, as we have argued in Chapter 5 and elsewhere, we believe engineers have a wider professional obligation. Engineers have a professional obligation to participate in democratic deliberation regarding risk by contributing their expertise to this debate. In doing so, they must be aware of alternative approaches and agendas in order to avoid serious confusion and undue dogmatism. In light of this, we propose the following guidelines for engineers in risk communication:[25]

1. Engineers, in communicating risk to the public, should be aware that the public's approach to risk is not the same as that of the risk expert. In particular, "risky" cannot be identified with a measure of the probability of harm. Thus, engineers should not say "risk" when they mean "probability of harm." They should use the two terms independently.

2. Engineers should be wary of saying, "There is no such thing as zero risk." The public often uses "zero risk" to indicate not that something involves no probability of harm but that it is a familiar risk that requires no further deliberation.

3. Engineers should be aware that the public does not always trust experts and that experts have sometimes been wrong in the past. Therefore, engineers, in presenting risks to the public, should be careful to acknowledge the possible limitations in their position. They should also be aware that laypeople may rely on their own values in deciding whether or not to base action on an expert's prediction of probable outcomes.

4. Engineers should be aware that government regulators have a special obligation to protect the public, and that this obligation may require them to take into account considerations other than a strict cost–benefit approach. Although public policy should take into account cost–benefit considerations, it should take into account the special obligations of government regulators.

5. Professional engineering organizations, such as the professional societies, have a special obligation to present information regarding technological risk. They must present information that is as objective as possible regarding probabilities of harm. They should also acknowledge that the public, in thinking about public policy regarding technological risk in controversial areas (e.g., nuclear power), may take into consideration factors other than the probabilities of harm.

A major theme in these guidelines is that engineers should adopt a *critical attitude* toward the assessment of risk. This means that they should be aware of the existence of perspectives other than their own. The critical attitude also implies that they should be aware of the limitations in their own abilities to assess the probabilities and magnitude of harms. In the next section, we consider an example of these limitations and the consequent need for the critical attitude even in looking at the mode of risk assessment characteristic of engineering.

An Example of Public Policy: Building Codes

One of the most immediate ways in which public policy must rely on engineering expertise and engineering is in turn affected by public policy is through local building codes. The local building codes specify factors of safety and construction steps (e.g., fireproofing or material requirements) that are required in the area. Building codes have the status of law and may not be changed without public hearings and legislative action. The legislature will often appoint a committee of experts to propose a new building code or necessary changes in an existing one. For example, following the collapse of the World Trade Center's Twin Towers, there was a major multiagency investigative effort to identify the causes of the collapses and to propose changes in New York City's building codes that would protect against future collapses.

One of the more important ways professional engineers show a concern for the general public (and their safety) is in carrying out the local building code requirements in designing such things as buildings, elevators, escalators, bridges, walkways, roads, and overpasses. When a responsible engineer recognizes a violation of a building code in a design and does not object to it, the engineer bears some responsibility for any injuries or deaths that result. Similarly, when an engineer learns of a proposed change in a building code that he or she is convinced creates danger for the public and does nothing to prevent this change, the engineer bears some responsibility for any harm done.

The Twin Towers case illustrates these issues. The New York City building codes in place in 1945 required that all stairwells be surrounded with heavy masonry and concrete structure. As a consequence, in 1945 the firemen were able to get to the area inside the Empire State Building immediately through the stairwells and put out the fire in 40 minutes.

However, when the planners and designers of the Twin Towers project examined their early designs, they saw that there would not be enough rentable space in their plans to make the project financially viable. The solution was to eliminate the masonry requirement around the stairwells to increase the amount of rental space in the building for financial planners. They secured permission from the city of New York to ignore this safety requirement, with the result that firemen could not access the upper stories of the towers and the inhabitants in the floors above the fire could not escape the fire.

The codes were also changed in the mid-1960s to cut in half the amount of fireproofing required to be sprayed on the structural steel in the Twin Towers. This reduction reduced the weight of the building, thus allowing for the extra height. It also provided more rentable space. Tragically, it also facilitated the collapse of the structure. The fireproofing was further reduced below the new lowered requirements because criminal elements in the structural firms pocketed the costs without using the required minimum amount of spray. The combination of these two changes in the New York City building codes directly contributed to the deaths of more than 2,000 people on the morning of September 11, 2001.

New York City building codes in 1945 and now include a requirement that each load-carrying, high-rise steel structural building component be tested in furnaces that subject the structure to the harsh conditions of a major fire. Even though a major fire in 1975 spread from the 9th to the 19th floor of the north tower and caused significant sagging of the steel floor trusses similar to the sagging of the floor structure observed just before the towers collapsed on September 11, 2001, the prescribed tests were never done. It may be that the omission of these code-required tests was a major contributor to the magnitude of the 2001 disaster.

As another example of a serious shortcoming of the New York City building codes, see the Citicorp building case in the appendix. In this case, William LeMessurier designed the building's main load-carrying steel structure to a code-specified worst-case wind condition that was incorrect. Fortunately, LeMessurier recognized the error in the code and modified the already built structure to correct for it. The codes were subsequently corrected.

Building codes are one of the aspects of public policy that both directly affect engineers and most clearly require information from engineers in their formulation. They illustrate one of the most concrete and specific ways in which engineering expertise is needed in the formulation of public policy and in which public policy in turn vitally affects engineering design.

7.6   DIFFICULTIES IN DETERMINING THE CAUSES AND LIKELIHOOD OF HARM: THE CRITICAL ATTITUDE

Estimating risk, no doubt defined in terms of the probabilities and magnitudes of harm, has been described by one writer as looking "through a glass darkly."[26] It would be highly desirable, of course, to be able to accurately predict the harm resulting from engineering work. Instead, engineers can only estimate the magnitude and probability of harm. To make matters worse, often engineers cannot even make estimates satisfactorily. In actual practice, therefore, estimating risk (or "risk assessment") involves an uncertain prediction of the probability of harm. In this section, we consider some of the methods of estimating risk, the uncertainties in these methods, and the value judgments that these uncertainties necessitate.

Limitations in Detecting Failure Modes

With respect to new technologies, engineers and scientists must have some way of estimating the risks that they impose on those affected by them. One of the methods for assessing risk involves the use of a fault tree. In a fault tree analysis, we begin with an undesirable event, such as a car not starting or the loss of electrical power to a nuclear power plant's safety system. We reason *back* to the events that might have caused this undesirable event. Fault trees are often used to anticipate hazards for which there is little or no direct experience, such as nuclear meltdowns. They enable an engineer to analyze systematically the various failure modes attendant to an engineering project. A *failure mode* is a way in which a structure, mechanism, or process can malfunction. For example, a structure can rip apart in tension, crumble to pieces in compression, crack and break in bending, lose its integrity because of corrosion (rusting), explode because of excessive internal pressure, or burn because of excessive temperature. Figure 7.1 illustrates how a fault tree analysis can be used to discover why an automobile will not start.

Another approach to a systematic examination of failure modes is the event tree analysis. Here, we reason *forward* from a hypothetical event to determine what

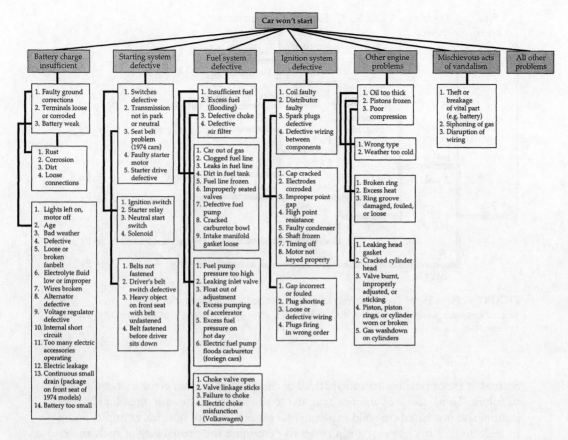

FIGURE 7.1 Fault Tree Analysis of Failure of an Automobile to Start
The failure appears at the top of the fault tree, and the possible causes of the failure appear as "branches" of the fault tree.

Source: This diagram is from B. Fischoff, P. Slovick, and S. Lichtenstein, "Fault Trees: Sensitivity and Estimated Failure Problem Representation," *Journal of Experimental Psychology: Human Perception and Performance, 4* (1978): 330–344. Used with permission.

consequences this hypothetical event might have and the probabilities of these consequences. Figure 7.2 illustrates in schematic form an event tree analysis. This simplified event tree for an accident involving a loss of coolant in a typical nuclear power plant begins with a failure and enumerates the various events to which this failure could lead. This event tree shows the logical relationships between the possible ways that a pipe break can affect the safety systems in a nuclear plant. If both a pipe and on-site power fail simultaneously, then the outcome will be an enormous release of radioactive coolant. If these two systems are independent, then the probability of this happening is the product of the two probabilities taken separately. For example, if there is one chance in 10^{-4} ($P_1 = 0.0001$) that the pipe will break and one chance in 10^{-5} ($P_2 = 0.00001$) that the on-site power will simultaneously fail, then the chance of a loss of a large release is 1 in 10^{-9} ($P = P_1 \times P_2$), or 1 in 1 billion.

Although engineers rightly believe that it is necessary to go through such analyses to ensure that they have taken into account as many failure modes as possible, the analyses have severe limitations. First, it is not possible to anticipate all of the mechanical, physical, electrical, and chemical problems that might lead to failure.

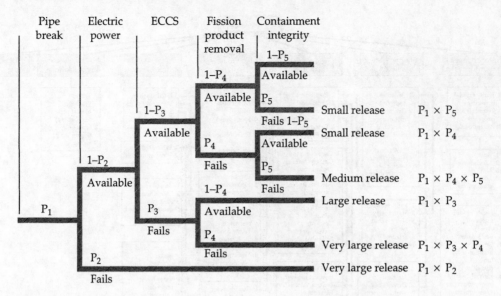

FIGURE 7.2 An Event Tree Analysis of a Pipe Break in a Nuclear Plant

Source: Reproduced, with permission, from the *Annual Review of Energy,* Volume 6, © 1981 by Annual Reviews, Inc. Courtesy N. C. Rasmussen.

Second, it is not possible to anticipate all of the types of human error that could lead to failure. Third, the probabilities assigned to the failure modes are often highly conjectural and not based on solid experimental testing. We are not, for example, going to melt down a nuclear reactor in order to determine the probability of such an occurrence leading to a chain reaction fission explosion. In many cases, we do not know the probability of the behavior of materials at extremely elevated temperatures. Fourth, we can never be sure we have all of the possible initiating events (even ones we know exist in different contexts) included on the event tree or placed in the right order.

Limitations Due to Tight Coupling and Complex Interactions

Sociologist Charles Perrow[27] confirms some of these problems by arguing that there are two characteristics of high-risk technologies that make them especially susceptible to accidents and allow us to speak of "normal accidents." These two features are the "tight coupling" and "complex interactions" of the parts of a technological system. These two factors make accidents not only likely but also difficult to predict and control. This, in turn, makes risk difficult to estimate.

In tight coupling, the temporal element is crucial. Processes are tightly coupled if they are connected in such a way that one process is known to affect another and will usually do so within a short time. In tight coupling, there is usually little time to correct a failure and little likelihood of confining a failure to one part of the system. As a result, the whole system is damaged. A chemical plant is tightly coupled because a failure in one part of the plant can quickly affect other parts of the plant. A university, by contrast, is loosely coupled because if one department ceases to function, then the operation of the whole university is usually not threatened.

In complex interaction, the inability to predict consequences is crucial. Processes can be complexly interactive in that the parts of the system can interact in unanticipated ways. No one dreamed that when X failed, it would affect Y. Chemical plants are complexly interactive in that parts affect one another in feedback patterns that cannot always be anticipated. A post office, by contrast, is not so complexly interactive. The parts of the system are related to one anther for the most part in a linear way that is well understood and the parts do not usually interact in unanticipated ways that cause the post office to cease functioning. If a post office ceases to function, it is usually because of a well-understood failure.

Examples of complexly interactive and tightly coupled technical systems include not only chemical plants but also nuclear power plants, electric power grid networks, space missions, and nuclear weapons systems. Being tightly coupled and complexly interactive, they can have unanticipated failures, and there is little time to correct the problems or keep them from affecting the entire system. This makes accidents difficult to predict and disasters difficult to avoid once a malfunction appears.

Unfortunately, it is difficult to change tightly coupled and complexly interactive systems to make accidents less likely or to make them easier to predict. To reduce complexity, decentralization is required to give operators the ability to react independently and creatively to unanticipated events. To deal with tight coupling, however, centralization is required. In order to avoid failures, operators need to have command of the total system and to be able to follow orders quickly and without question. It may not be possible, furthermore, to make a system both loosely coupled and noncomplex. Engineers know that they can sometimes overcome this dilemma by including localized and autonomous automatic controls to protect against failures due to complexity and couple them with manual overrides to protect against tight coupling failures. Nevertheless, according to Perrow, some accidents in complex, tightly coupled systems are probably inevitable and, in this sense, "normal."

The following is an example of an accident in a system that was complexly interactive and tightly coupled. In the summer of 1962, the New York Telephone Company completed heating system additions to a new accounting building in Yonkers, New York. The three-story, square-block building was a paradigm of safe design, using the latest technology.

In October 1962, after the building was occupied and the workers were in place, final adjustments were being made on the building's new, expanded heating system located in the basement. This system consisted of three side-by-side, oil-fired boilers. The boilers were designed for low pressures of less than 6.0 psi and so were not covered by the boiler and pressure vessel codes of the American Society of Mechanical Engineers. Each boiler was equipped with a spring-loaded safety relief valve that had been designed to open and release steam into the atmosphere if the boiler pressure got too high. Each boiler was also equipped with a pressure-actuated cutoff valve that would cut off oil flow to the boiler burners in the event of excessive boiler pressure. The steam pressure from the boilers was delivered to steam radiators, each of which had its own local relief valve. Finally, in the event that all else failed, a 1-foot-diameter pressure gauge with a red "Danger Zone" marked on the scale and painted on the face sat on the top of each boiler. If the pressure got too high, the gauge was supposed to alert a custodian who operated the boilers so he could turn off the burners.

On October 2, 1962, the following events transpired:[28]

1. The building custodian decided to fire up boiler 1 in the heating system for the first time that fall. The electricians had just wired the control system for the new companion boiler (boiler 3) and successfully tested the electrical signal flows.

2. The custodian did not know that the electricians had left the fuel cutoff control system disconnected. The electricians had disconnected the system because they were planning to do additional work on boiler 3 the following week. They intended to wire the fuel cutoffs for the three boilers in series (i.e., high pressure in any one would stop all of them).

3. The custodian mechanically closed the header valve because it was a warm Indian summer day and he did not want to send steam into the radiators on the floors above. Thus, the boiler was delivering steam pressure against a blocked valve, and the individual steam radiator valves were out of the control loop.

4. As subsequent testing showed, the relief valve had rusted shut after some tests the previous spring in which the boilers had last been fired. (Later, laws were enacted in New York state that require relief valves for low-pressure boiler systems to be operated by hand once every 24 hours to ensure that they are not rusted shut. At the time, low-pressure boiler systems were not subject to this requirement.)

5. This was on Thursday, the day before payday, and the custodian made a short walk to his bank at lunch hour to cash a check soon after turning on boiler 1.

6. The cafeteria was on the other side of the wall against which the boiler end abutted. Employees were in line against the wall awaiting their turn at the cafeteria serving tables. There were more people in line than there would have been on Friday because on payday many workers went out to cash their paychecks and eat their lunches at local restaurants.

7. Boiler 1 exploded. The end of the boiler that was the most removed from the wall next to the cafeteria blew off, turning the boiler into a rocket-like projectile. The boiler lifted off its stanchions and crashed into the cafeteria, after which it continued to rise at great velocity through all three stories of the building. Twenty-five people were killed and almost 100 seriously injured.

The events that led to this disaster were complexly interrelated. There is no possible way that fault tree or event tree analyses could have predicted this chain of events. If the outside temperature had been cooler, then the custodian would not have closed the header valve and the individual steam radiator valves in each upstairs room would have opened. If the relief valve had been hand-operated every day, its malfunction would have been discovered and probably corrected. If the time had not been noon and the day before payday, the custodian might have stayed in the basement and seen the high-pressure reading and turned off the burners. If it had not been lunch time, the unfortunate victims would not have been in the cafeteria line on the other side of the wall from the boiler.

The events were also tightly coupled. There was not much time to correct the problem once the pressure started to rise, and there was no way to isolate the boiler failure from a catastrophe in the rest of the building. There was one engineering design change that, if adopted, could have broken the chain of events and prevented the accident. It would have been a simple matter to include a fuel flow cutoff if the fuel cutoff system were in any way disabled. However, in complex interconnected systems such as this one, hindsight is always easier than foresight.

Normalizing Deviance and Self-Deception

Still another factor that increases risk and also decreases our ability to anticipate harm is increasing the allowable deviations from proper standards of safety and acceptable risk. Sociologist Diane Vaughn refers to this phenomenon as the *normalization of deviance.*[29]

Every design carries with it certain predictions about how the designed object should perform in use. Sometimes these predictions are not fulfilled, producing what are commonly referred to as *anomalies.* Rather than correcting the design or the operating conditions that led to anomalies, engineers or managers too often do something less desirable. They may simply accept the anomaly or even increase the boundaries of acceptable risk. Sometime this process can lead to disaster.

This process is dramatically and tragically illustrated by the events that led to the *Challenger* disaster.[30] Neither the contractor, Morton Thiokol, nor NASA expected the rubber O-rings that sealed the joints in the solid rocket booster (SRB) to be touched by the hot gases of motor ignition, much less to be partially burned. However, because previous shuttle flights showed damage to the sealing rings, the reaction by both NASA and Thiokol was to accept the anomalies without attempting to remedy the problems that caused the anomalies.

The following are examples of how deviance was normalized before the disaster:

1. In 1977, test results showed that the SRB joints would rotate open at ignition, creating a larger gap between the tang and clevis. According to NASA engineers, the gap was large enough to prevent the secondary seal from sealing if the primary O-ring failed late in the ignition cycle. Nevertheless, after some modifications, such as adding sealing putty behind the O-rings, the joint was officially certified as an acceptable risk, even though the joint's behavior deviated from design predictions.[31]

2. Another anomaly was discovered in November 1981 after flight STS-2, which showed "impingement erosion" of the primary O-ring in the right SRB's aft field joint.[32] The hot propellant gases had moved through the "blow holes" in the zinc chromate putty in the joints. The blow holes were caused by entrapped air introduced at the time the putty was installed. Even though this troubling phenomenon was not predicted, the joints were again certified as an acceptable risk.

3. A third anomaly occurred in 1984 with the launch of STS-41-B when, for the first time, two primary O-rings on two different joints were eroded.[33] Again, the erosion on two joints was termed an acceptable risk.[34]

4. Another anomaly occurred in 1985 when "blow-by" of hot gases had reached the secondary seal on a nozzle joint. The nozzle joints were considered safe because, unlike the field joints, they contained a different and quite safe secondary "face seal." The problem was that a similar malfunction could happen with the field joint with the danger much more serious, and these problems were not dealt with.

5. Perhaps the most dramatic example of expanding the boundaries of acceptable risk was in the area of the acceptable temperature for launch. Before the *Challenger*

launch, the lowest temperature of the seals at launch time was 53 degrees Fahrenheit. (At that time, the ambient temperature was in the high 60s.) On the night before the launch of the *Challenger,* however, the temperature of the seals was expected to be 29 degrees and its ambient temperature below freezing. Thus, the boundaries for acceptable risk were expanded by 24 degrees.

The result of (1) accepting these anomalies without making any adequate attempt to remedy the basic problem (poor seal design) and (2) lowering the temperature considered acceptable for launch was the tragic destruction of the *Challenger* and the loss of its crew. Vaughn argues that these kinds of problems cannot be eliminated from technological systems and that, as a result, accidents are inevitable. Whether or not this is the case, there is no question that technology imposes risk on the public and that these risks are often difficult to detect and eliminate.

The case also illustrates how the self-deception involved in normalizing deviance can limit the ability of engineers to correctly anticipate risk. Some of the engineers, and especially engineering managers, repeatedly convinced themselves that allowing still one more deviation from design expectations would not increase the chance of failure or was at least an acceptable risk. The result was a tragic disaster.

7.7   THE ENGINEER'S LIABILITY FOR RISK

We have seen that risk is difficult to estimate and that engineers are often tempted to allow anomalies to accumulate without taking remedial action, and even to expand the scope of acceptable risk to accommodate them. We have also seen that there are different and sometimes incompatible approaches to the definition of acceptable risk as exhibited by risk experts, laypeople, and government regulators.

Another issue that raises ethical and professional concerns for engineers regards legal liability for risk. There are at least two issues here. One is that the standards of proof in tort law and science are different, and this produces an interesting ethical conflict. Another issue is that in protecting the public from unnecessary risk, engineers may themselves incur legal liabilities. Let us consider each of these issues.

The Standards of Tort Law

Litigation that seeks redress from harm most commonly appeals to the law of torts, which deals with injuries to one person caused by another, usually as a result of fault or negligence of the injuring party. Many of the most famous legal cases involving claims of harm from technology have been brought under the law of torts. The litigation involving harm from asbestos is one example. In 1973, the estate of Clarence Borel,[35] who began working as an industrial insulation worker in 1936, brought suit against Fiberboard Paper Products Corporation:

> During his career he was employed at numerous places usually in Texas, until disabled from the disease of asbestosis in 1969. Borel's employment necessarily exposed him to heavy concentrations of asbestos generated by insulation materials. In a pretrial deposition Borel testified that at the end of the day working with insulation materials containing asbestos his clothes were usually so dusty that he could barely pick them up without shaking them. Borel stated, "You just move them a little bit and there is going to be dust and I blowed this dust out of my nostrils by the handfuls at the end of the day. I even used Mentholatum in my nostrils to keep some of

the dust from going down my throat, but it is impossible to get rid of all of it. Even your clothes just stay dusty continuously, unless you blow it off with an air hose." In 1964, doctors examined Borel in connection with an insurance policy and informed him that x-rays of his lungs were cloudy. The doctors told Borel that the cause could be his occupation as an installation worker and advised him to avoid asbestos dust as much as he possibly could. On January 19,1969, Borel was hospitalized and a lung biopsy was performed. Borel's condition was diagnosed as pulmonary asbestosis. Since the disease was considered irreversible Borel was sent home.... [His] condition gradually worsened during the remainder of 1969. On February 11, 1970, he underwent surgery for the removal of his right lung. The examining doctors determined that Borel had a form of lung cancer known as mesothelioma, which had been caused by asbestos. As a result of these diseases, Borel later died before the district case reached the trial stage.[36]

The federal district court in Texas decided in favor of the estate of Mr. Borel and the Fifth Circuit Court of Appeals upheld the decision.

The standard of proof in tort law is the preponderance of evidence, meaning that there is more and better evidence in favor of the plaintiff than the defendant. The plaintiff must show

(1) that the defendant violated a legal duty imposed by the tort law, (2) that the plaintiff suffered injuries compensable in the tort law, (3) that the defendant's violation of legal duty caused the plaintiff's injuries, and (4) that the defendant's violation of legal duty was the proximate cause of the plaintiff's injuries.[37]

The standard of proof that a given substance was the proximate cause of a harm is less stringent than that which would be demanded by a scientist, who might well call for 95 percent certainty. It is also less stringent than the standard of evidence in criminal proceedings, which calls for proof beyond reasonable doubt.

As an illustration of this lower standard of evidence, consider the case of *Rubanick v. Witco Chemical Corporation and Monsanto Co.* The plaintiff's sole expert witness, a retired cancer researcher at New York's Sloan-Kettering Cancer Center, testified that the deceased person's cancer was caused by exposure to polychlorinated biphenyls (PCBs). He based his opinion on

(1) the low incidence of cancer in males under 30 (the deceased person was 29), (2) the decedent's good dietary and nonsmoking habits and the absence of familial genetic predisposition to cancer, (3) 5 of 105 other Witco workers who developed some kind of cancer during the same period, (4) a large body of evidence showing that PCBs cause cancer in laboratory animals, and (5) support in the scientific literature that PCBs cause cancer in human beings.[38]

The court did not require the expert to support his opinion by epidemiological studies, merely that he demonstrate the appropriate education, knowledge, training, and experience in the specific field of science and an appropriate factual basis for his opinion.[39]

Other better known cases, such as that of Richard Ferebee, who alleged that he suffered lung damage as a result of spraying the herbicide paraquat, also accepted standards of evidence for causal claims that would not have been acceptable for research purposes.[40]

Some courts, however, have begun to impose higher standards of evidence for recovery of damages through tort—standards that are similar to those used in science. In the Agent Orange cases, Judge Jack B. Weinstein argued that epidemiological studies were the only useful studies having any bearing on causation, and that by this standard no plaintiff had been able to make a case. Bert Black,[41] a legal commentator, has taken a similar view. He believes that the courts (i.e., judges) should actively scrutinize the arguments of expert witnesses, demanding that they be supported by peer-reviewed scientific studies or at least have solid scientific backing. In some cases, he believes, they should even overrule juries who have made judgments not based on scientific standards of evidence.[42]

Even though this view represents a departure from the normal rules of evidence in tort law, it might in some cases be fairer to the defendants because some decisions in favor of plaintiffs may not be based on valid proof of responsibility for harm. The disadvantage is also equally obvious. By requiring higher standards of proof, the courts place burdens of evidence on plaintiffs that they often cannot meet. In many cases, scientific knowledge is simply not adequate to determine casual relationships, and this would work to the disadvantage of the plaintiffs. There are also problems with encouraging judges to take such an activist role in legal proceedings. The major ethical question, however, is whether we should be more concerned with protecting the rights of plaintiffs who may have been unjustly harmed or with promoting economic efficiency and protecting defendants against unjust charges of harm.

Protecting Engineers from Liability

The apparent ease with which proximate cause can be established in tort law may suggest that the courts should impose a more stringent standard of acceptable risk. But other aspects of the law afford the public less protection than it deserves. For example, the threat of legal liability can inhibit engineers from adequately protecting the public from risk. Engineers in private practice may face especially difficult considerations regarding liability and risk, and in some cases they may need increased protection from liability.

Consider, for example, the safety issues in excavating for foundations, pipelines, and sewers.[43] A deep, steep-sided trench is inherently unstable. Sooner or later, the sidewalls will collapse. The length of time that trench walls will stand before collapsing depends on several factors, including the length and width of the cut, weather conditions, moisture in the soil, composition of the soil, and how the trench was excavated. People who work in deep trenches are subjected to considerable risk, and hundreds of laborers are injured or killed each year when the walls collapse.

To reduce the risk, construction engineers can specify the use of trench boxes in their designs. A trench box is a long box with an upside-down U-shaped cross section that is inserted inside the trench to protect the laborers. As long as workers remain inside the trench boxes, their risk of death or injury is greatly reduced.

Unfortunately, the use of trench boxes considerably increases the expense and time involved in construction projects. The boxes must be purchased or rented, and then they must be moved as excavation proceeds, slowing construction work and adding further expense.

Engineers are placed in an awkward position with regard to the use of trench boxes, especially where the boxes are not required by building codes. If they do not specify the use of the boxes, then they may be contributing to a situation that subjects workers to a high risk of death and injury. If they do specify the use of boxes, then they may be incurring liability in case of an accident because of the use of trench boxes. With situations such as this in mind, the National Society of Professional Engineers has been actively lobbying the U.S. Congress to pass a law that specifically excludes engineers from liability for accidents where construction safety measures are specified by engineers but then are either not used or used improperly by others. This would enable engineers to more effectively protect the safety of workers. Unfortunately, the proposals have never become law.

The problem with trench boxes illustrates a more general issue. If engineers were free to specify safety measures without being held liable for their neglect or improper use, they could more easily fulfill one aspect of their responsibility to protect the safety of the public.

7.8 BECOMING A RESPONSIBLE ENGINEER REGARDING RISK

The development of new technology is intimately connected with risk. The obligation of engineers is to be ethically responsible with regard to risk. The first step in the process of becoming ethically responsible about risk is to be aware of the fact that risk is often difficult to estimate and can be increased in ways that may be subtle and treacherous. The second step is to be aware that there are different approaches to the determination of acceptable risk. In particular, engineers have a strong bias toward quantification in their approach to risk, which may make them insufficiently sensitive to the concerns of the lay public and even the government regulators. The third step is to assume their responsibility, as the experts in technology, to communicate issues regarding risk to the public, with the full awareness that both the public and government regulators have a somewhat different agenda with regard to risk.

We conclude with an attempt to formulate a principle of acceptable risk. To formulate this principle, let us consider further some of the legal debate about risk.

The law seems to be of two minds about risk and benefits. On the one hand, some laws make no attempt to balance the two. The Chemical Food Additives Amendments to the Food, Drug and Cosmetics Act, enacted in 1958, require that a chemical "deemed to be unsafe" not be added to food unless it can be "safely used."[44] *Safe use* was defined by the Senate Committee on Labor and Public Welfare as meaning that "no harm will result" from its addition to food.[45]

The well-known Delaney Amendment also prohibits the addition to food of any chemical known to cause cancer when ingested by animals.[46]

On the other hand, there is often an attempt to strike a balance between the welfare of the public and the rights of individuals. The Toxic Substances Control Act of 1976 authorized the EPA to regulate any chemical upon a finding of "unreasonable risk of injury to health or the environment."[47] But it is only "unreasonable risk" that triggers regulation, so some degree of risk is clearly tolerated. The report of the House Commerce Committee describes this balancing process as follows:

> Balancing the probabilities that harm will occur and the magnitude and severity of that harm against the effect of proposed regulatory action on the availability to society of the benefits of the substance or mixture, taking into account the availability of substitutes for the substance or mixture which do not require regulation, and other adverse effect which such proposed action may have on society.

Having said this, the report goes on to say that "a formal benefit–cost analysis under which monetary value is assigned to the risks ... and to the costs of society" is not required.[48]

The Atomic Energy Act of 1954 continually refers to the "health and safety of the public" but makes little attempt to define these terms. The Nuclear Regulatory Commission's rules, however, use the expression "without undue risk" and seem to suggest again a balancing of risks and benefits.[49] In the words of one legal commentator, in practice, especially in the earlier years,

> the acceptability of risk was measured largely in terms of the extent to which industry was capable of reducing the risk without jeopardizing an economic and financial environment conducive to continuing development of the technology.[50]

Again, we have an attempt to balance protection of individuals and promotion of the public welfare.

Sometimes the conflict between these two approaches is evident in a single debate. In a Supreme Court case involving exposure to benzene in the workplace, OSHA took an essentially respect for persons standpoint, arguing that the burden of proof should be on industry to prove that a given level of exposure to benzene was not carcinogenic. In its rebuke of OSHA, the Supreme Court argued that in light of the evidence that current standards did not lead to harm to workers, risk must be balanced against benefits in evaluating more stringent standards and that the burden of proof was on OSHA to show that the more stringent standards were justified.[51]

Given these considerations, we can construct a more general principle of acceptable risk, which may provide some guidance in determining when a risk is within the bounds of moral permissibility:

> People should be protected from the harmful effects of technology, especially when the harms are not consented to or when they are unjustly distributed, except that this protection must sometimes be balanced against (1) the need to preserve great and irreplaceable benefits, and (2) the limitation on our ability to obtain informed consent.

The principle does not offer an algorithm that can be applied mechanically to situations involving risk. Many issues arise in its use; each use must be considered on its own merits. We can enumerate some of the issues that arise in applying the principle.

First, we must define what we mean by "protecting" people from harm. This cannot mean that people are assured that a form of technology is free from risk. At best, "protection" can only be formulated in terms of probabilities of harm, and we have seen that even these are subject to considerable error.

Second, many disputes can arise as to what constitutes a harm. Is having to breathe a foul odor all day long harm? What about workers in a brewery or a sewage disposal plant? Here the foul odors cannot be eliminated, so the question of what harms should be eliminated cannot be divorced from the question of whether the harms can be eliminated without at the same time eliminating other goods.

Third, the determination of what constitutes a great and irreplaceable benefit must be made in the context of particular situations. A food additive that makes the color of frozen vegetables more intense is not a great and irreplaceable benefit. If an additive were found to be a powerful carcinogen, then it should be eliminated.

On the other hand, most people value automobiles highly, and they would probably not want them to be eliminated, despite the possibility of death or injury from automobile accidents.

Fourth, we have already pointed out the problems that arise in determining informed consent and the limitations in obtaining informed consent in many situations. From the standpoint of the ethics of respect for persons, informed consent is a consideration of great importance. However, it is often difficult to interpret and apply.

Fifth, the criterion of unjust distribution of harm is also difficult to apply. Some harms associated with risk are probably unjustly distributed. For example, the risks associated with proximity to a toxic waste disposal area that is not well constructed or monitored are unjustly distributed. The risks associated with coal mining might also be conceded to be unjustly distributed, but coal may also be considered a great and irreplaceable benefit. So the requirement to reduce risk in the coal industry might be that the risks of coal mining should be reduced as much as possible without destroying the coal industry. This might require raising the price of coal enough to make coal mining safe and more economically rewarding.

Sixth, an acceptable risk at a given point in time may not be an acceptable risk at another point in time. Engineers' responsibility to protect the health and safety of the public requires them to reduce risk when this can be done as a result of technological innovation. As new risk-reducing technology becomes available, the responsibility of engineers to reduce risk changes.

7.9   CHAPTER SUMMARY

Risk is a part of engineering and especially of technological progress. The concept of "factors of safety" is important in engineering. Virtually all engineering codes give a prominent place to safety. Engineers and risk experts look at risk in a somewhat different way from others in society. For engineers, risk is the product of the likelihood and magnitude of harm. An acceptable risk is one in which the product of the probability and magnitude of the harm is equaled or exceeded by the product of the probability and magnitude of the benefit, and there is no other option where the product of the probability and magnitude of the benefit is substantially greater. In calculating harms and benefits, engineers have traditionally identified harm with factors that are relatively easily quantified, such as economic losses and loss of life. The "capabilities" approach attempts to make these calculations more sophisticated by developing a more adequate way of measuring the harms and benefits from disasters to overall well-being, which it defines in terms of the capabilities of people to live the kind of life they value. A risk is acceptable if the probability is sufficiently small that the adverse effect of a hazard will fall below a threshold of the minimum level of capabilities attainment that is acceptable in principle.

The public does not conceptualize risk simply in terms of expected deaths or injury but, rather, considers other factors as well, such as whether a risk is assumed with free and informed consent and whether the risk is imposed justly. Government regulators take a still different approach to risk because they have a special obligation to protect the public from harm. Consequently, they place greater weight on protecting the public than on benefiting the public. In light of these different agendas, social policy must take into account a wider perspective than that of the risk expert.

Engineers, and especially professional engineering societies, have an obligation to contribute to public debate on risk by supplying expert information and by recognizing that the perspectives in the public debate will comprise more than the perspective of the risk expert. Debates over building codes illustrate some aspects of this public debate over risk.

Estimating the causes and likelihood of harm poses many difficulties. Engineers use various techniques, such as fault trees and event trees. However, the phenomena of "tight coupling" and "complex interactions" limit our ability to anticipate disasters. The tendency to accept increasing deviations from expected performance can also lead to disasters.

Engineers need to protect themselves from undue liability for risk, but this need sometimes raises important issues for social policy. One issue is the conflict between the standards of science and tort law. The standard of proof in tort law for whether something causes a harm is the preponderance of evidence, but the standard of evidence in science is much higher. The lower standard of tort law tends to protect the rights of plaintiffs who may have been unjustly harmed, and the higher standard of science tends to protect defendants and perhaps promote economic efficiency. The problems engineers have in protecting themselves from unjust liabilities while protecting the public from harm are illustrated by the use of trench boxes. Finally, a principle of acceptable risk provides some guidance in determining when a risk is within the bounds of moral permissibility.

NOTES

1. "B-25 Crashes in Fog," *New York Times,* July 29, 1945, p. 1.
2. Peter Glantz and Eric Lipton, "The Height of Ambition," *New York Times Sunday Magazine,* Sept. 8, 2002, Sec. 6, p. 32.
3. "Wider Inquiry into Towers Is Proposed," *New York Times,* May 2, 2002, p. B1; and "Bloomberg, Learning from 9/11, Seeks High Rise Safety Reforms," *New York Times,* Sept. 24, 2003, p. B2.
4. Glantz and Lipton, p. 32.
5. William W. Lowrance, "The Nature of Risk," in Richard C. Schwing and Walter A. Albers, Jr., eds., *Societal Risk Assessment: How Safe Is Safe Enough?* (New York: Plenum, 1980), p. 6.
6. The National Public Radio story was aired on "Morning Edition," Dec. 3, 1992. This account is taken from the *Newsletter* of the Center for Biotechnology Policy and Ethics, Texas A & M University, 2, no. 1, Jan. 1, 1993, p. 1.
7. For further discussion of the concept of capabilities and description of this approach to measuring harm in this section, see Amartya Sen, *Development as Freedom* (New York: Anchor Books, 1999); Martha Nussbaum, *Women and Human Development: The Capabilities Approach* (New York: Cambridge University Press, 2000); Colleen Murphy and Paolo Gardoni, "The Role of Society in Engineering Risk Analysis: A Capabilities-Based Approach," *Risk Analysis,* 26, no. 4, pp. 1073–1083; and Colleen Murphy and Paolo Gardoni, "The Acceptability and the Tolerability of Societal Risks: A Capabilities-based Approach," *Science and Engineering Ethics,* 14, no. 1, 2008, pp. 77–92. We are indebted to Professors Murphy and Gardoni for supplying the basis of this section.
8. S. Anand and Amartya Sen, "The Income Component of the Human Development Index," *Journal of Human Development,* 1, no. 1, 2000, 83–106.
9. Colleen Murphy and Paolo Gardoni, "Determining Public Policy and Resource Allocation Priorities for Mitigating Natural Hazards; A Capabilities-based Approach," *Science & Engineering Ethics,* 13, no. 4, 2007, pp. 489–504.

10. Colleen Murphy and Paolo Gardoni, "The Acceptability and the Tolerability of Societal Risks: A Capabilities-Based Approach," *Science and Engineering Ethics*, 14, no. 1, 2008, pp. 77–92.

11. Chauncey Starr, "Social Benefits versus Technological Risk," *Science*, 165, Sept. 19, 1969, 1232–1238. Reprinted in Theodore S. Glickman and Michael Gough, *Readings in Risk* (Washington, DC: Resources for the Future), pp. 183–193.

12. Paul Slovic, Baruch Fischhoff, and Sarah Lichtenstein, "Rating the Risks," *Environment*, 21, no. 3, April 1969, pp. 14–20, 36–39. Reprinted in Glickman and Gough, pp. 61–74.

13. Starr, pp. 183–193.

14. For this and several of the following points, see Paul B. Thompson, "The Ethics of Truth-Telling and the Problem of Risk," *Science and Engineering Ethics*, 5, 1999, 489–510.

15. Starr, op. cit., pp. 183–193.

16. D. Litai, "A Risk Comparison Methodology for the Assessment of Acceptable Risk," Ph.D. dissertation, Massachusetts Institute of Technology, Cambridge, MA, 1980.

17. John Rawls, *A Theory of Justice* (Cambridge, MA: Harvard University Press, 1971), p. 3.

18. From the *Charlotte* (N.C.) *Observer*, Feb. 10, 1980. Quoted in Carl F. Cranor, *Regulating Toxic Substances: A Philosophy of Science and the Law* (New York: Oxford University Press, 1993), p. 152.

19. Ralph L. Kenny, Ram B. Kulkarni, and Keshavan Nair, "Assessing the Risks of an LGN Terminal," in Glickman and Gough, pp. 207–217.

20. See William D. Ruckelshaus, "Risk, Science, and Democracy," *Issues in Science and Technology*, 1, no. 3, Spring 1985, pp. 19–38. Reprinted in Glickman and Gough, pp. 105–118.

21. Epidemiology is the study of the distribution of disease in human populations and the factors that causally influence this distribution.

22. See Carl F. Cranor, "Some Moral Issues in Risk Assessment, " *Ethics*, 101, Oct. 1990, pp. 123–143. See also Cranor's *Regulating Toxic Substances*, pp. 12–48.

23. Cranor, *Regulating Toxic Substances*, pp. 139–143.

24. Harold P. Green, "The Role of Law in Determining Acceptability of Risk," in *Societal Risk Assessment: How Safe Is Safe Enough?* (New York: Plenum, 1980), pp. 255–269.

25. This list was suggested by the four "dicta for risk communication" proposed by Paul Thompson, in Thompson, op. cit., pp. 507–508. Although some items in this list are the same as Thompson's, we have modified and expanded his list.

26. Cranor, *Regulating Toxic Substances*, p. 11.

27. Charles Perrow, *Normal Accidents: Living with High-Risk Technologies* (New York: Basic Books, 1984), p. 3.

28. See *New York Times*, Oct. 15, 1962, for an account of this tragic event. The engineering details are cited from an unpublished report by R. C. King, H. Margolin, and M. J. Rabins to the City of New York Building Commission on the causes of the accident.

29. Diane Vaughn, *The Challenger Launch Decision* (Chicago: University of Chicago Press, 1996), pp. 409–422.

30. See the Presidential Commission on the Space Shuttle *Challenger* Accident, *Report to the President by the Presidential Commission on the Space Shuttle Challenger Accident* (Washington, DC: Government Printing Office, 1986), vol. 1, p. 120. Quoted in Vaughn, p. 77.

31. See Vaughn, pp. 110–111. The following account is taken from Vaughn and from personal conversations with Roger Boisjoly. This account should be attributed to the authors, however, rather than to Diane Vaughn or Roger Boisjoly.

32. Ibid., pp. 121 ff.

33. Ibid., pp. 141 ff.

34. Ibid., pp. 153 ff.

35. *Borel v. Fiberboard Paper Products Corp. et al.*, 493 F.2d (1973) at 1076, 1083. Quoted in Cranor, *Regulating Toxic Substances*, p. 52.

36. Cranor, *Regulating Toxic Substances,* p. 58.

37. 576 A.2d4 (N.J. Sup. Ct. A.D.1990) at 15 (concurring opinion).

38. Cranor, *Regulating Toxic Substances,* p. 81. Summarized from "New Jersey Supreme Court Applies Broader Test for Admitting Expert Testimony in Toxic Case," *Environmental Health Letter,* Aug. 27, 1991, p. 176.

39. "New Jersey Supreme Court Applies Broader Test," p. 176.

40. *Ferebee v. Chevron Chemical Co.,* 736 F.2d 11529 (D.C. Cir 1984).

41. Bert Black, "Evolving Legal Standards for the Admissibility of Scientific Evidence," *Science,* 239, 1987, pp. 1510–1512.

42. Bert Black, "A Unified Theory of Scientific Evidence," *Fordham Law Review,* 55, 1987, pp. 595–692.

43. See R. W. Flumerfelt, C. E. Harris, Jr., M. J. Rabins, and C. H. Samson, Jr., *Introducing Ethics Case Studies into Required Undergraduate Engineering Courses, Final Report to the NSF on Grant DIR-9012252,* Nov. 1992, pp. 262–285.

44. Public Law No. 85-929, 72 Stat. 784 (1958).

45. Senate Report No. 85-2422, 85th Congress, 2d Session (1958).

46. c21 United States Code, sect. 348 (A) (1976).

47. Public Law No. 94-469, 90 Stat. 2003 (1976). The same criterion of "unreasonable risk" is found in the Flammable Fabrics Act. See Public Law No. 90-189, Stat. 568 (1967).

48. Public Law No. 83-703, 68 Stat. 919 (1954), 42 United States Code 2011 et. seq. (1976).

49. 10 CFR 50.35 (a) (4).

50. Harold P. Green, "The Role of Law in Determining Acceptability of Risk," in *Societal Risk Assessment: How Safe Is Safe Enough?* (New York: Plenum, 1980), p. 265.

51. *Industrial Union Department, AFL-CIO v. American Petroleum Institute et al.,* 448 U.S. 607 (1980).

CHAPTER EIGHT

Engineers in Organizations

Main Ideas in this Chapter

- The common law doctrine of employment at will has been modified by the courts and by statutes to give some protection to employees in the workplace.

- Employees should become familiar with the culture of the organization in which they are employed and take advantage of organizational resources in order to enhance their own integrity and independence.

- Engineers and managers have different perspectives, both legitimate, and it is useful to distinguish between decisions that should be made by managers or from the management perspective and decisions that should be made by engineers or from the engineering perspective.

- Sometimes organizational disobedience is necessary. There is disobedience by contrary action and disobedience by nonparticipation, but the most widely discussed type of organizational disobedience is whistleblowing. Richard DeGeorge and Michael Davis have developed theories of justified whistleblowing.

- Roger Boisjoly's attempt to stop the launch of the *Challenger* illustrates the conflict between management and engineering perspectives in decision making. His testimony before the Rogers Commission raises questions about when whistleblowing is justified.

PAUL LORENZ WAS A MECHANICAL engineer employed by Martin Marietta. He was laid off on July, 25, 1975, for allegedly failing to engage in acts of deception and misrepresentation concerning the quality of materials used by Martin Marietta in designing equipment for the National Aeronautics and Space Administration (NASA). The equipment was for the external tank of the space shuttle program. Before he was laid off, Lorenz was informed that he should "start playing ball with management." After being laid off, Mr. Lorenz filed a tort claim against Martin Marietta for wrongful discharge on the grounds that he was fired for refusing to perform an illegal act. Federal law does prohibit knowingly and willingly making a false representation to a federal agency. Lower courts rejected Lorenz's claim on the grounds that Colorado recognizes no claim of wrongful discharge against employers.

In 1992, the Colorado Supreme court concluded that "Lorenz did present sufficient evidence at trial to establish a prima facie case for wrongful discharge under the public-policy exception to the at-will employment doctrine." The Court directed a new trial in accordance with its findings, but the new trials never took place, probably because of an out-of-court settlement between Mr. Lorenz and his former employer.[1]

8.1   INTRODUCTION

The Lorenz case is an important case in the development of the law regarding the rights of professional employees in the workplace. The crucial idea in the case was the so-called "public-policy exception" to the traditional common law doctrine of "employment at will." Common law is the tradition of case law or "judge-made-law" that originated in England and is fundamental in U.S. law. It is based on a tradition in which a judicial decision establishes a precedent, which is then used by succeeding jurists as the basis for their decisions in similar cases. Common law is distinguished from statutory law, or laws made by legislative bodies.

Traditionally, U.S. law has been governed by the common law doctrine of "employment at will," which holds that in the absence of a contract, an employer may discharge an employee at any time and for virtually any reason. Recent court decisions, such as this one, have held that the traditional doctrine must be modified if there is an important interest at stake. Precisely how far the public policy exception extends is still being formulated by the courts, but it includes such things as a refusal to break the law (such as in the Lorenz case), performing an important public obligation (e.g., jury duty), exercising a clear legal right (e.g., exercising free speech or applying for unemployment compensation), and protecting the public from a clear threat to health and safety. In general, the public policy exception has not been invoked to protect an employee when there is a mere difference in judgment with the employer.[2] The courts have also given more weight to the codes of administrative and judicial bodies, such as state regulatory boards, than to the codes promulgated by professional societies.[3]

In addition to the judicial modification of at-will employment, dissenting employees have also received some statutory protection, primarily through whistleblower laws. The first such state law was passed in Michigan in 1981. If the employee is unfairly disciplined for reporting an alleged violation of federal, state, or local law to public authorities, the employee can be awarded back pay, reinstatement to the job, costs of litigation, and attorney's fees. The employer can also be fined up to $500.[4] New Jersey's Conscientious Employee Protection Act forbids termination for conduct undertaken for the sake of compliance with "a clear mandate of public policy concerning the public health, safety, or welfare."[5]

Many cases in the area of what might very generally be called "employee rights" involve nonprofessional employees, but our special interest is professional employees, especially engineers. Many of the cases, like the Lorenz case, involve a conflict between professional employees and managers. In fact, most of the classic cases in engineering ethics involve conflicts between engineers and managers. Therefore, this relationship bears close examination. It is the focus of this chapter. We begin with a very cynical and pessimistic picture of the relationship of engineers and managers—one that offers little prospect of a productive relationship between these two groups. Then we develop a more optimistic—and, we believe, more realistic—view of the relationship.

8.2 ENGINEERS AND MANAGERS: THE PESSIMISTIC ACCOUNT

Management theorist Joseph Raelin, reflecting the position of some students of management, says, "There is a natural conflict between management and professionals because of their differences in educational background, socialization, values, vocational interests, work habits, and outlook."[6] We can be somewhat more precise about the areas of conflict between engineers and managers.

First, although engineers may not always maintain as much identity with their wider professional community as some other professionals (e.g., research scientists), engineers do often experience a conflict between loyalty to their employer and loyalty to their profession.[7] Most engineers want to be loyal employees who are concerned about the financial well-being of their firms and who carry out instructions from their superiors without protest. In the words of many engineering codes, they want to be "faithful agents" of their employers. At the same time, as engineers they are also obligated to hold paramount the health, safety, and welfare of the public. This obligation requires engineers to insist on high standards of quality and (especially) safety.[8]

Second, many managers are not engineers and do not have engineering expertise, so communication is often difficult. Engineers sometimes complain that they have to use oversimplified language in explaining technical matters to managers and that their managers do not really understand the engineering issues.

Third, many engineers who are not managers aspire to the management role in the future, where the financial rewards and prestige are perceived to be greater. Thus, many engineers who do not yet occupy the dual roles of engineer and manager probably expect to do so at some time in their careers. This conflict can be internalized within the same person because many engineers have roles as both engineers and managers. For example, Robert Lund, vice president for engineering at Morton Thiokol at the time of the *Challenger* disaster, was both an engineer and a manager. Before the disaster, Lund was even directed by his superior to take the managerial rather than the engineering perspective.

This account of the differences between the perspectives of engineers and managers suggests the possibility of frequent conflicts. This prediction is confirmed by a well-known study by sociologist Robert Jackall. Although his study focuses only infrequently on the relationship between managers and professionals, his occasional references to the relationship of managers to engineers and other professionals make it clear that he believes his general description of the manager–employee relationship applies to the relationship of managers to professionals, including engineers. In his study of managers in several large U.S. corporations, Jackall found that large organizations place a premium on "functional rationality," which is a "pragmatic habit of mind that seeks specific goals." Jackall found that the managers and firms he studied had several characteristics that were not conducive to respecting the moral commitments of conscientious professionals.[9]

First, the organizational ethos does not allow genuine moral commitments to play a part in the decisions of corporate managers, especially highly placed ones. A person may have whatever private moral beliefs she chooses, as long as these beliefs do not influence behavior in the workplace. She must learn to separate individual conscience from corporate action. Managers, according to Jackall, prefer to think in terms of trade-offs between moral principles, on the one hand, and expediency, on the other hand. What we might think of as genuine moral considerations play little part in managerial decisions, according to Jackall. Faulty products are bad because they will ultimately harm the company's public image, and environmental damage is bad for business or will ultimately affect managers in their private role as consumers.

This attitude is in contrast to that of White, an employee who, according to Jackall, was concerned with a problem of excessive sound in his plant. White defined the issue of possible harm to employees as a moral concern instead of approaching it pragmatically. In another example, Jackall recounted the story of Brady, an accountant who found financial irregularities that were traced to the CEO. Whereas Brady saw the issue as a moral one, managers did not. In discussing the case, they held that Brady should have kept his mouth shut and dropped the matter. After all, the violations were small relative to the size of the corporation.[10]

Second, loyalty to one's peers and superiors is the primary virtue for managers. The successful manager is the team player, the person who can accept a challenge and get the job done in a way that reflects favorably upon himself and others.[11]

Third, lines of responsibility are deliberately blurred to protect oneself, one's peers, and one's superiors. Details are pushed down and credit is pushed up. Actions are separated from consequences insofar as this is possible so that responsibility can be avoided. In making difficult and controversial decisions, a successful manager will always get as many people involved as possible so he can point his finger at others if things go wrong. He should also avoid putting things in writing to avoid being held responsible. Protecting and covering for one's boss, one's peers, and oneself supersedes all other considerations.

According to this account of managerial decision making, the moral scruples of professionals have no place. In such an atmosphere, a principled professional would often appear to have no alternative to organizational disobedience. Such was the case with Joe Wilson, an engineer who found a problem with a crane that he believed involved public health and safety. Wilson wrote a memo to his boss, who replied that he did not need such a memo from Wilson and that the memo was not constructive. After Wilson was fired and went public, a *New York Times* investigation cited a corporate official's comment that Wilson was someone who "was not a team player."[12]

If engineers typically work in an organizational environment like the one Jackall described, their professional and ethical concerns will have little chance of being accorded respect. There is, however, a more constructive aspect of Jackall's study. He does suggest some characteristics of managerial decision making that are useful in analyzing the manager–engineer relationship:

1. Jackall's study implies that managers have a strong and probably overriding concern for the well-being of the organization. Well-being is measured primarily in financial terms, but it also includes a good public image and relatively conflict-free operation.

2. Managers have few, if any, loyalties that transcend their perceived obligations to the organization. They do not, for example, have professional obligations that they might consider to override or even counterbalance their obligations to the organization.

3. The managerial decision-making process involves making trade-offs among the relevant considerations. Ethical considerations are only one type of consideration. Furthermore, if we are to believe Jackall, managers tend not to take ethical considerations seriously, unless they can be translated into factors affecting the well-being (e.g., the public image) of the firm.

Jackall presents a very pessimistic picture of the place of a morally committed professional in an organization. In the next section, we suggest ways in which engineers can have a much more positive and productive relationship between them and their organizations.

8.3 BEING MORALLY RESPONSIBLE IN AN ORGANIZATION WITHOUT GETTING HURT

The Importance of Organizational Culture

In order to be morally responsible in an organization without suffering the fate of the employees in Jackall's study, engineers must first have some understanding of the organization in which they are employed. This knowledge helps engineers to understand (1) how they and their managers tend to frame issues under the influence of the organization and (2) how one can act in the organization effectively, safely, and in a morally responsible way.

The qualities of the organization we have in mind here often fall into the category of "organizational culture." It is generally agreed that organizational culture is set at the top of an organization—by high-level managers, by the president or chief executive officer of the organization, by directors, and sometimes by owners. If the organization values success and productivity over integrity and ethical principles, these values will powerfully influence the decisions of members of the organization. The values become, in the words of one writer, "a mindset, a filter through which participants view their world."[13] If this filter is strongly rooted in an organizational culture of which one is a part, it is an even more powerful influence on behavior.

Some writers use the term "organizational scripts" or "schemas" to refer to the way an organization conditions its members to view the world in a certain way, seeing some things and not seeing others. Dennis Gioia was a manager at Ford. He made the recommendation not to recall the Pinto, even though the car had been involved in the tragic deaths of passengers after relatively minor accidents. He describes his experience at Ford as follows:

> My own schematized . . . knowledge influenced me to perceive recall issues in terms of the prevailing decision environment and to *unconsciously overlook* key features of the Pinto case, mainly because they did not fit an existing script. Although the outcomes of the case carry retrospectively obvious ethical overtones, the schemas driving my perceptions and actions precluded considerations of the issues in ethical terms because the scripts did not include ethical dimensions.[14]

We have to be careful here not to allow an appreciation of the influence of organizational culture to completely override a belief in individual moral responsibility. Nevertheless, employees, including professional employees, do make decisions in the context of the organization in which they are employed, and one needs to understand the forces that bear upon his or her decision making.

With funding from the Hitachi Corporation, Michael Davis and his associates studied the position of engineers in engineering firms. Their published study, often called the Hitachi report, found that companies fall into one of three categories: engineer-oriented companies, customer-oriented companies, and finance-oriented companies. Understanding these three types of firms helps us understand the organizational cultures in which engineers work.

Three Types of Organizational Culture

Engineer-Oriented Companies. In these firms, there is general agreement that quality takes priority over other considerations, except safety. In the words of one manager, "We have overdesigned our products and would rather lose money than diminish our reputation."[15] Engineers often described their relationship to managers in these kinds of firms as one in which negotiation or arriving at consensus was prominent. Engineers often said that managers would rarely overrule them when there was a significant engineering issue, although they might make the final decision when primarily such issues as cost or marketing are involved. Managers in such companies said that they never withhold information from engineers, although they suspect engineers sometimes withhold information in order to cover up a mistake.

Customer-Oriented Companies. Decision making is similar to that of engineer-oriented firms, but with four significant differences. First, managers think of engineers as advocates of a point of view different from their own. Whereas managers must focus on such business factors as timing and cost, engineers should focus on quality and safety. Second, more emphasis is placed on business considerations than in engineer-oriented companies. Third, as with engineer-oriented companies, safety outranks quality. Sometimes quality can be sacrificed to get the product out the door. Finally, communication between engineers and managers may be somewhat more difficult than in engineer-oriented firms. Managers are more concerned about engineers' withholding information, even though consensus is highly valued.

Finance-Oriented Companies. Although possessing far less information about this category of firms, Davis conjectures, based on the information available, that these firms are more centralized and that this has important consequences. For example, engineers may receive less information for making decisions and consequently their decisions are given less weight by managers. Managers are less inclined to try to reach consensus, and engineers are seen as having a "staff" and advisory function.

Acting Ethically without Having to Make Difficult Choices

Acting in an ethical manner and with little harm to oneself is generally easier in engineer-oriented and customer-oriented companies than in finance-oriented companies. In the first two types of firms, more respect is given to the types of values with which engineers are typically concerned, especially safety and quality. Communication is better, and there is more emphasis on arriving at decisions by consensus rather than by the authority of the managers. All of this makes it much easier for an engineer to act in a professional and ethical manner. However, there are some additional suggestions that should make acting ethically easier and less harmful to the employee.

First, engineers and other employees should be encouraged to report bad news. Sometimes there are formal procedures for lodging complaints and warnings about impending trouble. If possible, there should be formal procedures for lodging complaints. One of the best known procedures is the Differing Professional Views and Differing Professional Opinions of the Nuclear Regulatory Commission.[16] Another procedure is the Amoco Chemical Hazard and Operability program.[17] In addition, many large corporations have "ombudsmen" and "ethics officers," who can promote ethical behavior as well as serve as a conduit for complaints. Some writers have suggested, however, that in-house ethics officers are too much the creatures of the organizations; instead, outside ethics consultants should be hired to handle complaints and internal disagreements. The argument is that in-house ethics officers have been nurtured in the organizational culture and are dependent on the organizations for their salaries, so they are not able to adopt a genuinely objective perspective.[18]

Second, companies and their employees should adopt a position of "critical" loyalty rather than uncritical or blind loyalty. *Uncritical loyalty* to the employer is placing the interests of the employer, as the employer defines those interests, above every other consideration. By contrast, *critical loyalty* is giving due regard to the interests of the employer but only insofar as this is possible within the constraints of the employee's personal and professional ethics. We can think of the concept of critical loyalty as a creative middle way that seeks to honor the legitimate demands of the organization but also to honor the obligation to protect the public.

Third, when making criticisms and suggestions, employees should focus on issues rather than personalities. This helps avoid excessive emotionalism and personality clashes.

Fourth, written records should be kept of suggestions and especially of complaints. This is important if court proceedings are eventually involved. It also serves to "keep the record straight" about what was said and when it was said.

Fifth, complaints should be kept as confidential as possible for the protection of both the individuals involved and the firm.

Sixth, provisions should be made for neutral participants from outside the organization when the dispute requires it. Sometimes, employees within the organization are too emotionally involved in the dispute or have too many personal ties to make a dispassionate evaluation of the issues.

Seventh, explicit provision for protection from retaliation should be made, with mechanisms for complaint if an employee believes he or she has experienced retaliation. Next to the fear of immediate dismissal, probably the greatest fear of an employee who is in disagreement with a superior is that he or she will suffer discrimination in promotion and job assignment, even long after the controversy is resolved. Protection from this fear is one of the most important of employee rights, although it is one of the most difficult to provide.

Eighth, the process for handling organizational disobedience should proceed as quickly as possible. Delaying resolution of such issues can be a method of punishing dissent. Sufficient delay often allows management to perform the actions against which the protest was made. Prolonging the suspense and cloud of suspicion that accompanies an investigative process also serves to punish a protesting employee, even if his or her actions were completely justifiable.

8.4 PROPER ENGINEERING AND MANAGEMENT DECISIONS

Functions of Engineers and Managers

How should we establish the boundary between decisions that should be made by engineers and those that should be made by managers? An answer to this question must begin with a delineation of the proper functions of engineers and managers in an organization and of the contrasting points of view associated with these differing functions.

The primary function of engineers within an organization is to use their technical knowledge and training to create structures, products, and processes that are of value to the organization and its customers. But engineers are also professionals, and they must uphold the standards that their profession has decided should guide the use of their technical knowledge. Thus, engineers have a dual loyalty—to the organization and to their profession. Their professional loyalties go beyond their immediate employer.[19]

These obligations include meeting the standards usually associated with good design and accepted engineering practice. The criteria embedded in these standards include such considerations as efficiency and economy of design, the degree of invulnerability to improper manufacturing and operation, and the extent to which state-of-the-art technology is used.[20] We summarize these considerations by saying that engineers have a special concern for quality.

Engineers also ascribe preeminent importance to safety. Moreover, they are inclined to be cautious in this regard, preferring to err on the conservative side in safety considerations. In the *Challenger* case, for example, the engineers did not have firm data on the behavior of the O-rings at low temperatures, even though their extrapolations indicated that there might be severe problems. So they recommended against the launch.

The function and consequent perspective of managers is different. Their function is to direct the activities of the organization, including the activities of engineers. Managers are not professionals in the strict sense. Rather than being oriented toward standards that transcend their organization, they are more likely to be governed by the standards that prevail within the organization and, in some cases, perhaps by their own personal moral beliefs. Both Jackall and the Hitachi report imply that managers view themselves as custodians of the organization and are primarily concerned with its current and future well-being. This well-being is measured for the most part in economic terms, but it also includes such considerations as public image and employee morale.

This perspective differs from that of engineers. Rather than thinking in terms of professional practices and standards, managers tend to enumerate all of the relevant considerations ("get everything on the table," as they sometimes say) and then balance them against one another to come to a conclusion. Managers feel strong pressure to keep costs down and may believe engineers sometimes go too far in pursuing safety, often to the detriment of such considerations as cost and marketability. By contrast, engineers tend to assign a serial ordering to the various considerations relevant to design so that minimal standards of safety and quality must be met before any other considerations are relevant.[21] Although they may also be willing to balance safety and quality against other factors to some extent, engineers are more likely to believe that they have a special obligation to uphold safety and quality standards in negotiations with managers. They will usually insist that a product or process must never violate accepted engineering standards and that changes be made incrementally.

These considerations suggest a distinction between what we call a *proper engineering decision* (PED), a decision that should be made by engineers or from an engineering perspective, and what we call a *proper management decision* (PMD), a decision that should be made by managers or from the management perspective. While not claiming to give a full definition of either PED or PMD in the sense of necessary and sufficient conditions, we can formulate some of the features that should ordinarily characterize these two types of decision procedures. We refer to the following descriptions as "characterizations" of proper engineering and management decisions. They are as follows:

PED: a decision that should be made by engineers or at least governed by professional engineering standards because it either (1) involves technical matters that require engineering expertise or (2) falls within the ethical standards embodied in engineering codes, especially those that require engineers to protect the health and safety of the public.

PMD: a decision that should be made by managers or at least governed by management considerations because (1) it involves factors relating to the well-being of the organization, such as cost, scheduling, and marketing, and employee morale or welfare; and (2) the decision does not force engineers (or other professionals) to make unacceptable compromises with their own technical or ethical standards.

We make three preliminary remarks about these characterizations of engineering and management decisions. First, the characterizations of the PED and PMD show that the distinction between management and engineering decisions is made in terms of the standards and practices that should predominate in the decision-making process. Furthermore, the PMD makes it clear that management standards should never override engineering standards when the two are in substantial conflict, especially with regard to safety and perhaps even quality. However, what is considered a "substantial conflict" may often be controversial. If engineers want *much* more than *acceptable* safety or quality, then it is not clear that the judgment of engineers should prevail.

Second, the PMD specifies that a legitimate management decision not only must not force engineers to violate their professional practices and standards but also must not force other professionals to do so either. Even though the primary contrast here is the difference between engineering and management decisions, the specification of a legitimate management decision must also include this wider prohibition against the violation of other professional standards. A complete characterization of a legitimate management decision should also include prohibitions against violating the rights of nonprofessional employees, but this would make the characterization even more complicated and is not relevant for our purposes.

Third, engineers may often be expected to give advice, even in decisions properly made by managers. Management decisions can often benefit from the advice of engineers. Even if there are no fundamental problems with safety, engineers may have important contributions with respect to such issues as improvements in design, alternative designs, and ways to make a product more attractive.

Paradigmatic and Nonparadigmatic Examples

Several terms in both characterizations are purposely left undefined. The characterization of the PED does not define "technical matters,' and it certainly does not define "health" and "safety." PMD does not fully specify the kinds of considerations that are typical management considerations, citing only "factors relating to the well-being of the company, such as cost, scheduling, marketing, and employee morale or welfare." The characterization of the PMD requires that management decisions not force engineers to make "unacceptable compromises with their own professional standards," but it does not define *unacceptable*. We do not believe that it is useful to attempt to give any general definition of these terms. The application of these terms will be relatively uncontroversial in some examples, and no attempts at definition can furnish a definitive clarification in all of the controversial cases.

It will be useful to employ the line-drawing technique in handling moral issues that arise in this area. We refer to the relatively uncontroversial examples of PEDs and PMDs as *paradigmatic*.[22] The characterizations of PED and PMD provided earlier are intended to describe such paradigms. These two paradigms can be thought of as marking the two ends in a spectrum of cases.

We can easily imagine a paradigmatic PED. Suppose engineer Jane is participating in the design of a chemical plant that her firm will build for itself. She must choose between valve A and valve B. Valve B is sold by a friend of Jane's manager, but it fails to meet minimum specifications for the job. It has, in fact, been responsible for several disasters involving loss of life, and Jane is surprised that it is still on the market. Valve A, by contrast, is a state-of-the-art product. Among other things, it has a quicker shutoff mechanism and is also much less prone to malfunctions in emergencies. Although it is 5 percent more expensive, the expense is one that Jane's firm can well afford. Valve A, therefore, is the clear and unequivocal choice in terms of both quality and safety. Table 8.1 illustrates this.

Here, the decision should be made by Jane or other engineers, or at least in accordance with engineering considerations. This is because (1) the decision involves issues related to accepted technical standards and (2) the decision relates in important ways to the safety of the public and therefore to the ethical standards of engineers. The choice between valves A and B is a paradigmatic PED.

We can modify the example to make it a paradigmatic PMD. Suppose valves A and B are equal in quality and safety, but valve B can be supplied much faster than valve A, is 15 percent cheaper, and is manufactured by a firm that is a potential customer for some of the products of Jane's firm. Valve A, however, is made by a firm that is potentially an even bigger customer for some of the products of Jane's firm,

TABLE 8.1 A Paradigmatic PED

Feature	PMD	Test	PED
Technical expertise	Not needed	——————— X	Needed
Safety	Not important	——————— X	Important
Cost	Important	——————— X	Not important
Scheduling	Important	——————— X	Not important
Marketing	Important	——————— X	Not important

TABLE 8.2 A Paradigmatic PMD

Feature	PMD	Test	PED
Technical expertise	Not needed	X ———————————	Needed
Safety	Not important	X ———————————	Important
Cost	Important	X ———————————	Not important
Scheduling	Important	X ———————————	Not important
Marketing	Important	X ———————————	Not important

although cultivating a relationship with this firm will require a long-term commitment and be more expensive. If there are no other relevant considerations, the decision as to whether to purchase valve A or valve B should be made by managers, or at least made in accordance with management considerations. Comparing the decision by the two criteria in the PMD, we can say that (1) management considerations (e.g., speed of delivery, cost, and the decision as to which customers should be cultivated) are important, and (2) no violation of engineering considerations would result from either decision. Table 8.2 illustrates this case.

Many cases will lie between the two extremes of paradigmatic PEDs and paradigmatic PMDs. Some cases may lie so near the center of the imaginary spectrum of cases that they might be classified as either PED or PMD. Consider another version of the same case in which valve A has a slightly better record of long-term reliability (and is therefore somewhat safer), but valve B is 10 percent cheaper and can be both delivered and marketed more quickly. In this case, rational and responsible people might well differ on whether the final decision on which valve to buy should be made by engineers or managers. Considerations of reliability and safety are engineering considerations, but considerations of cost, scheduling, and marketing are typical management considerations. Table 8.3 illustrates this situation.

Would ordering valve B be an "unacceptable" compromise of engineering standards of safety and quality? Are the cost, scheduling, and marketing problems significant enough to overbalance the engineering considerations? Here, rational people of good will might differ in their judgments. In considering a case such as this, it is important to remember that, as in all line-drawing cases, *the importance or moral "weight" of the feature must be considered*. One cannot simply count the number of features that fall on the PMD or PED side or where the "X" should be placed on the line.

TABLE 8.3 PED/PMD: A Nonparadigmatic Case

Feature	PMD	Test	PED
Technical expertise	Not needed	——————X——	Needed
Safety	Not important	——————X——	Important
Cost	Important	—X————————	Not important
Scheduling	Important	————X—————	Not important
Marketing	Important	————X—————	Not important

Many issues regarding pollution also illustrate the problematic situations that can arise in the interface between proper engineering and proper management decisions. Suppose process A is so much more costly than process B that the use of process A might threaten the survival of the company. Suppose, furthermore, that process B is more polluting, but it is not clear whether the pollution poses any substantial threat to human health. Here again, rational people of good will might differ on whether management or engineering considerations should prevail.

8.5 RESPONSIBLE ORGANIZATIONAL DISOBEDIENCE

Sometimes engineers attempting to be both loyal employees and responsible professionals and citizens encounter difficulties. The engineer finds herself in a position of having to oppose her managers or her organization. Jim Otten finds the expression "organizational disobedience" appropriate as a generic term to cover all types of actions taken by an employee that are contrary to the wishes of her employer. Given the similarities between this kind of action and civil disobedience, the term seems appropriate.[23] We do not follow Otten's definition exactly, but we use his expression and define *organizational disobedience* as a protest of, or refusal to follow, an organizational policy or action.

It is helpful to keep the following two points about organizational disobedience in mind. First, the policy that a professional employee disobeys or protests may be either specific or general. It may be a specific directive of a superior or a general organizational policy, either a single act or a continuing series of actions.

Second, the employer may not intend to do anything morally wrong. For example, when an engineer objects to the production of a faulty type of steel pipe, he is not necessarily claiming that his firm intends to manufacture a shoddy product. Rather, he is objecting to a series of actions that would probably result in unfortunate consequences, however unintended.

There are at lest three distinct areas in which responsible engineers might be involved in organizational disobedience:

1. Disobedience by contrary action, which is engaging in activities contrary to the interests of the company, as perceived by management.
2. Disobedience by nonparticipation, which is refusing to carry out an assignment because of moral or professional objections.
3. Disobedience by protest, which is actively and openly protesting a policy or action of an organization.

What guidelines should the responsible engineer use in deciding when to engage in organizational disobedience in these areas, and how should he or she carry out this disobedience? We consider the first two types of organizational disobedience in this section and the third type in the next.

Disobedience by Contrary Action

Engineers may sometimes find that their actions outside the workplace are objectionable to managers. Objections by managers are usually in one of two areas. First, managers may believe that a particular action or perhaps the general lifestyle of an employee reflects unfavorably on the organization. For example, an engineer might be a member of a political group that is generally held in low esteem by the community. Second, managers may believe that some activities of employees are contrary to the interests of the organization in a more direct way. For example, an engineer may be a member of a local environmental group that is pressuring his or her company to install antipollution equipment that is not required by law or is lobbying to keep the company from purchasing some wetland area that it intends to drain and use for plant expansion. In an actual case, Mr. Novosel publicly opposed his employer in a public debate and referendum and was dismissed.[24] How should an engineer handle such delicate situations?

Although we cannot investigate all of the issues fully here, a few observations are essential. Disobedience by contrary action is not a paradigm case of harm to the organization (compared, for example, with theft or fraud), and its restriction by the organization is not a paradigm case of restriction of individual freedom (compared, for example, with a direction to do something the employee thinks is seriously immoral). Nevertheless, they are examples of harm to the individual and the organization. Let us consider some of the arguments that might be offered to confirm this claim.

On the one hand, there is no doubt that an organization can be harmed in some sense by the actions of employees outside the workplace. A company that has a reputation for hiring people whose lifestyles are offensive to the local community may not be able to hire highly desirable people, and it may lose business as well. The harm that an organization may suffer is even more obvious when employees engage in political activities that are directly contrary to the interests of the organization. A manager can argue with some persuasiveness that the simplistic assertion that nothing the employee does after 5 o'clock affects the organization does not do justice to the realities of business and community life. On these grounds, a manager might assert that the organization's right to the loyalty of its employees requires the employee not to harm the organization in these ways.

On the other hand, an employee's freedom suffers substantial curtailment if organizational restrictions force her to curtail activities to which she has a deep personal commitment. Nor can the manager persuasively argue that employees should simply resign if management finds their activities outside the workplace objectionable because the same activities might harm other organizations in the same way. Thus, consistently applying the argument that employees should never do anything that harms the organization results in the conclusion that employees should never engage in lifestyles or political activities that are controversial. This amounts to a substantial limitation of an employee's freedom.

In surveying these arguments, we believe that a good case can be made that organizations should not punish employees for disobedience by contrary action. Punishing employees for disobedience by contrary action amounts to a considerable infringement on individual freedom. Moreover, employees may not be able to avoid this type of harm to organizations simply by changing jobs. Many organizations might be harmed by an engineer's political views or efforts on behalf of the environment. Thus, allowing this type of harm to count as justification for organizational control permits organizations to exert considerable influence over an employee's life outside the workplace. In a society that values individual freedom as much as ours does, such a substantial abridgement of individual freedom is difficult to justify.

Despite these considerations, however, many managers will act strenuously when they believe they or their organizations are threatened by actions of employees outside the workplace. Therefore, two observations may be appropriate.

First, some actions by employees outside the workplace harm an organization more directly than others. An engineer's campaign for tighter restrictions on her own company's environmental pollution will probably have a more direct effect on her company than an engineer's private sexual life. Employees should be more careful in areas in which the harm to their organization is more direct.

Second, there can be a major difference in the degree to which curtailment of an employee's activities outside the workplace encroaches on his freedom. Curtailment of activities closely associated with one's personal identity and with strong moral or religious beliefs is more serious than limitation of activities that are associated with more peripheral beliefs. Therefore, employees should allow themselves more freedom in areas that are closely related to their basic personal commitments than in areas more peripheral to their most important concerns.

Disobedience by Nonparticipation

In one of the most famous legal cases that falls in this category, Dr. Grace Pierce, a physician, strongly objected to some impending tests on humans of a drug for diarrhea. Dr. Pierce had not actually refused to participate in the conduct of the tests, but the firm assumed that she would refuse and transferred her to another area. She eventually resigned.[25] Engineers are most likely to engage in disobedience by nonparticipation in projects that are related to the military and in projects that may adversely affect the environment. Engineer James, a pacifist, may discover that the underwater detection system that his company has contracted to build has military applications and thereupon request to be relieved of an assignment to the project. Engineer Betty may request not to be asked to design a condominium that will be built in a wetland area.

Disobedience by nonparticipation can be based on professional ethics or personal ethics. Engineers who refuse to design a product that they believe is unsafe can base their objections on their professional codes, which require engineers to give preeminence to considerations of public safety, health, and welfare. Engineers who refuse to design a product that has military applications because of their personal objections to the use of violence must base their refusal on personal morality because the codes do not prohibit engineers from participating in military projects. The basis of objections to participating in projects that engineers believe are harmful to the environment is more controversial. Some of the engineering codes have statements about the environment and some do not; when present, the statements are usually very general and not always easy to interpret.

Several things should be kept in mind about disobedience by nonparticipation. First, it is possible (although perhaps unlikely) for an employee to abuse the appeal to conscience, using it as a way to avoid projects he finds boring or not challenging or as a way to avoid association with other employees with whom he has personal difficulties. An employee should be careful to avoid any behavior that would support this interpretation of his or her actions. Second, it is sometimes difficult for employers to honor a request to be removed from a work assignment. For example, there may be no alternative assignments, there may be no other engineer who is qualified to do the work, or the change may be disruptive to the organization. These problems are especially severe in small organizations.

Nevertheless, we believe an organization, when it can do so, should honor most requests for nonparticipation in a project when the requests are based on conscience or a belief that the project violates professional standards. Common morality holds that a violation of one's conscience is a serious moral matter. Employers should not force employees to make a choice between losing their job or violating personal or professional standards. Sometimes employers may not have any alternative work assignments, but many organizations have found ways to respect employees' views without undue economic sacrifice.

8.6 DISOBEDIENCE BY PROTEST

We have saved this third type of organizational protest for a separate section because it is the best known and most extensively discussed form of organizational disobedience. In some situations, engineers find the actions of the employer to be so objectionable that they believe mere nonparticipation in the objectionable activity is insufficient. Rather, some form of protest, or "whistleblowing," is required. We begin by making some general comments about whistleblowing and then consider two important theories of whistleblowing.

What Is Whistleblowing?

The origin and exact meaning of the metaphor of whistleblowing are uncertain. According to Michael Davis, there are three possible sources of the metaphor: a train sounding a whistle to warn people to get off the track, a referee blowing a whistle to indicate a foul, or a police officer blowing a whistle to stop wrongdoing.[26] One problem with all of these metaphors, as Davis points out, is that they depict whistleblowers as outsiders, whereas a whistleblower is more like a team player who calls a foul play on his own team. This "insider" aspect is suggested by the *American Heritage Dictionary*'s definition of a whistleblower as "one who reveals wrongdoing within an organization to the public or to those in positions of authority." This suggests two characteristics of whistleblowing: (1) One reveals information that the organization does not want revealed to the public or some authority, and (2) one does this out of approved channels.

An important distinction is between internal and external whistleblowing. In *internal whistleblowing*, the alarm about wrongdoing stays within the organization, although the whistleblower may bypass his immediate superiors, especially if they are involved in the wrongdoing. In *external whistleblowing*, the whistleblower goes outside the organization, alerting a regulatory organization or the press. Another important distinction is between open and anonymous whistleblowing. In *open whistleblowing*, the whistleblower reveals his identity, whereas in *anonymous whistleblowing* the whistleblower attempts to keep his identity secret. Whether internal or external, open or anonymous, however, a whistleblower is usually defined as a person who is an insider, one who is a part of the organization. For this reason, the question of loyalty always arises. A whistleblower's actions are acts of disloyalty to his or her organization.[27] Therefore, whistleblowing needs a justification. Let's look at the two major approaches to the justification of whistleblowing. One uses primarily utilitarian considerations and the other employs considerations more appropriate to the standpoint of respect for persons.

Whistleblowing: A Harm-Preventing Justification

Richard DeGeorge has provided a set of criteria that must be satisfied before whistleblowing can be morally justified.[28] DeGeorge believes that whistleblowing is morally *permissible* if

1. the harm that "will be done by the product to the public is serious and considerable";
2. the employees report their concern to their superiors, and;
3. "getting no satisfaction from their immediate superiors, they exhaust the channels available" within the organization.

DeGeorge believes that whistleblowing is morally *obligatory* if

1. the employee has "documented evidence that would convince a responsible, impartial observer that his view of the situation is correct and the company policy is wrong"; and
2. the employee has "strong evidence that making the information public will in fact prevent the threatened serious harm."

Notice that the criteria involve a balancing of harms and benefits in a typical utilitarian fashion. There is potential harm to the public, and this is what initiates the considerations that whistleblowing might be justified. The public will benefit if these harms are eliminated. There is also potential harm to the organization, and the prospective whistleblower must attempt to minimize this harm by first trying to use available channels within the organization. There is also potential harm to the whistleblower, and the risk of harm must only be undertaken when there is some assurance that others would be convinced of the wrong and the harm might be prevented. There is no reason, DeGeorge seems to believe, to risk one's career if there is little chance the whistleblowing will have the desired effect. Taken as general tests for justified or required whistleblowing, however, DeGeorge's criteria are subject to criticisms.[29]

1. The first criterion seems too strong. DeGeorge seems to assume that the employee must *know* that harm will result and that the harm must be *great*. Sometime an employee is not in a position to gather evidence that is totally convincing. Perhaps just believing on the basis of the best evidence available that harm will result is sufficient.

2. It should not always be necessary for employees to report their criticisms to their superiors. Often, one's immediate superiors are the cause of the problem and cannot be trusted to give unbiased evaluation of the situation.

3. It should not always be necessary to exhaust the organizational chain of command. Sometimes there is not time to do this before a disaster will occur. Also, sometimes employees have no effective way to make their protests known to higher management except by going public.

4. It is not always possible to get documented evidence of a problem. Often, organizations deprive employees of access to the vital information needed to make a conclusive argument for their position. They deprive protesting employees of access to computers and other sources of information necessary to make their case.

5. The obligation to make the protest may not always mean there will be strong evidence that a protest will prevent the harm. Just giving those exposed to a harm the chance to give free and informed consent to the potential harm is often a sufficient justification of the protest.

6. Some have argued that if the whistleblower does not have evidence that would convince a reasonable, impartial observer that her view of the situation is

correct (criterion 4), her whistleblowing could not prevent harm and would not even be morally permissible, much less obligatory. Thus, if criterion 4 is not fulfilled, whistleblowing might not even be permissible.[30]

Whistleblowing: A Complicity-Avoiding View

Michael Davis has proposed a very different theory of the justification of whistleblowing: "We might understand whistleblowing better if we understand the whistleblower's obligation to derive from the need to avoid *complicity* in wrongdoing rather than from the ability to prevent harm."[31] Davis formulates his "complicity theory" in the following way.

You are morally required to reveal what you know to the public (or to a suitable agent or representative of it) when

(C1) what you will reveal derives from your work for an organization;

(C2) you are a voluntary member of that organization;

(C3) you believe that the organization, though legitimate, is engaged in a serious moral wrong;

(C4) you believe that your work for that organization will contribute (more or less directly) to the wrong if (but *not* only if) you do not publicly reveal what you know;

(C5) you are justified in beliefs C3 and C4; and

(C6) beliefs C3 and C4 are true.[32]

According to complicity theory, the moral motivation for blowing the whistle is to avoid participating in an immoral action, not to prevent a harm to the public. Thus, it is more in agreement with the basic ideas of the respect for persons tradition. One blows the whistle to avoid violating moral precepts, not to prevent harm to the public.

Davis' approach to the moral justification of whistleblowing has several distinct advantages. First, since preventing harm to the public is not a motivation for whistleblowing, one does not have to know that harm would result if he does not blow the whistle. Second, since preventing harm to the organization is not a motivation for blowing the whistle, one does not have to first work through organizational channels. Third, since preventing harm to oneself is not a motivation for whistleblowing, one does not have to be sure that blowing the whistle will prevent the harm before one risks one's career. Nevertheless, there are problems with Davis' theory as well.[33]

First, the requirement that what one reveals must derive from one's work in the organization (C1) and must contribute to the wrongdoing (C4) seems much too restrictive. Suppose engineer Joe is asked to review a design for a structure submitted to a customer by another member of the organization in which he is employed. Joe finds the design highly defective and, in fact, that it would be a serious threat to public safety if the structure were to be built. According to Davis, Joe would not have any obligation to blow the whistle because the design had nothing to do with Joe's work with the organization. Yet this seems implausible. Joe may well have an obligation to blow the whistle if the design poses a serious threat because of its potential for harm to the public regardless of his own involvement in the design.

Second, Davis also requires that a person be a voluntary member of an organization. But suppose Michael, an Army draftee, discovers a situation that poses a serious threat to his fellow soldiers. Michael has a moral obligation to blow the whistle, and the fact that he was drafted seems to have little relevance.

Third, Davis believes that one is only justified in blowing the whistle if in fact one believes that serious wrongdoing by the organization has occurred. But it seems more reasonable to say that one is justified in blowing the whistle if one has good reason to believe that wrongdoing will occur. Even if one turned out to be mistaken, one would still be justified in blowing the whistle, especially from the standpoint of the ethics of respect for persons. Otherwise, one's moral integrity would be compromised because one would be involved in activities that at least one *believes* to be wrong. To be doing something one believes to be wrong is still a serious compromise of one's moral integrity, even if by some more objective standard one is not actually involved in wrongdoing.

Finally, Davis does not take sufficient account of what many people would consider to be a clear—and perhaps the most important—justification of whistleblowing, namely that it is undertaken to prevent harm to the organization or (more often) to the public. Although avoiding complicity in wrongdoing is a legitimate and important justification for blowing the whistle, at the very least, it need not be the only one.

Despite the criticisms of both theories, there does seem to be truth in both. For Davis, whistleblowing must be justified because the whistleblower violates the obligation of loyalty. He justifies blowing the whistle to keep himself from complicity in wrongdoing. For DeGeorge, whistleblowing must be justified because of the harm it can produce to the organization and to the whistleblower. These harms can sometimes be outweighed by the harm to the public that would otherwise occur. All of these considerations seem valid. From a practical standpoint, they are all important to consider when thinking about blowing the whistle.

Some Practical Advice on Whistleblowing

We conclude this section with some practical considerations on protesting organizational wrongdoing.

First, take advantage of any formal or informal processes your organization may have for making a protest. Your organization may have an "ethics hotline" or an ombudsman. The Nuclear Regulatory Commission has a formal process for registering what it calls "Differing Professional Opinions."[34] Many managers have an "open door" policy, and there may be other informal procedures for expressing to a superior a different assessment of a situation.

Second, determine whether it is better to keep your protest as confidential as possible or to involve others in the process. Sometimes the most effective way to work within an organization is to work confidentially and in a nonconfrontational way with superiors and colleagues. At other times, it is important to involve your peers in the process so that a manager cannot justify disregarding your protest by assuming that it is the result of one disgruntled employee.

Third, focus on issues, not personalities. People get defensive and hostile when they are personally attacked, whether these people are your superiors or your peers. Therefore, it is usually a better tactic to describe the issues in impersonal terms insofar as this is possible.

Fourth, keep written records of the process. This is important if court proceedings are eventually involved. It also serves to "keep the record straight" about what was said and when it was said.

Fifth, present positive suggestions in association with your objection. Your protest should have the form, "I have a problem that I want to bring to your attention, but I also think I have a way to solve it." This approach keeps your protest from being wholly negative and suggests a positive solution to the problem you have identified. Positive suggestions can be helpful to managers, who must deal with the problem in a practical way.

8.7 ROGER BOISJOLY AND THE *CHALLENGER* DISASTER

Two events in the professional life of engineer Roger Boisjoly, both related to the *Challenger* disaster, illustrate several themes in this chapter. One of these events is the teleconference between Morton Thiokol and NASA the night before the launch of the *Challenger*. This dramatic event illustrates the conflict between engineers and management in decision making. The second experience is Boisjoly's testimony before the Presidential Commission on the Space Shuttle *Challenger* Accident. Boisjoly's testimony raises the issue of whistleblowing and the extent of the legitimacy of loyalty of an engineer to the organization in which he or she is employed.

Proper Management and Engineering Decisions

Robert Lund, vice president of engineering at Morton Thiokol, was both an engineer and a manager. In the teleconference on the evening before the fateful launch, he, in concert with other engineers, had recommended against launch. The recommendation was based on a judgment that the primary and secondary O-rings might not seal properly at the low temperatures at which the vehicle would be launched. NASA officials expressed dismay at the no-launch recommendation, and Thiokol executives requested an interruption in the teleconference to reassess their decision. During the 30-minute interruption, Jerald Mason, senior vice president of Morton Thiokol, turned to Lund and told him to take off his engineering hat and put on his management hat. Afterwards, Lund reversed his no-launch recommendation.

In admonishing Lund to take off his engineering hat and put on his management hat, Mason was saying that the launch decision should be a management decision. Testifying before the Rogers Commission, which investigated the *Challenger* accident, Mason gave two reasons for this belief. First, the engineers were not unanimous: "[W]ell, at this point it was clear to me we were not going to get a unanimous decision."[35] If engineers disagreed, then there was presumably not a clear violation of the technical or ethical standards of engineers; thus, it could be argued that neither requirement of the PMD was being violated.

There are reasons to doubt the factual accuracy of Mason's claim, however. In his account of the events surrounding the *Challenger* given at the Massachusetts Institute of Technology (MIT) in 1987, Roger Boisjoly reported that Mason asked if he was "the only one who wanted to fly."[36] This would suggest that he did not have evidence at this point that other engineers wanted to fly. Whatever validity Mason could give to his argument that some engineers supported the launch (and therefore that the opposition of the engineers to the launch was not unanimous) was apparently based on conversations with individual engineers after the teleconference. So Mason probably had little justification at the time of the teleconference for believing that the nonmanagement engineers were not unanimously opposed to the launch.

Nevertheless, Mason may be correct in maintaining that there was some difference of opinion among those most qualified to render judgment, even if this information was not confirmed until after the event. If engineers disagreed about the technical issues, then the engineering considerations were perhaps not as compelling as they would have been if the engineers had been unanimous. Thus, the first part of the PED criterion may not have been fully satisfied. Those who did not find a technical problem probably would not find an ethical problem either. So the second criterion of the PED may also not have been fully satisfied.

Mason's second reason was that no numbers could be assigned to the time required for the O-rings to seal at various temperatures:

Dr. Keel: Since Mr. Lund was your vice president of engineering and since he presented the charts and the recommendations not to launch outside of your experience base—that is, below a temperature of 53 degrees for the O-rings—in the previous 8:45 Eastern Standard Time teleconference, what did you have in mind when you asked him to take off his engineering hat and put on his management hat?

Mr. Mason: I had in mind the fact that we had identified that we could not quantify the movement of that, the time for movement of the primary [O-ring]. We didn't have the data to do that, and therefore it was going to take a judgment rather than a precise engineering calculation, in order to conclude what we needed to conclude.[37]

This might also be a reason for holding that the decision to launch did not violate criterion 2 of the PMD and did not clearly satisfy criterion 1 of the PED. However, the fact that no calculations could be made to determine the time it would take the O-rings to seal at various temperatures does not necessarily justify the conclusion that a management decision should be made. Surely the fact that failure of the O-rings to seal could destroy the *Challenger* implies that the engineering considerations were of paramount importance even if they could not be adequately qualified. The engineer's concern for safety is still relevant.

Nevertheless, Mason's comment may make a valid observation. Given that engineers generally prefer to make judgments on the basis of quantitative calculations, they may well have been uncomfortable with the fact that there were no precise numbers for the degree of degradation of the O-rings at lower temperatures. As a result, the engineering judgment did not have the same degree of decisiveness that it would have had otherwise. All that Roger Boisjoly could argue was that the degree of degradation seemed to be correlated with temperature, and even the data he used to back up this claim were limited.

Mason's arguments, taken together, might be seen as an attempt to meet criterion 2 of the PMD. If the decision to recommend launch is not a clear violation of engineering practice, then an engineer would not violate his technical practices by recommending launch. Thus, Mason's argument could be seen as a claim that the decision whether to launch was at the very least not a paradigm instance of a PED. A paradigm PED would be one in which (among other things) the experts clearly agree and there are quantitative measures that unambiguously point to one option rather than another. Thus, the recommendation to launch was at the very least not a paradigm case of a violation of technical engineering practices.

Mason might also have argued that criterion 1 of the PMD was satisfied. A renewed contract with NASA was not assured, and failure to recommend launch might have been the decisive factor that persuaded NASA officials not to renew the contract with Morton Thiokol. Thus, the well-being of the company might have been substantially harmed by a no-launch recommendation.

Despite these arguments, we believe that the launch decision was properly an engineering decision, even though it perhaps was not a paradigm case of such a decision.

First, criterion 1 of the PMD was not as compelling a consideration as Mason may have supposed. There was no evidence that a no-launch decision would threaten the survival of Morton Thiokol, or even that it would in any fundamental way jeopardize Thiokol's well-being. In any case, engineering considerations should have had priority.

Second, criterion 2 of the PED was relevant because the decision to launch violated the engineer's propensity to modify or change criteria only in small increments. The temperature on the launch day was more than 20 degrees below that of any previous launch day. This was an enormous change, which should have given an engineer good reason to object to the launch.

Third, criterion 1 of the PED was relevant. Even though the quantitative data were limited and clearly did not give conclusive evidence that there would be a disaster, the data did seem to point in that direction so that the engineering need for quantitative measures was satisfied to some extent. Engineers, furthermore, are alert to the fact that composites, such as the ones the O-rings are made of, are temperature sensitive and that one could reasonably expect substantially lower temperatures to produce substantially greater blow-by problems.

Fourth, criterion 2 of the PED was also relevant because life was at stake. Engineers are obligated by their codes of ethics to be unusually cautious when the health and safety of the public are involved. This should be particularly important when those at risk do not give informed consent to special dangers. This was the case with the astronauts, who did not have any knowledge of the problems with the O-rings.

The importance of the safety issue was further highlighted because of the violation of the practice of requiring the burden of proof to be borne by anyone advocating a launch decision rather than a no-launch decision. In testimony before the Rogers Commission, Robert Lund recounts this all-important shift in the burden of proof:

Chairman Rogers: How do you explain the fact that you seemed to change your mind when you changed your hat?

Mr. Lund: I guess we have got to go back a little further in the conversations than that. We have dealt with Marshall for a long time and have always been in the position of defending our position to make sure that we were ready to fly, and I guess I didn't realize until after that meeting and after several days that we had absolutely changed our position from what we had before. But that evening I guess I had never had those kinds of things come from the people at Marshall that we had to prove to them that we weren't ready....And so we got ourselves in the thought process that we were trying to find some way to prove to them it wouldn't work, and we were unable to do that. We couldn't prove absolutely that the motor wouldn't work.

Chairman Rogers: In other words, you honestly believed that you had a duty to prove that it would not work?

Mr. Lund: Well that is kind of the mode we got ourselves into that evening. It seems like we have always been in the opposite mode. I should have detected that, but I did not, but the roles kind of switched.[38]

This last-minute reversal of a long-standing policy, requiring the burden of proof to rest with anyone recommending a no-launch rather than a launch decision, was a serious threat to the integrity of the engineering obligation to protect human life.

Although hindsight no doubt benefits our judgment, it does seem that the decision whether to recommend launch was properly an engineering decision rather than a management decision, even though it may not have been a paradigm case of a proper engineering decision. There is insufficient reason to believe that the case diverged so much from the paradigm engineering decision that management considerations should have been allowed to override the engineering constraints. Engineers, not managers, should have had the final say on whether to launch. Or, if the person making the recommendation wore both an engineering hat and a management hat—as Robert Lund did—he should have kept his engineering hat on when he made the decision. The distinction between paradigmatic engineering and management decisions and the attendant methodology developed here help to confirm this conclusion.

Whistleblowing and Organizational Loyalty

Boisjoly's attempt in the teleconference to stop the launch was probably not an instance of whistleblowing. It certainly was not an instance of external whistleblowing because Boisjoly made no attempt to alert the public or officials outside Thiokol and NASA. His actions on the night before the launch were probably not even internal whistleblowing because (1) they did not involve revealing information that was not known (rather, they made arguments about the information already available) and (2) he did not go out of approved channels. His testimony before the Rogers Commission, however, might be considered a case of whistleblowing because it did fulfill these two criteria. His testimony revealed information that the general public did not know, and it used channels outside the organization, namely the Rogers Commission. Was his testimony a case of justified whistleblowing?

First, let us look at DeGeorge's criteria. Since his criteria are utilitarian in orientation and focus on preventing harm, our first response might be to say that Boisjoly's testimony before the Rogers Commission could not be an instance of whistleblowing because the tragedy had already occurred. One writer has argued, however, that Boisjoly thought his testimony might contribute to the safety of future flights. He cites as his evidence a speech Boisjoly made at MIT, during which he reminded the audience that, as professional engineers, they had a duty "to defend the truth and expose any questionable practice that may lead to an unsafe product."[39] Whether or not Boisjoly actually believed his testimony might prevent future disasters, we can ask whether his testimony is in fact justified as a possible way to prevent future disasters. Certainly the harm of future disasters is serious and considerable (criterion 1). We can probably agree that, given his past experience,

Boisjoly had reason to believe that reporting his concerns to his superiors would not give satisfaction (criteria 2 and 3). If this is correct, his testimony, considered as a case of whistleblowing, would be justified. Given the facts of the *Challenger* disaster, his testimony would probably convince a responsible, impartial observer that something should be done to remedy the O-ring problems (criterion 4). Whether he had strong evidence for believing that making this information public would prevent such harms in the future (criterion 5) is probably much more doubtful.

We can probably conclude, therefore, that from the standpoint of DeGeorge's criteria, Boisjoly's whistleblowing was justified but not required. In any case, it is clear that—as one would expect from criteria that adopt a utilitarian standpoint—the major issue has to do with the legitimacy of our beliefs about the consequences of certain courses of action.

Now let us consider Boisjoly's testimony from the standpoint of Davis' criteria for justified whistleblowing. Unlike DeGeorge's criteria, where concern for preventing future harms must be the primary consideration, here we must be concerned with Boisjoly's need to preserve his own moral integrity. Was he complicit enough in the wrongdoing so that whistleblowing was necessary to preserve his own moral integrity? To review the criteria, his whistleblowing was certainly related to his work in the organization (criterion 1). Furthermore, he was a voluntary member of that organization (criterion 2). Also, he almost certainly believed that Morton Thiokol, though a legitimate organization, was engaged in a serious moral wrong (criterion 3). The central issue is raised by the fourth criterion, namely whether he believed that his work for Thiokol contributed (more or less directly) to the disaster so that if (but *not* only if) he failed to publicly reveal what he knew he would be a contributor to the disaster (criterion 4). Following on this are the questions of whether he was justified in a belief that continued silence would make him complicit in wrongdoing (criterion 5) and whether in fact this belief was true (criterion 6).

In order to better focus on the question of what it means to say that one's work contributes to wrongdoing, A. David Kline asks us to consider the following two examples.[40] In the first example, Researcher 1 is directed by his tobacco company to provide a statistical analysis that shows that smoking is not addictive. He knows that his analysis is subject to serious criticism, but his company nevertheless uses his work to mislead the public. In the second example, Researcher 2 is directed by his tobacco company to study the issue of smoking and addiction. He concludes that there is strong evidence that smoking is addictive, but his firm ignores his work and makes public claims that smoking is not addictive. According to Kline, Researcher 1 is complicit in the deception of the public, and Researcher 2 is not complicit. However, Boisjoly's situation, according to Kline, is closer to that of Researcher 2 than that of Researcher 1. Since the claim that Boisjoly was complicit in wrongdoing is false, Kline believes that Davis cannot justify Boisjoly's blowing the whistle by his criteria. Boisjoly is not required to blow the whistle in order to preserve his own moral integrity.

However, let us modify Davis' criteria so that the question becomes whether remaining silent would make Boisjoly complicit in *future* wrongdoing by Thiokol. Here, there are two questions: whether blowing the whistle would prevent future wrongdoing (a factual question) and whether silence would make Boisjoly complicit in wrongdoing (an application question). If the answer to both of these questions is in the affirmative, Boisjoly should blow the whistle.

We shall leave it to the reader to more fully explore these questions, but only point out that both theories of whistleblowing add useful dimensions to the study of the moral dimensions of the issue. It is important to ask whether blowing the whistle will prevent wrongdoing and to ask whether and to what extent our own moral integrity is compromised by silence. In practical deliberation, both questions are important.

A final issue raised by Boisjoly's testimony is whether he violated the obligation of loyalty to his firm. His action was probably a violation of uncritical loyalty, but it was not a violation of critical loyalty, at least if his blowing the whistle was justified. In this situation, these two questions cannot be divorced.

8.8 CHAPTER SUMMARY

Common law on the rights of employees in the workplace has been governed by the common law doctrine of employment at will, which holds that in the absence of a contract, an employer may discharge an employee at any time and virtually for any reason. Some recent court decisions have modified this doctrine by an appeal to "public policy," which gives protection to some actions by employees. Some statutory law also accords employees some rights against their employers.

Conflicts between employees, including engineers, and managers often occur in the workplace. Sociologist Robert Jackall gives a pessimistic account of the moral integrity of managers, implying that it may be difficult for an employee to preserve his integrity in the workplace. Other writers, however, have contradicted this account, implying that employees can usually be morally responsible without sacrificing their careers. In order to preserve their careers and their integrity, employees should educate themselves in the "culture" of their organization. They should also adopt some common-sense techniques for minimizing the threats to their careers when making a legitimate protest.

Given that engineers and managers have different perspectives, problems can be avoided if organizations make a distinction between decisions that should be made by managers and decisions that should be made by engineers. In general, engineers should make the decision when technical matters or issues of professional ethics are involved. Managers should make the decision when considerations related to the well-being of the organization are involved and the technical and ethical standards of engineers are not compromised. Many decisions do not neatly fall into either category, and the line-drawing method can be useful in deciding who should make a decision.

Sometimes organizational disobedience is necessary. One type of organizational disobedience is engaging in activities (typically outside the workplace) contrary to the interest of the organization, as these interests are defined by managers. Another type of organizational disobedience is refusing to participate, or asking to be relieved of an obligation to participate, in some task in the organization. A third type of organizational disobedience is protesting a policy or action of an organization. The most widely discussed example of this third type of disobedience is whistleblowing. Richard DeGeorge's theory of justified whistleblowing focuses on the weighing of the relevant harms and benefits. Michael Davis' theory of justified whistleblowing focuses on the question whether whistleblowing is required in order to relieve one of complicity in wrongdoing.

Roger Boisjoly's attempt to stop the launch of the *Challenger* illustrates the conflict between the prerogatives of managers and engineers in decision making. In this case, the recommendation on whether to launch should probably have been made by engineers. Boisjoly's testimony before the Rogers Commission was a case of whistleblowing because (1) his testimony revealed information not generally known and (2) he went outside of approved channels. We can usefully apply both DeGeorge's and Davis' criteria to Boisjoly's testimony before the Rogers Commission. Boisjoly's testimony illustrates critical, rather than uncritical, loyalty to an organization.

NOTES

1. 823 P.2d 100 (Colo. 1992).
2. See (no author) "Protecting Employees at Will against Wrongful Discharge: The Public Policy Exception," *Harvard Law Review*, 96, no. 8, June 1983, pp. 1931–1951.
3. Genna H. Rosten, "Wrongful Discharge Based on Public Policy Derived from Professional Ethics Codes," 52 *American Law Reports*, 5th 405.
4. *Wall Street Journal*, April 13, 1981.
5. NJ Stat Ann at 34:19-1 to 19-8.
6. Joseph A. Raelin, *The Clash of Cultures: Managers and Professionals* (Boston: Harvard Business School Press, 1985), p. xiv.
7. Ibid., p. 12.
8. Ibid., p. 270.
9. Robert Jackall, *Moral Mazes: The World of Corporate Managers* (New York: Oxford University Press, 1988), p. 5.
10. Ibid., p. 105–107.
11. Ibid., p. 69.
12. Ibid., p. 112.
13. Christopher Meyers, "Institutional Culture and Individual Behavior: Creating the Ethical Environment," *Science and Engineering Ethics*, 10, 2004, p. 271.
14. Patricia Werhane, *Moral Imagination and Management Decision Making* (New York: Oxford University Press, 1999), p. 56. Quoted in Raelin, op. cit., p. 271.
15. Michael Davis, "Better Communication between Engineers and Managers: Some Ways to Prevent Many Ethically Hard Choices," *Science and Engineering Ethics*, 3, 1997, p. 185. See also pp. 184–193 for the ensuing discussion.
16. Stephen H. Unger, *Controlling Technology: Ethics and the Responsible Engineer*, 2nd ed. (New York: Wiley, 1994), pp. 122–123.
17. See the discussion in Davis, op. cit., p. 207.
18. Meyers, op. cit., p. 257.
19. Raelin also points out the importance of professional loyalties that transcend the organization, contrasting the "local" orientation of managers with the "cosmopolitan" orientation of most professionals. While describing engineers as more locally oriented than most professionals, he does not deny that engineers have loyalties to professional norms that transcend loyalties to their own organization. See Raelin, pp. 115–118, for a description of the local–cosmopolitan distinction.
20. State-of-the-art technology may not always be appropriate. If an engineer is designing a plow for use in a less industrialized country, then simplicity, ease of repair, and availability of repair parts may be more important than the use of the most advanced technology.
21. We are indebted to Michael Davis for this and several other insights in this section. Davis uses John Rawls's term *lexical ordering* to describe the assigning of priorities. Rawls, however, seems to equate serial ordering with lexical ordering. He defines a lexical order as "an order which requires us to satisfy the first principle in the ordering before we can move on to the second, the second before we consider the third, and so on. A principle

does not come into play until those previous to it are either fully met or do not apply. A serial ordering avoids, then, having to balance principles at all; those earlier in the ordering have an absolute weight, so to speak, with respect to the later ones, and hold without exception." See John Rawls, *A Theory of Justice* (Cambridge, MA: Harvard University Press, 1971), p. 43. See also Michael Davis, "Explaining Wrongdoing," *Journal of Social Philosophy,* 20, Spring–Fall 1988, pp. 74–90.

22. For purposes of this analysis, we are assuming that all decisions are either PEDs or PMDs—that is, they should be made by either managers or engineers rather than by anyone else.

23. Jim Otten, "Organizational Disobedience," in Albert Flores, ed., *Ethical Problems in Engineering* (Troy, NY: Center for the Study of the Human Dimensions of Science and Technology, 1980), pp. 182–186.

24. *Novosel v. Nationwide Insurance* 721 F 2d 894.

25. *Pierce v. Ortho Pharmaceutical* 319 A.2d. at 178. This case occurred in 1982.

26. Michael Davis, "Whistleblowing," in Hugh LaFollette, ed., *The Oxford Handbook of Practical Ethics* (New York: Oxford University Press, 2003), p. 540.

27. Heins Luegenbiehl, "Whistleblowing," in Carl Mitcham, ed., *Encyclopedia of Science, Technology and Ethics* (Detroit, MI: Thomson, 2005), pp. 2062–2064.

28. Richard T. DeGeorge, *Business Ethics* (New York: Macmillan, 1982), p. 161. This account is taken directly from Richard T. DeGeorge, "Ethical Responsibilities of Engineers in Large Organizations," *Business and Professional Ethics Journal,* 1, no. 1, Fall 1981, pp. 1–14.

29. Several of these criticisms are suggested by Gene G. James, "Whistle-Blowing: Its Moral Justification," reprinted in Deborah G. Johnson, *Ethical Issues in Engineering* (Englewood Cliffs, NJ: Prentice Hall, 1991), pp. 263–278.

30. David Theo Goldberg, "Turning in to Whistle Blowing," *Business and Professional Ethics Journal,* 7, 1988, pp. 85–99.

31. Michael Davis, "Whistleblowing," op. cit., p. 549.

32. Ibid., pp. 549–550.

33. Several of these criticisms are taken from A. David Kline, "On Complicity Theory," *Science and Engineering Ethics,* 12, 2006, pp. 257–264.

34. See Unger, pp. 122–123.

35. *Report of the Presidential Commission on the Space Shuttle Challenger Accident,* vol. IV, Feb. 26, 1986 to May 2, 1986, p. 764.

36. Roger Boisjoly, "The *Challenger* Disaster: Moral Responsibility and the Working Engineer," in Deborah Johnson, *Ethical Issues in Engineering* (Englewood Cliffs, NJ: Prentice Hall, 1991), p. 6.

37. *Presidential Commission on the Space Shuttle Challenger Accident,* pp. 772–773.

38. Ibid., p. 811.

39. Roger Boisjoly, "The *Challenger* Disaster: Moral Responsibility and the Working Engineer," op. cit., p. 14. Quoted in A. David Kline, "On Complicity Theory," *Science and Engineering Ethics,* 12, 2006, p. 259.

40. Kline, op. cit., p. 262.

CHAPTER NINE

Engineers and the Environment

Main Ideas in this Chapter

- Engineering codes are increasingly including provisions about the environment, but their implications for many environmental issues are not clear.

- In the United States, environmental law, especially federal law, sets the context in which many engineering decisions about the environment are made. For the most part, environmental law focuses on making the environment "clean."

- There are various criteria for "clean," and this book favors a "degree-of-harm" criterion.

- A "progressive" attitude toward the environment goes beyond what the law requires, and this attitude can apply to corporate policy as well as to the career commitments of individual engineers. The progressive attitude can have several grounds, but the nonanthropocentric orientation is one of them.

- Despite the arguments made for a progressive attitude toward the environment, the authors continue to believe that the mandatory obligations of engineers toward the environment should be limited to protecting the environment insofar as it protects human health.

- However, when possible, organizations should allow engineers to refuse to participate in environmental projects (or other projects) they find objectionable and to voice responsible objections in and outside the workplace.

MARK HOLTZAPPLE, PROFESSOR OF CHEMICAL engineering at Texas A & M University, is a paradigm of an environmentally conscious engineer. Early in his career, Holtzapple decided to commit his research agenda to developing energy-efficient and environmentally friendly technologies. To this end, he is pursuing the following four areas of research:

- *Biomass conversion.* He is developing a process that converts biological materials into useful fuels and chemicals. Feedstocks to the process include municipal solid waste, sewage sludge, agricultural residues (e.g., sugarcane bagasse, corn stover, or manure), and energy crops (e.g., energy cane or sweet sorghum). He has a pilot plant near the university that will generate data needed to help commercialize the process. The process is sustainable and reduces greenhouse gas emissions.

- *A water-based air conditioner.* This air conditioner does not employ environmentally destructive refrigerants. Furthermore, it is substantially more energy-efficient than conventional air conditioners.
- *StarRotor engine.* This rotary engine is highly efficient and emits virtually no pollution. It can use a wide variety of fuels, including gasoline, diesel, methane, alcohol, and even vegetable oil. Holtzapple believes the engine should last for as long as 1 million miles when employed in automobiles.
- *Water desalinization.* He is currently working on a water-desalinization process that he believes will cost-effectively convert seawater into freshwater suitable for human consumption.

Among his numerous awards, Professor Holtzapple won the 1996 Green Chemistry Challenge Award given by the president and vice president of the United States.

9.1   INTRODUCTION

Engineers have a complex relationship to the environment. On the one hand, they have helped to produce some of the environmental problems that plague human society. Projects designed by engineers produce toxic chemicals that pollute the land, air, and rivers. Engineers also design projects that flood farmlands, drain wetlands, and destroy the forests. On the other hand, engineers can design projects, products, and processes that reduce or eliminate these same threats to environmental integrity. If engineers have contributed to our environmental problems (as have most of the rest of us), then they are also an essential part of their solution.

What obligations should the engineering profession and individual engineers assume with regard to the environment? We can begin to answer this question by examining the engineering codes. Then we discuss the laws that set the context in which much engineering work affecting the environment must be conducted. Next, we look at some examples of organizations that have adopted a progressive attitude toward the environment that goes beyond what the law requires. Then we consider some reasons for this more progressive approach, arguing that a nonanthropocentric orientation should be a part of the foundation for this progressive attitude. Finally, we consider the obligations regarding the environment that we believe should be mandatory for engineers.

9.2   WHAT DO THE CODES SAY ABOUT THE ENVIRONMENT?

Many engineering codes make no reference to the environment at all, but increasingly they are adopting some environmental provisions. The codes of the American Society of Civil Engineers (ASCE), the Institute of Electrical and Electronics Engineers (IEEE), the American Society of Mechanical Engineers (ASME), the American Institute of Chemical Engineers, and the Association for Computing Machinery have environmental provisions. The latest addition to the list is the National Society of Professional Engineers (NSPE). The ASCE code, however, still has the most extensive references to the environment. The 1977 code included for the first time the statement that "Engineers should be committed to improving the environment to enhance the quality of life" (section 1.f). The code as revised since then contains many more references to the environment. The code's environmental statements fall into two categories, which we may refer to as *requirements* and *recommendations*. By using the expression "engineers shall," the code *requires* engineers to "strive to comply with the principles of sustainable development," to inform their clients or employers of the possible consequences of ignoring the principles of sustainable development, to present information regarding the failure to comply with the principles of sustainable development to the proper authority in writing, and to "cooperate with the proper authority in furnishing such further information or assistance as may be required." By using the expression "engineers should," the code merely *recommends* that engineers seek opportunities to work for the "protection of the environment through the practice of sustainable development" and that they be "committed to improving the environment by adherence to the principles of sustainable development so as to enhance the quality of the life of the general public."

In another ASCE document, "The Role of the Engineer in Sustainable Development," sustainable development is defined as follows:

> Sustainable development is a process of change in which the direction of investment, the orientation of technology, the allocation of resources, and the development and functioning of institutions [is directed] to meet present needs and aspirations without endangering the capacity of natural systems to absorb the effects of human activities, and without compromising the ability of future generations to meet their own needs and aspirations.[1]

It is noteworthy that even though the provisions regarding sustainable development are requirements, they are only requirements to "strive to comply" with the principles of sustainable development or to inform clients or employers of the possible consequences of violating the principles of sustainable development. Even the strongest requirements, therefore, are rather weak. A look at the other codes will reveal similar weaknesses in the statements about the environment. In addition, the professional codes have no power to compel compliance. Nevertheless, professional codes provide some guidance for decision making by engineers. In some cases, individuals, companies, and even whole industry groups have gone further than the professional codes require.

9.3 THE ENVIRONMENT IN LAW AND COURT DECISIONS: CLEANING UP THE ENVIRONMENT

Most environmental laws are directed toward producing a cleaner environment. Environmental pollution, however, was not the subject of serious federal regulation until the late 1960s.[2] Until that time, an individual who wanted to combat pollution was usually forced to appeal to the common law. If no single individual was sufficiently harmed by pollution to be motivated to bring suit against a polluter, then no action was taken. The states were equally ineffective in controlling pollution. This situation opened the way for federal intervention.

Federal Laws on the Environment

In 1969, Congress passed the National Environmental Policy Act, which may well be the most important and influential environmental law in history. It has served as a model for legislation not only in the particular states but also in many other countries. The act declared "a national policy which will encourage productive and enjoyable harmony between man and his environment." The act attempts to "assure for all Americans safe, healthful, productive, and aesthetically and culturally pleasing surroundings."[3] One of its best known mandates is the environmental impact statement, which is now required of federal agencies when their decisions affect the environment. Congress then created the Environmental Protection Agency (EPA) to enforce its mandates.

Although directly concerned with worker health, the Occupational Safety and Health Act (1970) has important implications for the more general control of toxic substances. It authorizes the Secretary of Labor to set standards for "toxic materials or harmful physical agents." The standard for a given substance must be one that

> most adequately assures, to the extent feasible, on the basis of the best available evidence, that no employee will suffer material impairment of health or functional capacity even if such employee has regular exposure to the hazard dealt with by such standards for the period of his working life.[4]

The act seems to impose a strict standard: The employee must be protected from "material impairment of health or functional capacity" for his entire working life. But it also allows that the feasibility of the protection must be considered and that it need last only for the "working life" of the employee.

In 1970, Congress passed the Clean Air Act, amending it in 1977 and 1990. The act places health considerations ahead of balancing costs and benefits when dealing with hazardous pollutants.[5] It set a goal of 90 percent reduction in auto emissions. It also permitted the EPA to consider economic and technological feasibility in deciding when the goals were to be met but not in setting the goals themselves.

The Federal Water Pollution Control Act (FWPCA) was first enacted in 1948. In 1972, Congress passed a comprehensive revision of the FWPCA and amended it again in 1977. Some authorities attach the popular designation "Clean Water Act" to the 1972 amendments and others to the 1977 amendments. As stated in the 1972 amendments, the act is "designed to restore and maintain the chemical, physical, and biological integrity of the Nation's waters." It makes it unlawful for any person, business, or governmental body to discharge any pollutant into navigable waters without a permit. The act mandated pollution-control measures in two stages: (1) By 1977, all plants were to have installed water pollution-control devices that represented the best practicable pollution-control technology, and (2) by 1989 all plants were to have installed equipment that met more stringent standards. Plants discharging conventional pollutants must apply the best conventional pollution-control technology. Plants discharging toxic or unconventional pollutants were to apply the best available technology that was economically achievable. The act requires polluters to do their best to stop polluting regardless of the cost.[6]

In 1976, Congress passed the Resource Conservation and Recovery Act, which was designed to control the transportation, storage, treatment, and disposal of hazardous wastes. The act requires the producer of a hazardous waste to complete a "manifest," a form that describes the nature of the hazardous waste and its method of disposal. The transporter and the operator of the disposal site must both sign the manifest and return it to the producer of the waste. This procedure is supposed to provide a complete record of the disposal of the waste. The EPA is also required to regulate the disposal of the sites. The act requires that standards that regulate hazardous waste be based solely on the protection of public health and the environment.[7]

The Pollution Prevention Act of 1990 established pollution prevention as a national objective. The act requires the EPA to develop and implement a strategy to promote reduction of the pollutant's source. This policy is in sharp contrast to most environmental protection laws, which simply attempt to manage pollutants. The act established pollution prevention as the most desirable practice, followed by recycling, treatment, and disposal, in descending order or preference.

There are many other important pieces of environmental legislation. The federal Insecticide, Fungicide and Rodenticide Act took its contemporary form in 1972 and has had five amendments. The Safe Drinking Water Act was passed in 1974 and was substantially amended in 1996. The Toxic Substances Control Act was passed in 1976 and amended three times. The act, commonly referred to as " Superfund," was passed in 1980. In 1990, in response to the *Exxon Valdez* accident, the oil spill provisions of the Clean Water Act were amended to form what is usually called the Oil Pollution Act. The Residential Lead-Based Paint Hazard Reduction Act was passed in 1992, and the Mercury-Containing and Rechargeable Battery Management Act was passed in 1996.

This short list by no means enumerates all of the environmental and health-related legislation passed by Congress in the past several decades. It does illustrate, however, the range of positions taken on the proper criterion for a clean environment, from the refusal to allow cost to play any part in the determination to the clear acceptance of cost considerations. None of these acts mandates cost–benefit analysis, although some allow cost to be considered in their implementation.

The Courts on the Environment

Critics still maintain, however, that congressional legislation is often unrealistic in the extent to which it ignores cost considerations. The courts, they argue, must face even more directly the costs of pollution control to industry and governmental agencies, as well as the technological limits to our ability to control pollution. In the process, they might provide a more useful guide to a criterion for a clean environment.

In *International Harvester v. Ruckelshaus,* the District of Columbia Circuit Court ruled in 1973 that EPA regulations might have been congruent with the Clean Air Act but were defective in their rulings because they failed to consider the feasibility and practicality of the technology required.[8] The District of Columbia Court of Appeals rendered a decision in 1973 with similar import. It interpreted a relevant section of the Clean Air Act as permitting the EPA to consider cost but not to impose a cost–benefit test.[9] In the famous "Benzene" decision of 1980, a plurality of justices on the U.S. Supreme Court found that "safe" does not entail "risk free." Justice Stevens argued that the Occupational Safety and Health Administration (OSHA) could not regulate a chemical simply because it posed *some* risk; OSHA would also have to show that the risk was "significant."[10]

In 1986, a tribunal for the Circuit Court in the District of Columbia reviewed a decision by the EPA to set a standard for vinyl chloride emissions at levels more strict than industry could have satisfied at the time even if it had devoted great effort and expense to the task. The court ruled that when the EPA cannot determine a "safe" threshold for a pollutant, it may take not only health but also technological and economic factors into account in establishing emission standards that industry can achieve without paying costs that are "grossly disproportionate" to the level of safety achieved.[11]

In an earlier decision on asbestos, the District of Columbia Circuit Court of Appeals noted that Congress recognized that "employees would not be protected if employers were put out of business." It also called attention, however, to the fact that "standards do not become infeasible simply because they may impose substantial costs on an industry, force the development of new technology, or even force some employers out of business."[12]

Carl Cranor summarizes the implications of the Circuit Court's decisions in the following way:

> The implicit principles embodied in the D.C. Circuit Court's decisions suggest the following. On the one hand OSHA may set standards more stringent than existing ones in pursuit of better health for workers, unless they threaten the economic viability of an entire industry; that is too steep a price to pay for improved health. On the other hand, even the court interprets Congress as being willing to tolerate the loss of some jobs, and even some firms in an industry, if failure to impose health regulations would materially impair the health or functional capacity of workers in that industry.[13]

Any rational criterion for a clean environment must take into account both the need to protect the health of workers and the need to protect the financial viability of industries on which workers and the general public depend. Yet the balance suggested by Cranor's summary may not be the correct one because it appears to allow serious violations of the health of individuals if this is necessary to protect a whole "industry." According to Cranor's summary, we may impose stricter health regulations, even if the result is the closing of some firms, but we may not impose regulations that force the closing of a whole "industry." There are also conceptual and application issues that have to do with how we determine what constitutes an "industry." As Cranor asks, "Are plastic container and metal container manufacturers part of the same industry, or are they two different industries?"[14] Suppose that protecting human health requires that we impose regulation on plastic production that would put all plastic container manufacturers out of business. If plastic container manufacturers are considered an "industry" in themselves, then we may not impose these severe regulations because an entire industry would be eliminated. This limitation on our ability to protect the public would presumably apply, regardless of the severity of the health risks to workers or the public.

9.4   CRITERIA FOR A "CLEAN" ENVIRONMENT

Most environmental laws focus on making the environment "clean"—that is, free from various pollutants. The project of making the environment clean naturally raises the question of what criterion for "clean" underlies these regulations. Table 9.1 lists various criteria for a clean environment that can be found in the discussions of this topic. For the reasons listed in the table, we find all of the criteria inadequate.

In attempting to construct a more adequate criterion, we must begin with the assumption that we are trying to balance the goals of increasing job opportunities

TABLE 9.1 Inadequate Definitions of "Clean"

Criterion	Objections	Application
1. *Comparative criterion:* The environment is clean if it imposes no greater threat to human life or health than do other risks.	The levels of pollution currently accepted may be too high.	Workers should not expect working conditions to be safer than the drive to and from work.
2. *Normalcy criterion:* The environment is clean if the pollutants present in it are normally present in nature to the same degree.	The levels of pollution in nature vary and may sometimes be accepted only because they are unavoidable.	Radiation as high as the level of ultraviolet radiation in Denver is acceptable.
3. *Optimal-pollution reduction criterion:* The environment is clean if funds required to reduce pollution further could be used in other ways that would produce more overall human well-being.	Cost and benefits may be unfairly distributed.	The funds required to reduce a pollutant further would save more lives if used elsewhere.
4. *Maximum protection criterion:* The environment is clean only if any identifiable risk from pollution that poses a possible threat to human health has been eliminated, up to the capacity of available technology and legal enforcement to do so.	This criterion would require the elimination of many substances whose toxicity is doubtful or extremely limited.	A new chemical is assumed to be harmful unless shown to be harmless.
5. *Demonstrable harm criterion:* The environment is clean if every pollutant that is demonstrably harmful to human health has been eliminated.	It is often difficult to prove a substance is harmful, even when it is. Also, eliminating a pollutant completely may be too costly, as well as unnecessary if the pollutant is harmless at low levels.	Eliminate anything that can be proven to be a pollutant; leave everything else as it is.

and income, on the one hand, with protecting the health of individuals, on the other hand. Here, we have a classic conflict between utilitarian and respect for persons ethical considerations. From the utilitarian standpoint, we want to increase income, job opportunities, and even overall public health.[15] An increase in income produces utility, whether it is the income of workers or owners. Similarly, an increase in the number and the desirability of jobs also increases utility. Finally, good health is a precondition for achieving most other goods and so is also desirable from a utilitarian standpoint. Utilitarians, however, permit trade-offs between any of these goods, if the trade-off produces a net increase in overall utility. Because utilitarians consider the well-being of individuals only insofar as it affects overall utility, minor benefits to many might outweigh severe harms to a few. Thus, we might be justified in reducing health protection for some people in exchange for a net increase in overall utility.

Some environmental laws and (especially) some recent court decisions have attempted to guard against this unfortunate tendency of utilitarianism to forget the individual in the effort to promote greater overall utility; this often involves an appeal to considerations that are more compatible with the respect for persons point of view. The ethics of respect for persons attempts to take account of the distribution of goods and harms and of the special weight that some goods (e.g., health) have. From this respect for persons standpoint, an individual's health should not be sacrificed, even to increase the general welfare of everyone.

We believe that the engineer's obligation to hold paramount the health of the public should not be interpreted in purely utilitarian terms. However, the need to consider the economic effects of regulations that seek to protect human health must not be forgotten. The proper criterion for evaluating what is clean must cover a spectrum of cases, with two extremes. This spectrum is delineated in the following criterion, which we call the *degree-of-harm criterion:*

> When pollutants pose a clear and pressing threat to human health, they must be reduced below any reasonable threshold of harm. Cost should not be considered a significant factor. Insofar as substances pose an uncertain (but possible) risk to health or when the threshold of danger cannot be determined, economic factors may be considered. If a harm is irreversible, its prevention should be given higher priority.

According to this criterion, the task of protecting the environment oscillates between two extremes. At one extreme, where the risk of causing harm to the public is grave, the imperative of protecting human health should be made primary. In some cases, this imperative might require the elimination of virtually all pollution. At the other extreme, where the risk to human health appears to be small or indeterminate, cost–benefit considerations are more appropriate. Although cost–benefit analysis cannot be used to determine to what extent serious threats must be eliminated, it may be used—within limits that cannot be precisely specified—to determine the extent to which suspected but undetermined threats must be eliminated.

We believe that this policy should guide the engineer's interpretation of the injunction in most engineering codes to protect the health of the public, but that the need to protect the environment should be guided by requirements that are more extensive and that are discussed in the next section. Before we consider these requirements, however, let us review some cases to which this criterion can be applied.

We can begin with a case in which the criterion has been violated. Suppose engineer Vivian is employed by Shady Chemical. The firm has a long history of producing pesticides that not only damage the environment but also pose a threat to the workers who manufacture them, the farmers who apply them, and the consumers who eat the food to which they have been applied. When one of its products is banned by the government, the usual procedure of Shady Chemical is to slightly modify the formula of the product so that it no longer falls under the ban. By the time the new chemical is banned, a still newer product is ready to be marketed.

Vivian has been asked to participate in the development of an alternative to one of Shady Chemical's most successful products. The firm has learned on good authority that the product will soon be banned because it has been found to be a virulent carcinogen. Following its usual policy, Shady wants to find a substitute for the active ingredient in the pesticide that is as close to the old product as possible.

Although one can never be sure that the modified product has similar toxic properties to the old one until extensive testing has been done, Vivian has good reason to believe that the proposed substitute may even be worse. Shady Chemical has violated the degree-of-harm criterion. In fact, this is a paradigmatic violation.

Consider another example. The plant that employs engineer Bob has just discovered that its discharge into the atmosphere includes a new chemical that comes from one of its new product lines. The chemical is structurally similar to a class of chemicals that has been declared safe by the governmental regulatory agency. There is no reason to suspect that the chemical is dangerous, although its effects on humans have not been extensively tested. The plant's environmental affairs department is monitoring the new chemical and it is prepared to eliminate the chemical from the plant's discharge if any reason to suspect it is found, even if it is not banned by the government. In this case, Bob's firm is probably showing sufficient regard for human health and not violating the degree-of-harm criterion.

Many intermediate cases are more difficult to decide. Suppose engineer Melinda is employed at a plant that has identified a new chemical in its discharge into the local river. The chemical is not a regulated substance, although its chemical structure is similar to substances that have been found to be carcinogenic in large concentrations. Eliminating the substance would be expensive, but it would be economically feasible. In this situation, the degree-of-harm criterion would probably require that the plant at least begin making preparations to eliminate the substance from its discharge. Melinda would need more information, however, before she could be sure about the implications of the degree-of-harm criterion in this case. The importance of a thorough analysis of the facts, especially in nonparadigmatic line-drawing cases, cannot be overemphasized.

9.5 THE PROGRESSIVE ATTITUDE TOWARD THE ENVIRONMENT

Three Attitudes toward the Environment

In 1987, Joseph Petulla surveyed the attitudes taken by industry toward environmental law. Petulla found that there were three fundamentally different positions, all of which can apply to the attitudes of managers and individual engineers, as well as to corporate policy.[16] The first attitude is what we call the *sub-minimal attitude*. Industries in this group do as little as is possible—and sometimes less than is required—in meeting environmental regulations. They often have no full-time personnel assigned to environmental concerns, devote as few financial resources as possible to environmental matters, and fight environmental regulations. If it is cheaper to pay the fines than make the mandated changes, this is what they will do. Industries in this group generally hold that the primary goal of the company is to make money, and environmental regulations are merely an impediment to this goal.

The second attitude is what we call the *minimalist* or *compliance attitude*. Firms adopting this orientation accept governmental regulation as a cost of doing business but often without enthusiasm or commitment. There is often a great deal of skepticism about the value of environmental regulation. Nevertheless, these firms usually have established company policies that regulate environmental matters and have established separate units devoted to them.

A third attitude is what we call the *progressive attitude*. In these companies, being responsive to environmental concerns has the complete support of the CEO. The companies have well-staffed environmental divisions, use state-of-the-art equipment, and generally have good relationships with governmental regulators. The companies generally view themselves as good neighbors and believe that it is probably in their long-term interests to go beyond legal requirements because doing so generates good will in the community and avoids lawsuits. More than this, however, they may be genuinely committed to environmental protection and even environmental enhancement.

Two Examples of the Progressive Attitude toward the Environment

Although the progressive attitude has been adopted by individual engineers, such as Mark Holtzapple, we discuss two examples of this attitude on the corporate level. The number of industries in this third group seems to be increasing. In some cases, an entire industry group has developed a program to increase environmental responsibility. One example is an initiative by the Chemical Manufacturers Association (CMA). For many years, the chemical industry had received a considerable amount of public criticism about such issues as safety and pollution. In response to these criticisms, the CMA established a program called "Responsible Care: A Public Commitment." On April 11, 1990, more than 170 member firms of CMA published a set of guiding principles in the *New York Times* and the *Wall Street Journal*. The principles commit the industry to such policies as

- promoting the safe manufacture, transportation, use, and disposal of chemicals;
- promptly giving notice of safety and environmental hazards to the public and others who are potentially affected;
- operating plants in an environmentally safe manner;
- promoting research to improve chemicals with regard to their effects on health, safety, and the environment;
- participating with government in creating responsible laws regulating chemicals; and
- sharing with others information useful in promoting these goals.

To meet one major objective of the Responsible Care initiative, namely responding to public concerns, the CMA established a public advisory panel consisting of 15 non-industry representatives of the public. The CMA has also made commitment to Responsible Care a condition of membership.

Minnesota Mining and Manufacturing (3M) has not only subscribed to the Responsible Care guidelines but also adopted policies that go beyond their requirements. In the past, 3M was one of the nation's major emitters of air pollutants. In the early 1990s, 3M initiated a vigorous environmental program to reduce its emissions to levels below those allowed by environmental regulations. For example, it installed more than $175 million worth of equipment to recycle and burn off solvents, even though the plants were already meeting EPA standards for air emissions. This resulted in a reduction of volatile organic air emissions by 93 percent between 1990 and 2000, normalized to net sales.

3M has built its environmental strategy into all layers of management and production. It also helps customers reduce their waste problems by recycling some of the company's packaging. According to 3M, its thousands of 3P projects have saved the company $975 million since 1975.

A second example is represented by the CERES Principles, formerly called the Valdez Principles. (Ceres was the Roman Goddess of agriculture and fertility.) After the oil spill from the *Exxon Valdez,* a number of oil companies voluntarily adopted a set of principles that embody a progressive attitude toward the environment. We strongly suggest that the reader review this admirable set of principles for protecting the environment in their complete form at www.iisd.org/educate/learn/ceres.htm. The following is our summary, in abbreviated form, of the 10 principles:

1. *Protection of the biosphere.* Reduce and make progress toward the elimination of any environmentally damaging substance and safeguard habitats and protect open spaces and wilderness, while preserving biodiversity.
2. *Sustainable use of natural resources.* Make sustainable use of renewable natural sources, such as water, soils, and forests, and make careful use of nonrenewable resources.
3. *Reduction and disposal of wastes.* Reduce and, if possible, eliminate waste, and handle and dispose of waste through safe and responsible methods.
4. *Energy conservation.* Conserve energy and improve the energy efficiency of all operations, and attempt to use environmentally safe and sustainable energy sources.
5. *Risk reduction.* Strive to minimize the environmental, health, and safety risks to employees and surrounding communities, and be prepared for emergencies.
6. *Safe products and services.* Reduce and, if possible, eliminate the use, manufacture, or sale of products and services that cause environmental damage or health or safety hazards, and inform customers of the environmental impacts of our products or services.
7. *Environmental restoration.* Promptly and responsibly correct conditions we have caused that endanger health, safety, or the environment, redress injuries, and restore the environment when it has been damaged.
8. *Informing the public.* Inform in a timely manner everyone who may be affected by the actions of our company that affect health, safety, or the environment, and refrain from taking reprisals against employees who report dangerous incidents to management or appropriate authorities.
9. *Management commitment.* Implement these principles in a process that ensures that the board of directors and chief executive officer are fully informed about environmental issues and fully responsible for environmental policy, and make demonstrated environmental commitment a factor in selecting members of the board of directors.
10. *Audits and reports.* Conduct an annual self-evaluation of progress in implementing these principles, and complete and make public an annual CERES report.

9.6 GOING BEYOND THE LAW

How Far Does the Progressive View Go Beyond the Law?

Using the previous discussion as a base, we can summarize some of the ways in which the progressive view goes beyond the law.

First, with the exception of the Pollution Prevention Act of 1990, the law as we have summarized it says nothing about sustainable development and recycling, although the ASCE code and the CERES Principles do so. The progressive attitude encourages developing these types of technology.

Second, the CERES Principles and the 3P programs emphasize not only recycling but also avoiding the production of waste in the first place. Most of the laws, with the notable exception of the Pollution Prevention Act, do not address the issue of reducing waste.

Third, whereas the law does not require promoting research into the creation of more environmentally friendly chemicals, processes, and products, the progressive attitude does, as illustrated by the CERES Principles and the work of Mark Holtzapple.

Fourth, the progressive attitude requires the active participation with government in the creation of effective and responsible regulation. This kind of proactive cooperation with regulatory agencies is not required by the law.

Fifth, the progressive attitude encourages aggressive attempts to reduce emissions and other sources of pollution below those levels required by the law. Even eliminating altogether those substances known to be environmentally harmful is often considered a desirable aim.

Sixth, progressive companies attempt to reduce their consumption of energy, even though this is rarely, if ever, required by the law. Individual engineers who adopt the progressive attitude attempt to develop such technologies.

Seventh, progressive principles require firms to promptly report to the public when their actions produce threats to the health or safety of the public or to the environment.

Eighth, progressive firms pledge to make demonstrated environmental commitments one of the criteria for selecting officers and members of the board of directors.

Ninth, progressive firms commit to annual environmental audits in which they hold themselves accountable for environmentally responsible operation. These audits are to be made available to the public.

What Reasons Support Adopting the Progressive Attitude?

What reasons can we cite for adopting the progressive attitude, whether on the part of individuals or corporations?

First, there is the motive of individual and corporate self-interest. Many firms and industry groups, such 3M and the chemical industry generally, have instituted progressive policies only after legal problems or strong and persistent public criticism. No doubt one of the motivations for these policies is the desire to regain public trust and to avoid still more bad publicity. Progressive environmental policies also keep business out of trouble with regulators. In addition, progressive policies may result in the creation of products and processes that will be profitable. It is reasonable to assume that environmental standards will become increasingly stringent, and that products and processes that are the most environmentally friendly will find a market—if not today, then tomorrow. Progressive industries are the ones most likely to develop these new environmentally advanced products and will capture the markets for them. On the individual level, an engineer may be stimulated by a progressive environmental attitude to create products and processes that could advance one's professional career and perhaps be personally profitable. Mark Holtzapple's projects, if they turn out to be commercially successful, could be quite profitable for him.

Second, progressive policies generally promote human well-being, especially the well-being of future generations. If reducing emissions to the levels required by law generally promotes human health, reducing harmful emissions further, in some cases at least, probably promotes human health even more. Probably a more important consideration is that progressive policies, by increasing efficiency and reducing waste, can conserve resources for future generations.

Third, a powerful motive for the progressive attitude is a respect for nature for its own sake, independently of its contribution to human well-being. There is something immediately compelling about the grandeur and majesty of nature. In the words of one environmentally conscious engineer, "Nature is awesome." The awe and wonder that we often have when experiencing the natural world can be a powerful motivator for action. In particular, they can—and always have—inspired the belief that there is something wrong about "desecrating" nature and that the natural world should be respected, preserved, and nurtured.

Respect for nature for its own sake is probably the only way we can justify certain actions that most people probably feel intuitively are right. For example, blind salamanders live in the aquifers underneath the city of Austin, Texas. Most people probably believe—as do the citizens of Austin—that there is something wrong, or at least repugnant, about draining the aquifers and destroying these forms of life, unless there are very good reasons for doing it. Yet it is difficult to justify these feelings on any other grounds than a respect for nature for its own sake. In the next section, we develop the idea of respect for nature, which we have increasingly come to believe is an important, perhaps even crucial, foundation for the progressive environmental attitude.

9.7 RESPECT FOR NATURE

Some Essential Distinctions

Let us begin with some important distinctions. As we have seen, individual engineers and firms can be concerned for environmental harm if it poses a direct and clear threat to human health. We can call this a *health-related concern* for the environment. Engineers and firms can also be concerned about the environment even when human health is not directly affected. We can call this a *non-health-related concern* for the environment.

When engineers are concerned for environmental protection because polluting the air or water introduces carcinogens, this is an example of a health-related concern. Engineering projects often have an impact on the environment, however, even when human health is not directly affected. An engineer may be asked to design a dam that will destroy a wild river or flood thousands of acres of farmland. She may be asked to design a sawmill that will be located in the middle of an old-growth forest or to design a condominium that will be built on wetlands. If an engineer objects to these projects for reasons having to do with the environment, then the objection is based on a non-health-related concern.

A second distinction is between the intrinsic and instrumental value of nature. Some people believe that trees, rivers, animals, mountains, rivers, and other natural objects have *intrinsic value*—that is, value in themselves—apart from human use or appreciation of them. Another way to make the same point is to say that natural objects (or at least some natural objects) are "morally considerable." Others believe that natural objects have only *instrumental value*—that is, value insofar as they are used or appreciated by human beings. To these people, natural objects are not morally considerable.

If we do not believe that forests or lakes or mountains—or even animals—have value in themselves, then we can still justify attributing instrumental value to them, even if they are not directly related to human health. Destruction of forests can affect the supply of wood and the availability of recreational opportunities. Destruction of plant and animal species can damage the ecosystem and damage recreational opportunities. Flooding farmlands can reduce the supply of food for present and future generations. Draining wetlands can damage the ecosystem in ways that ultimately affect human beings.

These distinctions enable us to understand the most important distinction, namely that between anthropocentric and nonanthropocentric ethics. An *anthropocentric* ethics holds that only human beings have intrinsic value. Nonhuman natural objects, including other animals, have value only as they contribute to human well-being. A *nonanthropocentric* ethics holds that at least some natural objects other than human beings (animals, plants, and perhaps even inanimate things such as rivers and mountains) have intrinsic value.

Although the codes are not clear enough to make a decisive determination, it appears that most of their provisions have an anthropocentric basis, although some may suggest nonanthropocentric commitments. Most engineering codes already implicitly commit engineers to health-related environmental concerns, whether or not they use the term "environment." Most codes commit engineers to holding paramount the safety, health, and welfare of the public. Insofar as protecting the environment is necessary to protecting human health and safety, commitment to the environment is already present by implication. Any commitments to non-health-related concerns and to the intrinsic value of the environment can only be sought in the codes that explicitly refer to the environment, where the precise interpretation is controversial. For example, the IEEE code requires members to disclose factors that could endanger "the public or the environment." Concern for "the public" probably refers to health-related issues, but concern for the environment could refer to non-health-related issues. These non-health-related concerns could be based on the intrinsic value of the environment, but they could also be based on considerations related to human welfare, such as preserving forests.

The ASCE code's commitment to sustainable development is justified as a way to "enhance the quality of life of the general public." Therefore, it does not appear to contain a commitment to the intrinsic value of the environment. On the other hand, sustainable development does involve more than health-related concerns. To take another example, the ASME canon referring to "environmental impact" could suggest both health-related and non-health-related concerns, and these concerns might be based on the intrinsic value of the environment. The duty of chemical engineers to "protect the environment" could have similar implications, but the implications are not clear. It is difficult to find any clear nonanthropocentric commitment in the engineering codes.

Aldo Leopold's Nonanthropocentric Ethics

Aldo Leopold, an important figure in the contemporary environmental movement, represents the nonanthropocentric perspective. Leopold used the expression "biotic community" to refer to the living and nonliving aspects of the natural world. According to many proponents of environmentalism, contemporary technologically advanced civilization is guilty of massive assaults on the biotic community.

Western society in particular has tended to conceive of nature as passive and thus as the fit object of human manipulation and control. This view of nature as passive is amply reflected in our language about the natural world. Land is to be "developed." "Raw" land is to be "improved." Natural resources are to be "exploited" and "consumed." Trees are to be "harvested." Rivers are to be "harnessed" to produce electrical power. Wilderness must be "managed." The nonhuman world is to be made subservient to human purposes.

In *A Sand County Almanac,* Leopold presents a very different view of nature:

> We abuse land because we regard it as a commodity belonging to us. When we see land as a community to which we belong we may begin to use it with love and respect.... Perhaps such a shift in values can be achieved by reappraising things unnatural, tame, and confined in terms of things natural, wild, and free.[17]

Leopold's view interprets nature as something to which we belong rather than something that belongs to us. It is something "wild" and "free" rather than a passive object on which we work our purposes. Nature is "a fountain of energy flowing through a circuit of soils, plants, and animals."[18]

Viewing nature as an interdependent biotic community, Leopold believed that nature elicits an ethical response. He called this ethical response the "land ethic" and stated its moral standard in these words: "A thing is right when it tends to preserve the integrity, stability, and beauty of the biotic community. It is wrong when it tends otherwise."[19]

A Modified Nonanthropocentric Ethics

Although we believe that a respect for nature for its own sake is an important and perhaps crucial foundation for the progressive attitude toward the environment we have been advocating, engineers may still believe they have special obligations to the health, safety, and welfare of human beings. Thus, although they may believe that the blind salamanders in aquifers beneath the city of Austin, Texas, deserve to be protected even when they do not contribute to human welfare, environmental protection that promotes human welfare deserves special consideration. Furthermore, there may be circumstances in which nature can justifiably be harmed for the sake of promoting human well-being. Therefore, we might construct a moral standard that has a stronger anthropocentric element than Leopold's but has a clear nonanthropocentric component:

> An action is right if it preserves and protects the natural world, even if it is not necessary to promote human welfare, but it is justifiable to take actions that harm the environment if the production of a human good is sufficiently great.

Thus, the blind salamanders should be protected for their own sake, but if draining the underground aquifers and thereby destroying the salamanders is the only way to save the lives of large numbers of people in Austin, the action would still be justified.

Two things about this standard should be noted. First, there is probably no way to define the term "sufficiently" in a general way. One must determine in particular situations what the term means. Second, the way we have balanced respect for nature and human well-being is much like the way we have balanced rights and utility in other contexts. Placing the previous standard in terms of the concepts of rights and utility, we might say that the natural world has a right to protection and preservation even apart from human well-being, but this right may be overridden if the utility to humans is of sufficient magnitude. As we have already seen, utility to humans includes not only health but also economic well-being.

9.8 THE SCOPE OF PROFESSIONAL ENGINEERING OBLIGATIONS TO THE ENVIRONMENT

Should Engineers Have Environmental Obligations?

Because engineers create much of the technology that is involved in environmental degradation and also environmental improvement, they should have a special professional obligation to the environment. People are responsible for something when they knowingly bring it about or cause it to exist or happen. If I turn out the lights while friends are walking up the stairs, knowing full well that they may fall, then I am responsible if they fall; that is, I can be blamed for their fall. If I did not know that anyone was on the stairs and had no reason to believe that they were, then I am not responsible; that is, I cannot be blamed.

According to this argument, engineers should share in the responsibility for environmental concerns because they are often causal agents in projects and activities that affect the environment for good or ill. Engineers design dams that flood farmlands and wild rivers. They design chemical plants that pollute the air and water. They also design solar energy systems that make hydroelectric projects unnecessary and pollution-control systems that eliminate the discharge of pollutants into the air and water. Furthermore, they usually are (or should be) aware of the effects of their work on the environment. If engineers are morally responsible agents in issues that affect the environment, this argument goes, then they should also be required as professionals to promote environmental integrity.

There are, of course, objections to imposing responsibilities on engineers regarding the environment. One objection is that many judgments regarding the environment fall outside of the professional expertise of engineering, often finding their basis in the biological sciences. If Mary objects to the building of a condominium on a drained wetland because she believes it will cause unacceptable damage to the ecology of the area, she is making a judgment outside the area of her professional competence. However, Mary may be acting on the testimony of experts in the area, or the knowledge to which she appeals may be so common and generally accepted that it is no longer the exclusive property of the expert.

Another objection is that imposing substantial environmental obligations on engineers can cause problems of conscience for engineers who disagree with them, and it can cause problems for engineers with their employers if they work in organizations that do not accept these environmental obligations.

This objection has merit, but we believe it can be accounted for by new provisions in engineering codes. Just as medical codes have provisions for excusing physicians from performing procedures to which they have moral objections (e.g., abortions), so engineering codes should have similar provisions.[20] Although engineering codes currently do not have such provisions, there is some recognition of the right of conscience in the engineering profession. For example, NSPE's Board of Ethical Review (BER) has recognized the appeal to conscience. In Case 82-5 (i.e., case 5 in 1982), the board defended the right of an engineer to protest what he believed were excessive costs and time delays in a defense contract. The BER concluded that although the engineer did not have an ethical obligation to continue to protest his employer's tactics, he had "a right to do so as a matter of personal conscience." We believe that the right of conscience should be formalized in engineering codes. In the next section, we make specific recommendations.

Two Modest Proposals

We believe that professional engineering obligations regarding non-health-related issues can best be handled in terms of two proposals:

1. Although engineers should be required to hold paramount human health in the performance of their engineering work (including health issues that are environmentally related), they should not be required as professionals (i.e., required by the codes) to inject non-health-related environmental concerns into their engineering work.
2. Engineers should have the right to organizational disobedience with regard to environmental issues, as this is required by their own personal beliefs or their own individual interpretations of what professional obligation requires.

The first proposal embodies the idea that a minimal conception of professional obligation to safeguarding the environment should be incorporated into professional codes. Despite advocating the progressive attitude in previous sections, we do not believe that this attitude should be mandated by professional codes. The second proposal assumes that individual engineers may have a conception of what it is for them to act as professionals that is not a consensus view, or that they may have personal beliefs about the environment unconnected with their professional obligations. They might, for example, hold strongly to the progressive attitude, and they might (or might not) think it should be part of their professional obligation. It further holds that these views should be respected.

With regard to the second proposal, suppose an engineer says, "I know that all engineers do not agree with me here, but I believe it is unprofessional to participate in a project involving draining a wetland, even if this project is not illegal." Here, the engineer would be holding a view about what a professional obligation to the environment entails.

The three types of organizational disobedience discussed in Chapter 8 are relevant to this second proposal. First, engineers should have the right to disobedience by contrary action with regard to environmental issues; that is, they should have the right to promote their personal beliefs or their own individual interpretations of what professional obligation requires, including their beliefs about the environment, outside the workplace. For example, an engineer should be able to join an environmental group devoted to saving wetlands, even when her employer wants to drain a wetland to build a new plant. An engineer should be able to speak out against the building of a dam that will destroy a wild river, even when his firm may profit from construction of the dam.

Second, engineers should have the right to disobedience by nonparticipation with regard to environmental issues; that is, they should have the right to refuse to carry out assignments they believe are wrong, including environmentally related assignments. An engineer should be able to refuse to participate in the design of a plant that will adversely affect human health or that will be built on a reclaimed wetland. Similarly, she should have the right to refuse to design a dam that will destroy a wild river.

Third, engineers should have the right to disobedience by protest with regard to environmental issues; that is, they should have the right to protest employer actions they believe to be wrong, including actions they believe are harmful to human health or the environment. Within the bounds of discretion and due regard for the employer, an engineer should be able to protest an employer's plan to design or build a dam that will destroy a wild river or a project that will involve draining a wetland.

To make these rights clear, we believe the following provision regarding the rights of engineers should be incorporated into engineering codes:

> Where responsible assessment indicates it is feasible, organizations should not compel engineers to participate in projects that violate their professional obligations as determined by the codes, their own interpretations of professional responsibility, or their personal beliefs. Engineers shall have the right to voice responsible objections to engineering projects and products that they believe are wrong, without fear of reprisal. Engineers shall also have the right to support programs and causes of their own choosing outside the workplace.

We can offer the following arguments in favor of this proposed code provision. First, as we have previously noted, medicine already has a nonparticipation policy, and the BER has recognized a right of conscience on the part of engineers. Second, the proposal accepts the fact that some organizations may not be able to honor such a provision. Third, the proposal provides a way for engineers who are not in management positions and have no power to direct or change policy to avoid violating their conscience on issues they consider important. Fourth, the engineers who do not share a concern for environmental issues where human health is not at stake may follow their own beliefs as well.

The question of the nature and extent of the rights and obligations of engineers regarding environmental issues is still controversial. The discussion of this question is still in an early stage. The proposals offered in this chapter are intended to contribute to the discussion of this question as it takes place in both the engineering community and the larger public arena.

9.9   CHAPTER SUMMARY

The engineering codes mandate holding paramount the health and welfare of the public, so they indirectly mandate protecting the environment, insofar as it is necessary to promote this goal, even if the environment is not mentioned. The ASCE code has the most detailed environmental provisions, but no engineering code unambiguously advocates concern for nature for its own sake.

Most federal laws concerning the environment focus on making the environment "clean." Court decisions attempt to balance consideration for the environment with cost and economic well-being. The discussion of environmental matters includes many definitions of "clean," but this book advocates the degree-of-harm criterion.

Corporations and individuals can adopt one of three attitudes toward the environment: sub-minimal, minimalist, and progressive. The progressive attitude is increasingly being adopted by industry. It goes beyond the law in advocating sustainable development, recycling, eliminating the production of waste, promoting environmentally friendly technology, promoting active participation with government in environmental regulation, reducing pollution below governmental requirements, reducing energy consumption, selecting managers and members of boards of directors who are environmentally friendly, and making annual reports to the public regarding their actions affecting the environment.

In addition to self-interest and concern for human welfare (including future generations), respect for nature for its own sake can motivate and justify the progressive attitude. Aldo Leopold's "land ethic" represents such an attitude. We believe, however, that this respect for nature should be balanced against considerations of human well-being.

Despite advocating the progressive attitude, we believe that requirements of engineering codes should be limited to protecting the environment where human health is concerned but that, where responsible assessments of organizational constraints indicate feasibility, engineers should not be required to participate in projects that violate their professional obligations as determined by the codes, their individual interpretations of the codes, or personal beliefs. Furthermore, engineers should have the right to voice responsible objections to projects they find objectionable, without fear of reprisal, and to support progressive causes of their own choosing outside the workplace.

NOTES

1. The document is accessible on the ASCE website at http://www.asce.org/pressrom/news/policy_details.cfm?hdlid=60.
2. This section utilizes several sources, both for legal citations and for ideas. See Mark Sagoff, "Where Ickes Went Right, or Reason and Rationality in Environmental Law," *Ecology Law Quarterly,* 1987, pp. 265–323. See also Al H. Ringleb, Roger E. Meiners, and Frances L. Edwards, *Managing in the Legal Environment* (St. Paul, MN: West, 1990), pp. 553–583; Carl F. Cranor, *Regulating Toxic Substances: A Philosophy of Science and the Law* (New York: Oxford University Press, 1993), especially pp. 160–163; and James Salzman and Barton H. Thompson, Jr., *Environmental Law and Policy* (New York: Foundation Press, 2003). We are also indebted to Dr. Roy W. Hann, Department of Civil Engineering, Texas A & M University, for conversations about environmental law.
3. 42 United States Code [USC] sect. 4331 (1982), note 20.
4. 29 USC sect. 655(b)(5) (1976).
5. 42 USC sect. 7412(b)(1)(B) (1982), note 21.
6. See Thomas F. P. Sullivan, ed., *Environmental Law Handbook* (Rockdale, MD: Government Institutes, 1997); and Vicki R. Patton-Hulce, *Environment and the Law: A Dictionary* (Santa Barbara, CA: ABC Clio, 1995).
7. 42 USC sections 6901–6986 (1982 & Sup. 1985), note 21.
8. 478 F.2d 615 (D.C. Cir. 1973).

9. *Portland Cement Association v. Ruckelshaus,* 486 F.2d 375, 387 (D.C. Cir. 1973), note 197.

10. *Industrial Union Dept. AFL-CIO v. American Petroleum Institute,* 448 U.S. 607, 642 (1980).

11. *Natural Resources Defense Council v. EPA,* 804 F.2d 719 (D.C. Cir. 1986).

12. *Industrial Union Dept. AFL-CIO v. Hodgson* 162 U.S. App. D.C. at 342, 499 F.2d at 467, 477–78 (D.C. Cir. 1974).

13. Cranor, *Regulating Toxic Substances,* p. 161.

14. Ibid., pp. 161–162.

15. For suggestions for this utilitarian argument, see Cranor, *Regulating Toxic Substances,* pp. 163–168.

16. Joseph M. Petulla, "Environmental Management in Industry," *Journal of Professional Issues in Engineering,* 113, no. 2, April 1987, pp. 167–183.

17. Aldo Leopold, *A Sand County Almanac* (New York: Oxford University Press, 1994), pp. vii, ix.

18. Ibid., p. 216.

19. Ibid., pp. 224–225.

20. Article VI of the "Principles of Medical Ethics" of the American Medical Association states, "A physician shall, in the provision of appropriate medical care, except in emergencies, be free to choose whom to serve, with whom to associate, and the environment in which to provide medical services." Available at the American Medical Association's website at http://www.ama-assn.org/ama/pub/category/2512.html.

CHAPTER TEN

International Engineering Professionalism

Main Ideas in this Chapter

- Economic, cultural, and social differences between countries sometimes produce "boundary-crossing problems" for engineers. Solutions to these problems must avoid absolutism and relativism and should find a way between moral rigorism and moral laxism.

- Some actions, such as exploitation and bribery, can rarely, if ever, be justified, but some situations are susceptible to creative middle way solutions, as long as the solutions do not violate several familiar moral standards.

- Boundary-crossing problems are produced by such factors in host countries as low levels of economic development, extended family situations, the practice of building business relationships on personal relationships and cementing these relationships with gifts, low levels of scientific and technical sophistication, the practice of negotiating tax rates, and differing environmental and safety standards.

- These factors can, in turn, give rise to moral problems related to such issues as exploitation, bribery, extortion and grease payments, nepotism, excessively large gifts, and paternalism.

THE CLOTHING INDUSTRY IS PERHAPS the most competitive in the world.[1] Clothing manufacturing has been the first level of industrialization in most countries: Hong Kong, South Korea, Taiwan, China, Myanmar, Bangladesh, Sri Lanka, the Maldives, Laos, Vietnam, Bahrain, Indonesia, El Salvador, Honduras, and the Dominican Republic. Many factories in these countries employ young women in sweatshop conditions. Yet some critics argue that sweatshops (and even perhaps child labor) are a necessary part of economic development. David Landauer, Wellesley economist and World Bank consultant, remarks:

> We know of no case where a nation developed a modern manufacturing sector without first going through a "sweatshop" phase. How long ago was it that children could be found working in the textile factories of Lowell, Massachusetts; or Manchester, England; or Osaka, Japan?[2]

Similarly, a workers' rights advocate in Bangladesh argues that throwing children out of work and onto the streets "would be a serious violation of their human rights."[3]

Harwell & James (H&J) is a small clothing manufacturer, with sales that equal only 1.5 percent of Levi Strauss, the industry leader. It owns and operates a plant in Country X whose employees are mostly young women from the countryside. The young women live in company dormitories and work for $0.80 per day, producing garments that are at the low end of the price spectrum. They work 12-hour days in a clean, safe, and well-lit factory. The young women describe the work as hard but say they still prefer it to village life. Some of the young women are probably the sole wage earners in their families and without these jobs might well be forced into begging or prostitution. H&J does not employ children under the age of 14, and there are no serious health or safety problems at the plant. Some critics have argued, however, that H&J should leave Country X. A manager for H&J responds that if his firm left Country X, another firm would take its place. "The message from business," he maintains, "is to follow the dollar and learn to effect changes from within."[4]

Hanna is an engineer whose company has been asked to design and supervise the manufacture of some new equipment for the H&J plant in Country X. Hanna will be asked to spend 1 year in Country X, supervising the installation of the equipment and training plant personnel in its use. The new equipment should improve efficiency and safety in the plant. Nevertheless, some of Hanna's engineering colleagues argue that she should not take the assignment because it makes her a party to the exploitation the young women.

10.1 INTRODUCTION

Many engineers in the United States are accepting assignments overseas or are engaged in the design or manufacture of products for other countries. Furthermore, engineers and engineering students are coming to the United States for study or work. Similar crossings of national and cultural boundaries are occurring throughout the world. Let us refer to *boundary-crossing problems* as ethical problems that are produced by entering countries or regions with different cultural, social, or economic conditions. We can refer to the country that one leaves as the *home country* and the country that one enters as the *host country*. Problems in moving from home to host country are especially severe when the host country is not fully industrialized. Let us call these countries *lesser industrialized countries*. As we shall see, there are other differences that can cause problems as well.

Two simple solutions to boundary-crossing problems usually prove unsatisfactory. The *absolutist solution* follows the rule that the laws, customs, and values of the home country should always be followed. Home-country standards, however, may pose serious, if not insurmountable, problems if applied in host countries. Customs regarding giving large gifts and requests for small fees might be so pervasive and deeply entrenched in a host country, for example, that it might not be possible to do business in the host country without following the customs. Also, host county values might be as good as, or better than, home country standards, just different.

The other extreme is the *relativist solution,* which follows the rule, "When in Rome, do as the Romans do." In other words, home country citizens should simply follow host country laws, customs, and values, even if they are contrary to home country standards. This solution also has severe problems. It might sometimes lead to illegal actions. For example, the Foreign Corrupt Practices Act, passed by the U.S. Congress in 1977, makes it illegal for U.S. citizens to engage in practices such as paying some kinds of bribes and making some kinds of extortion payments, although these may be common practices in the host country. Another problem is that certain practices in the host country might be so morally repugnant that a home country engineer might have trouble following them. For example, the health and safety standards might be so low that they are clearly endangering the health and safety of host country workers.

In this chapter, we explore various ways of handling boundary-crossing problems that avoid the simplistic solutions suggested by absolutism and relativism. First, however, we discuss the creative middle way approach to resolving boundary-crossing problems and some of the constraints and limitations under which this solution must operate. Then, we consider several common types of boundary-crossing problems and how they might be resolved.

10.2   ETHICAL RESOURCES FOR SOLVING BOUNDARY-CROSSING PROBLEMS

There are a number of resources for resolving boundary-crossing problems. The following enumeration of resources should be considered a "tool kit." Although creative middle ways are often especially satisfying solutions, they may sometimes be impossible to implement or even ethically unacceptable. Therefore, we need other resources to determine whether creative middle ways are ethically acceptable. If they are not acceptable, we need ways of determining what should be done. Fortunately, we shall find that the ethical methods that we have already discussed are sufficient. As with any problem in practical ethics, however, one chooses what resources one needs for resolving a particular problem. In other words, one chooses the tool that is most appropriate for a particular ethical task.

Creative Middle Ways

The most obvious and, in many situations, the most useful resource for resolving boundary-crossing problems is a creative middle way solution, in which both the host country and the home country customs are honored in some form. In a creative middle way solution, we find a solution that honors competing moral demands (in this case, usually the demands of home and host country ethical and cultural considerations), assuming that neither home country or host country customs violate fundamental moral considerations. We often suggest this approach in the following discussions.

In using creative middle way solutions, especially in the international context, it should always be kept in mind that two extremes must be avoided. One extreme is *moral laxism,* which holds that in some situations, when moral principles cannot be strictly applied, we are justified in applying them so loosely that moral constraints are almost completely abandoned. Thus, the laxist allows solutions to moral problems that involve serious violations of moral standards. The laxist often argues that violations of moral principles are simply part of the price of living in the "real" world, especially in the world of another and very different culture.[5] The reasoning here might be as follows: "Because there is no option in this situation that allows me to act in an ideal moral way, I will simply abandon any moral considerations and act in a way that is most compatible with my self-interest or with the self-interest of my firm." This option involves an abandonment of ethical and professional considerations and may in some cases even lead an engineer to embrace choices that are illegal.

Another response is to be so concerned with doing the right thing in an ideal sense that one may become excessively moralistic and adopt a position known as moral rigorism. According to our interpretation, *moral rigorism* holds that moral principles must be strictly applied in every situation.[6] The moral rigorist is unwilling to accept the fact that although a given course of action is not the ideal, it may be the best that one can do in the situation, morally speaking, and it may not involve any serious violation of moral principles.

Consider the following situation:

> Laura's plant operates in Country X and produces fertilizer in an area where farmers live at almost a subsistence level. The plant produces relatively inexpensive fertilizer that the local farmers can afford, but it also produces pollution. However, although it would violate U.S. standards, the pollution produced does not violate the standards of Country X. In order to remedy the pollution problems, the plant would have to raise the price of the fertilizer so much that the farmers could not afford it. Laura has been asked by management to give her advice as to what the plant should do.

The competing moral obligations are to reduce pollution while enabling the local farmers to buy the fertilizer. Although the ideal solution would be to reduce the pollution from the plant, economic conditions may not allow this solution without pricing the fertilizer out of the reach of local farmers. Due to the pressing economic restraints, probably in this case the plant should continue to produce the fertilizer at the cheap price, even though it involves pollution, while attempting to find a cheaper solution to its environmental problems. Although a moral rigorist might find this solution unacceptable, it does not exemplify moral laxism either. The solution does not violate any fundamental moral principles, and it does not involve a wholesale abandonment of moral considerations or a simple self-interested solution to the problem. It may be the best solution possible in the situation. This kind of solution—the best possible in the circumstances that does not violate fundamental moral principles—is the very essence of a creative middle way solution.

Sometimes there may be such serious moral problems with one of the options that a creative middle way solution is not appropriate and even a person who is not a moral rigorist could not accept it. To take an extreme example, suppose you find out that the cheapest supplier of one of the parts used in your plant in Country X employs slave labor. Your friend suggests what he calls a creative middle way solution: Tell the supplier that you will use his products only if he does not use child slaves. This is not an acceptable creative middle way solution because slavery is still immoral, even if the slaves are all adults. Accepting the solution would be a form of moral laxism. It would abandon moral standards altogether.

How do we determine, then, when a creative middle way solution, or for that matter any other solution, is so far outside the moral boundaries that it cannot be accepted? How do we identify those tests, standards, or considerations that would help us determine when a solution to a boundary-crossing problem is or is not acceptable? The standards we employ should meet two criteria. First, they should be as nearly transcultural or universal as possible. That is, insofar as it is possible, they should apply to all cultures. At a minimum, they should apply to many cultures other than our own. Otherwise, they would not be useful in solving boundary-crossing problems. Second, the tests or standards should have an immediate plausibility. It should not require a complicated argument to convince people that the standards are relevant for evaluating a solution to a boundary-crossing problem. If the standards do not have an immediate intuitive appeal, they may not be convincing to many people.

We believe that the tests and standards already elaborated in this book meet these two criteria. These standards are useful not only in determining whether a creative middle way solution is acceptable but also in testing the moral acceptability of solutions that do not involve creative middle ways. The following are standards that we believe are especially useful.

First Standard: The Golden Rule

As we have already shown, the Golden Rule is embraced by most of the major religions and ethical philosophies. Using the Golden Rule, I can ask, "Would I be willing to accept the effects of this practice?" This question is especially difficult to answer when it requires putting myself in the position of a host country citizen, whose culture, economic status, living conditions, and values may be very different from my own. The classic problems in applying the Golden Rule present themselves in especially acute forms when using the rule to resolve boundary-crossing problems. Nevertheless, there are times when the answer seems rather clear. It is difficult to imagine, for example, that anyone would want to be exploited, be forced to violate deeply held moral beliefs, or have one's own person not respected.

Second Standard: Universal Human Rights

People in many cultures, including many non-Western cultures, now appeal to human rights in making their case for everything from minimal standards of living to protection from torture or political oppression. We have already seen that rights can be justified by the ethics of respect for persons because rights help protect the moral agency of individuals. Utilitarians also often argue that respecting the rights of individuals promotes human happiness or well-being. People live happier lives when their fundamental rights are respected.

"Rights talk" has become a near-universal vocabulary for ethical discourse. One measure of the cross-cultural nature of rights talk is the United Nation's International Bill of Human Rights, adopted in 1948, and two later documents—the International Covenant on Economic, Social, and Cultural Rights and the International Covenant on Civil and Political Rights.[7] These documents ascribe to all human beings the rights to

- life,
- liberty,
- security of person,
- recognition before the law,
- an impartial trial,

- marriage,
- property ownership,
- freedom of thought,
- peaceful assembly and participation in government,
- social security and work,
- education,
- participate in and form trade unions,
- nondiscrimination, and
- a minimal standard of living.

They also affirm the rights to freedom from slavery, torture, and inhuman or degrading punishment and from marriage without free consent.

This is a long list, and some theorists might argue that it is simply a "wish list," given the conditions that actually prevail in many countries. Notice also that some of the rights are what we have called "positive" rights: That is, they are not simply rights to noninterference from others, such as the rights not to be held in slavery or tortured. By contrast, rights to certain advantages, such as education, social security, and work, require from others not only a negative duty to noninterference but also a positive duty to help others achieve these rights. Most of us, however, would consider all of these rights highly desirable. The question is whether they should be considered as rights rather than simply as desirable things to have. For example, should we say that one has a right to a minimal standard of living? Which of these rights do professionals in the international arena have an obligation to respect?

James Nickel has proposed three criteria for determining when a right is what we shall call an *international right*—that is, a right that every country should, if resources and conditions permit, grant to its citizens. In terms of generality and abstraction, an international right falls between the very abstract rights discussed in Chapter 4 and the more specific rights guaranteed by laws and constitutions of individual governments. Nickel's conditions most relevant to our discussion are the following:

1. The right must protect something of very general importance.
2. The right must be subject to substantial and recurrent threats.
3. The obligations or burdens imposed by the right must be affordable in relation to the resources of the country, the other obligations the country must fulfill, and fairness in the distributions of burdens among citizens.[8]

Judged by these criteria, some of the United Nations' list of rights might not be applicable. Some countries may not have the economic resources to support the claims to a minimal education and subsistence, however desirable these may be. Perhaps we should say that these rights are desirable, insofar as a country can afford to provide them.

Third Standard: Promoting Basic Human Well-Being

Another test for determining whether a solution to a boundary-crossing problem is satisfactory is whether the solution promotes the well-being of host country citizens. If an option does not promote their well-being, this is a strong argument against it. The most important way in which engineering and business can promote well-being is through economic development. How do we measure economic development? As we noted in the chapter on risk, economist Amartya Sen and philosopher Martha

Nussbaum have addressed this issue. In particular, Nussbaum has derived a set of what she calls "basic human functional capabilities"—that is, basic capabilities that a person needs to be able to satisfy in order to live a reasonable quality of life.[9] From the standpoint of utilitarianism, we can consider these to be capabilities that should be increased in order to promote human well-being. The following is our summary of Nussbaum's 10 functional capabilities:

1. Being able to live a human life of normal length.
2. Being able to enjoy good health, nourishment, shelter, sexual satisfaction, and physical movement.
3. Being able to avoid unnecessary and nonbeneficial pain and to have pleasurable experiences.
4. Being able to use the senses, imagine, think, and reason.
5. Being able to form loving attachments to things and persons.
6. Being able to form a conception of the good and to engage in critical reflection about the planning of one's life.
7. Being able to show concern for others and to engage in social interaction.
8. Being able to live with concern for and in relation to animals, plants, and the world of nature.
9. Being able to laugh, play, and enjoy recreational activities.
10. Being able to live one's own life and nobody else's.

It is important to note that engineering is involved, either directly or indirectly, in many of these factors, which, according to Nussbaum, contribute to human well-being. By providing clean water and sanitation, engineering makes an enormous contribution to health and longevity. Production of fertilizer and other aids to farming increases the ability of a host country to feed its citizens. Technological development contributes to the level of wealth in a country and thereby plays an important part in promoting almost all of the other capabilities mentioned by Nussbaum.

Fourth Standard: Codes of Engineering Societies

Many of the major engineering codes are clearly intended to apply to their members wherever they live, even if they practice engineering in host countries. The Institute of Electrical and Electronics Engineers is explicitly an international organization. Its code opens with an acknowledgment of "the importance of our technologies in affecting the quality of life throughout the world." To take another example, the American Society of Mechanical Engineers (ASME-International) makes similar references to the international environment. A 1996 decision by the National Society of Professional Engineers (NSPE) Board of Ethical Review (Case 96-5) held that an NSPE member is bound by the NSPE's code of ethics, even in another country. In this case, the issue was whether a U.S. engineer could ethically retain a host country engineer who would then offer bribes to a host country official in order to get a contract. The board held that the practice would violate the NSPE code and it would be unethical for a U.S. engineer to be a party to such a practice. Professional codes give important guidance for engineers in the international arena as well as for engineers in their home country. Much of this guidance is in the form of prohibitions (against bribery, conflicts of interest, etc.), but it can be important in determining when a creative middle way solution is acceptable.

Now we are ready to discuss some areas in which boundary-crossing problems can be especially troublesome. We consider some representative cases and when creative middle way solutions are acceptable.

10.3   ECONOMIC UNDERDEVELOPMENT: THE PROBLEM OF EXPLOITATION

Exploitation, especially of the weak and vulnerable, is a serious moral problem, and it is particularly likely to occur in economically underdeveloped countries, where workers have few options for jobs. According to Robert E. Goodin, the risk of exploitation arises when the following five conditions are present:[10]

- There is an imbalance of (usually economic) power between the dominant and subordinate or exploited party.
- The subordinate party needs the resources provided by the dominant party to protect his or her vital interests.
- For the subordinate party, the exploitative relationship is the only source of such resources.
- The dominant party in the relationship exercises discretionary control over the needed resources.
- The resources of the subordinate party (natural resources, labor, etc.) are used without adequate compensation.

Consider the following case:

Joe's firm, Coppergiant, is the most powerful copper mining and copper smelting company in the world. It controls world prices and keeps competitors away from some of the most lucrative sources of copper. Joe works for Coppergiant in Country X, the firm's most lucrative source of copper. In Country X, Coppergiant buys copper at prices considerably below the world market and pays the workers the lowest wages for mining and smelting work in the world. As a result, Coppergiant makes enormous profits. Because the company pays off government officials and has so much control over the world market in copper, no other mining and smelting company is allowed into the country. Country X is desperately poor, and copper is virtually its only source of foreign currency.

This case meets all five of Goodin's criteria for exploitation. There is an asymmetrical balance of power between Country X and Jim's firm. Country X desperately needs the foreign currency provided by Jim's firm. The revenues through Jim's firm are the only source of the currency. Jim's firm, through its control of the market, exercises discretionary control over these revenues. Finally, the natural and labor resources of Country X are used without adequate compensation.

Exploitation is usually wrong because it violates several of the moral standards mentioned previously. For example, it violates the Golden Rule. It is difficult to imagine that anyone in any culture would, in normal circumstances, want to be the victim of exploitation. It violates the right to a minimal standard of living, and it keeps citizens of Country X from realizing many of the capabilities mentioned by Nussbaum. It is possible to argue that the exploitation is justified on utilitarian grounds if it is the only way Country X can undergo economic development, but this argument is implausible because economic development could almost certainly occur without this exploitation.

Since the exploitation in this case cannot be justified, we must conclude that the current situation should be changed. It may be that raising wages and copper prices to market levels would still provide workers in Country X and the general economy of Country X with revenues that would be less than desirable. At this point, a creative middle way solution might justify this condition because any further increase in wages might result in the economic collapse of Coppergiant or its leaving the country. This might leave workers and the economy in worse shape than before.

The case with which this chapter began is not a paradigm case of exploitation. Even if we concede that the first three conditions are met, the last two are more problematic. H&J may not exercise discretionary control over the resources because raising the price of their product even slightly might price its garments out of the highly competitive clothing market. If H&J cannot raise wages without raising the price of its garments, it does not exercise discretionary control over the resources.

Whether the compensation is "adequate" may also raise difficult conceptual, application, and factual issues. Although the wages are low by U.S. standards, they may not be low by the standards of the country. Furthermore, the company provides other benefits that must be considered in deciding whether the wages are adequate.

It is possible to view this situation as representing a creative middle way. The ideal situation would be one in which the young women were paid more, but an even worse situation would be one in which they did not have even this opportunity for advancement. If it is an acceptable creative middle way, one would have to argue that there is no violation of the Golden Rule, individual rights, or professional codes. Finally, one might argue that the situation increases the well-being of the young women and, indirectly, of the society in which they live.

10.4 PAYING FOR SPECIAL TREATMENT: THE PROBLEM OF BRIBERY

Bribery is one of the most common issues faced by U.S. engineers when they practice in host countries. In response to this problem, the U.S. Congress passed the Foreign Corrupt Practices Act in 1977. The act, however, only prohibits bribery of government officials. Typically, a bribe is made to a government official in exchange for violating some official duty or responsibility. The payment might result, for example, in an official's not making a decision to buy a product on its merits. The following is a typical or paradigm case of bribery:

> An executive of Company A hopes to sell 25 airplanes to the national airline of County X. The deal requires the approval of the head of the ministry of transportation in Country X. The executive knows that the official, who has a reputation for honesty, can make a better deal elsewhere, but he is also experiencing personal financial difficulties. So the executive offers the official $300,000 to authorize the purchase of the planes from Company A. The official accepts the bribe and orders the planes to be purchased.[11]

On the basis of this paradigm case, we can give the following definition of a bribe:

> A bribe is a payment of money (or something of value) to another person in exchange for his giving special consideration that is incompatible with the duties of his office, position, or role.[12]

A bribe also induces one person (the person given the bribe) to give to another person (the person giving the bribe) something that he does not deserve. Keep in mind that bribes presuppose an agreement that the bribe must be in exchange for a certain type of conduct. If this agreement is not present, then it is difficult to distinguish bribes from gifts or rewards.

Giving and receiving bribes are both forbidden by professional engineering codes. There are several good reasons for this. First, if an engineer takes a bribe, she is creating a situation that will most likely corrupt her professional judgment and tarnish the reputation of the engineering profession. Second, if she offers a bribe, then she engages in activity that will also tarnish the reputation of her profession if discovered and probably violate her obligation to promote the well-being of the public. Third, bribery induces the person who takes the bribe to act immorally by violating the obligation to act on behalf of the interests of his client or employer. For example, it can induce a government official to break the obligation to act on behalf of the best interests of the citizenry. Fourth, bribery can undermine the efficiency of the market by inducing someone to buy products that are not the best for the price. Fifth, bribery can give someone an unfair advantage over his competitors, thus violating the standards of justice and fair play.

John T. Noonan, jurist and authority on the history of morality, argues that the opposition to bribery is becoming stronger throughout the world.[13] There is massive popular discontent with bribery in Japan, Italy, and other countries. The antibribery ethic is increasingly embodied in the law. Even campaign contributions, which have many similarities with bribery, are becoming increasingly suspect.

Although there are many points of dissimilarity between bribery and slavery, there is some basis for saying that just as slavery was once accepted and is now universally condemned, so too bribery is increasingly held to be morally unacceptable, even if not universally condemned. Bribery, then, is something that should simply be avoided. In most cases, at least, no creative middle way is acceptable. We shall leave it to the reader to find cases in which it might be justified.

10.5 PAYING FOR DESERVED SERVICES: THE PROBLEM OF EXTORTION AND GREASE PAYMENTS

Extortion

Many actions that might appear to be bribery are actually cases of extortion. Consider again the case of the executive of Company A described previously. Suppose that he knows he is offering the best deal on airplanes to the official of Country X who has the authority to authorize purchases for his national airlines. The executive knows, however, that his bid will not even be considered unless he offers the official a large cash payment. The payment will not guarantee that Company A will get the contract—only that his bid will be considered. This is extortion rather than bribery.

It is more difficult to construct a definition of extortion than bribery. Here is a proposed, but inadequate, definition: "Extortion is the act of threatening someone with harm (that the extorter is not entitled to inflict) to obtain benefits to which the extorter has no prior right."[14] This definition is inadequate because some actions not covered by the definition are still extortion. For example, it would be extortion if one threatened to expose the official misconduct of a government official unless he pays a large sum of money—even though exposing the official would be both morally and legally permissible. We find it impossible, however, to give a completely adequate definition of extortion. All we can say is that the definition offered previously gives a sufficient, although not a necessary, condition of extortion.

Sometimes it is difficult to know whether one is paying bribery or extortion. An inspector who demands a payoff to authorize a shipment of a product may claim that the product does not meet his country's standards. It may be difficult to know whether he is lying—and too expensive to find out. In this case, if the company decides to make the payment, it may not know whether it is paying a bribe or extortion. Of course, it may be irresponsible for the company to make no effort to find the truth.[15]

Many of the most famous cases of corruption seem to border on both bribery and extortion. Between 1966 and 1970, for example, the Gulf Oil Corporation paid $4 million to the ruling Democratic Republican Party of South Korea. Gulf was led to believe that its continued flourishing in South Korea depended on these payments. If the payments gave Gulf special advantages over its competitors, the payments were bribes. If they would have been required of any competitor as a condition of operating without undeserved reprisals or restrictions, the payments might better be classified as extortion.[16]

The moral status of paying extortion is different from the moral status of paying and accepting bribes for the following reasons. First, paying extortion will not usually corrupt professional judgment. Second, although paying extortion can tarnish one's professional reputation, it will probably not do so as much as paying a bribe. Second, paying extortion will not cause one to act contrary to the best interests of one's employer or client by, for example, selecting an inferior product, but it does involve the use of a client's or employer's money. Third, paying extortion does not undermine the efficiency of the market by promoting the selection of inferior or more expensive products, but it does divert funds from their most efficient use. Fourth, paying extortion does not give one an unfair advantage over others, except insofar as others do not or cannot pay the extortion. The main problem with paying extortion is that it perpetuates a practice that is a type of theft.

Given these considerations, it may sometimes be morally permissible to pay extortion. A moral rigorist might find paying extortion morally unacceptable because it involves secret payments that give one an advantage over those persons and corporations that cannot afford to pay the extortion. However, extortion does enable one to do business in the host country. Assuming that the business activity is good for the home and host country and there are no serious violations of other moral standards, it may be justifiable.

Grease Payments

Grease payments are offered to facilitate routine bureaucratic decisions, such as hastening the passage of goods through customs. They usually involve relatively small amounts of money compared to ordinary bribery or extortion. They are probably most commonly forms of petty extortion, and they often do not give an unfair advantage over others, assuming that others make the payments too. Furthermore, they are often tacitly condoned by governments. For example, in many countries customs officials may not be paid an adequate salary, and the government may assume that officials will receive grease payments to supplement their salary, just as employers assume that waiters will supplement their salary with tips.

Again, a moral rigorist might hold that making grease payments is impermissible. It would surely be better if they were eliminated and replaced by more adequate salaries. In this way, the payments would not have to be kept somewhat clandestine, as most grease payments are. Furthermore, sometimes grease payments are more like bribes because they enable the payer to get special considerations that he does not deserve. If a grease payment allows the passage of inferior goods through customs, it is a petty bribe. Paying a fee to get telephone service or a permit in 3 weeks rather than 3 months is probably best considered a bribe because it puts one at the "head of the line" and gives him an advantage over others that he does not deserve. However, grease payments are often extortion. One simply does not get through customs or get one's goods through customs, or get telephone service at all, unless one pays the fee. If doing business in the country promotes the well-being of the host and home countries, and if the other moral tests are not seriously violated, making grease payments might be considered acceptable.

10.6 THE EXTENDED FAMILY UNIT: THE PROBLEM OF NEPOTISM

In many areas of the world, the primary unit of society is not the individual, as it is in the modern West. Rather, the primary unit is some larger group of humans. The larger group might be an extended family, which includes brothers and sisters and their families, aunts, uncles, cousins, and so forth. The group might even be a larger unit, such as a tribe. The relationship of the members of the group is one of mutual support. If a member of the group has fallen on bad times, the other members have an obligation to care for him. Similarly, if a member of the group has good fortune, he has an obligation to share the fruits of this good fortune with other members of the group. If a member of an extended family finds an unusually good job in a new industrial plant, for example, he has an obligation to find jobs for his relative—perhaps his brother or sister, or their spouses or children. This custom, however, may produce problems for the firms involved. Consider the following example, which is modeled on a real case:[17]

> You work for a steel company in India, which has the policy of partially compensating its employees with a promise to hire one of the employee's children. This policy is extremely popular with employees in a country where there is a tradition of providing jobs for one's children and the members of one's extended family. But to you, the policy is nepotism and in conflict with the more desirable policy of hiring the most qualified applicant. What should you do?

There are good arguments that this is an acceptable creative middle way solution to the problem. The policy of hiring the most qualified applicant is probably the most desirable one, so it is clearly one option. On the other hand, the policy of hiring many members of an employee's family would probably be morally unacceptable because it would interfere too much with economic efficiency by allowing to many people to be hired who are not the best candidates for the job. It would also be too severe a violation of considerations of justice and the right to nondiscrimination. The policy of hiring only one other family member, by contrast, seems acceptable. It makes a concession to the deeply held convictions of many people in a tradition-oriented culture, and it promotes harmony in the workplace (and perhaps economic efficiency in this way). This solution also shows the need to steer a middle way between moral rigorism and laxism.

10.7 BUSINESS AND FRIENDSHIP: THE PROBLEM OF EXCESSIVE GIFTS

In many cultures, an exchange of gifts is a way to cement personal friendships. Furthermore, in some cultures, one is expected to show favoritism toward friends, even when one acts in an official capacity. For people in many societies, the impersonal nature of Western business transactions, separated as they are from personal friendships and family ties, is unnatural and even offensive. The rule "Don't mix business and pleasure" is unacceptable.

For many in the West, however, large personal gifts look too much like bribes. Is there a creative middle way solution to this problem? Jeffrey Fadiman has suggested an answer: Give the gifts to the community, not to individuals. In one of his examples, a firm planted a large number of trees in a barren area. In another example, a firm gave vehicles and spare parts to a country that was having trouble enforcing the laws against killing animals in national parks. These gifts created goodwill, without constituting bribes to individuals. To some, of course, these gifts still have too much in common with bribes, even though they are certainly not paradigmatic bribes. Like bribes, they curry influence by bestowing favors. Unlike bribes, however, they are public rather than secret, and they are not given to individuals. Unless one is a moral rigorist, they may sometimes be a minimally acceptable solution. They are creative middle ways between the need to avoid bribery and the desirable goal of doing business in the host country. Since the option has so much in common with bribery, however, we would not consider it a completely satisfactory solution.

This solution does not solve the problem that sometimes gifts of substantial size are given to individuals. An "excessive" gift in the United States might not be excessive in another part of the world, so the norm regarding bribes and gifts must be adapted to fit the cultural conditions in a given society. Suppose affluent members of Country X routinely give gifts of substantial size to one another as tokens of friendship and esteem. Because the gifts are routinely given and received by everyone, they do not command any special favors. Is this practice acceptable for an engineer doing business in Country X?

The following considerations are relevant. First, we must examine the gift-giving practices in a given culture and determine whether a gift should be considered "excessive" by the standards of the host country culture. Since these gifts are routinely given and received by everyone, they do not command any special favors. We can call these substantial gifts "normal" and distinguish them from still larger gifts that would be necessary to command special favors and that should be classified as "excessive." Second, we can look at the intent of the objection to giving "excessive" gifts. The intent of the prohibition of "excessive" gifts is to prevent the giving of gifts that would curry favor and thus be a form of unfair competition. However, gifts that would be termed "excessive" in the home country are considered "normal" gifts in Country X, and they do not curry special favor. Thus, in Country X, giving "normal" gifts by that county's standards would not violate the intent of the norm against giving excessive gifts. Texas Instruments has set a policy on gift-giving in countries other than the United States that seems to follow this way of thinking:

> TI generally follows conservative rules governing the giving and receiving of gifts. However, what we consider to be an excessive gift in the U.S. may differ from what local customs dictate in other parts of the world. We used to define gift limits in terms of U.S. dollars, but this is impractical when dealing internationally. Instead, we emphasize following the directive that gift-giving should not be used in a way that exerts undue pressure to win business or implies a quid-pro-quo.[18]

We consider this policy to be morally acceptable. It is a creative middle way between merely rejecting the practices of the host country and being able to do business in the country and engaging in something closely akin to bribery.

10.8 THE ABSENCE OF TECHNICAL–SCIENTIFIC SOPHISTICATION: THE PROBLEM OF PATERNALISM

Because of lower educational levels and the general absence of exposure to technology in their daily lives, citizens in some host countries can easily misunderstand many issues, especially those having to do with risk, health, and safety. This can lead to *paternalism,* which is overriding the ability of another person to decide what he or she should do (or should not do) for the recipient's own "good." The *paternalist* is the one who decides for another. The *recipient* is the person who is the object of the paternalistic action. Although the paternalist imposes his will on the recipient out of a benign motive (to "help" the recipient), he nevertheless deprives the recipient of the freedom to direct his own life in a particular situation.

Paternalism is in some ways the opposite of exploitation. If exploitation is imposing my will on another for my good, paternalism is imposing my will on another for the other's good. Both have in common depriving another person of the freedom to direct his own life, although the reasons for doing this are very different. The following is an example of paternalism:

> Robin's firm operates a large pineapple plantation in Country X. The firm has been having what it considers excessive problems with maintaining the health of its workers. It has determined that a major reason for the health problems of its workers is the unsanitary conditions of the traditional villages in which they live. In order to remedy this problem, it has required the workers to leave their traditional villages and live in small, uniform houses on uniformly laid-out streets. Managers believe that the workers can be "educated" to appreciate the cleaner conditions and the aesthetic qualities of the new villages, but the workers have strongly objected. They protest that the new accommodations are boring and have destroyed much of their traditional way of life.

In order to discuss the moral status of this action, we must distinguish between weak and strong paternalism. In *weak paternalism,* the paternalist overrides the decision-making powers of the recipient when there is reason to believe the recipient is not exercising his moral agency effectively anyhow. In *strong paternalism,* the paternalist overrides the decision-making powers of the recipient, even when there is no reason to believe the recipient is not exercising his moral agency effectively. The paternalist overrides the decision-making powers of the recipient simply because he believes the recipient is not making the "right" decision—that is, a decision that is really for the recipient's own "good." Of course, the paternalist interprets what this "good" is.

From both utilitarian and respect for persons standpoints, there are several situations in which weak paternalism might be justified. They all involve situations in which there is reason to believe (or at least to suspect) that a person's moral agency is impaired. Thus, in exercising paternalistic control over the recipient, the paternalist is really protecting the moral agency of the recipient, not destroying it. If any one of the following conditions is present, a person may not be able to exercise his moral agency effectively, so any one of them is sufficient to justify weak paternalism:

- A person may be under undue emotional pressure, so she is unable to make a rational decision.
- A person may be ignorant of the consequences of her action, so she is unable to make a genuinely informed decision.
- A person may be too young to comprehend the factors relevant to her decision, so she is unable to make a rational and informed decision.
- Time may be necessary for the paternalist to determine whether the recipient is making a free and informed decision, so the paternalist may be justified in intervening to keep the recipient from making any decision until it is clear that the recipient is indeed making one that is free and informed.

In strong paternalism, we assume that the recipient is making a free and informed decision, but the presumption is that the recipient is not making the "right" decision, from the standpoint of the paternalist. Strong paternalism, then, can be justified only from a utilitarian standpoint. The argument has to be that the recipient is not making a decision that will maximize her own good (or overall good), even though she may think that she is making the correct decision.

Now let us return to the example. From the short description given previously, it is not clear whether the managers were exercising weak or strong paternalism. If the workers do not fully understand the health risks associated with their traditional village life, the managers were exercising weak paternalism in forcing them to move into the more sanitary villages. If the workers did understand the consequences but still preferred more disease and perhaps even less health care for the disease, in order to preserve their traditional way of life, the managers were exercising strong paternalism. Since strong paternalism is more difficult to justify, the burden of proof on the managers to show that their action was justified would be much greater.

Citizens of lesser industrialized countries are particularly likely to experience the conditions that might justify weak paternalism, or even strong paternalism in some cases. A lower level of education and technological sophistication can render citizens in those countries less able to make responsible decisions that affect their well-being. In such cases, a rational person might consent to be treated paternalistically, and in a few cases the overall good might even justify strong paternalistic action.

The following is an example in which weak paternalism is probably justified:

John is employed by a large firm that sells infant formula in Country X. The firm is also the only one that markets infant formula in Country X. Many mothers mix the formula with contaminated water because they do not understand the health dangers to their infants. They also dilute the formula too much in order to save money, unaware that this leads to malnutrition in their babies. John recommends that his firm stop selling the product in Country X. Management agrees and stops the sale of the product in Country X.

In this case, at least one of the conditions sufficient to justify weak paternalism (ignorance of the consequences of actions) is satisfied, so the action was probably justified. Of course, in stopping the sale of the infant formula, John's firm deprives the mothers in Country X of the ability to feed their babies with infant formula. There is ample evidence, however, that the mothers (or at least many of them) were not able to exercise their moral agency in a free and informed way.

10.9   DIFFERING BUSINESS PRACTICES: THE PROBLEM OF NEGOTIATING TAXES

Sometimes the business practices in other countries cause dilemmas for U.S. engineers. Consider the following case:

> James works for a U.S. firm in Country X, where it is customary for the government to assess taxes at an exorbitant rate because it expects firms to report only half their actual earnings. If a firm reported its actual earnings, the taxes would force it out of business. James' firm wonders whether it is morally permissible to adopt the local practice of dishonestly reporting its profits, even though it would be illegal to do this in the United States. It would report it profits honestly to the U.S. tax office.

The practice in question is probably not the most desirable way to collect taxes. It opens the way to bribery in the negotiating process and unfairness in the assessment of taxes since some firms may negotiate a lower tax rate (especially if they use bribery) than others.

Thus, it would probably be morally permissible for James' firm to report only half of its profits to the government of Country X, as long as the practice does not violate the firm's own code of ethics and the firm does not report its profits inaccurately to the U.S. government.[19] The practice does not appear to violate the Golden Rule since the firm would be willing for other firms to do the same thing and for it to use the practice if it were the government of Country X. The practice also does not violate the rights of anyone, and it probably produces more overall good than the alternatives, assuming the firm benefits its employees and the citizens of Country X. Furthermore, although the tax practice may not be the most desirable, it finances the legitimate activities of the government of Country X. Finally, the practice is not secret since other firms follow the same practice.

10.10   CHAPTER SUMMARY

The differing economic, social, and cultural conditions in various countries often raise ethical issues that we call boundary-crossing problems. The relativist solution to this problem is to follow the standards of the host country, and the absolutist solution is to follow the standards of the home country, assuming that these are the correct ones. The best way to resolve boundary-crossing problems is often to find a creative middle way. However, the creative middle way must not be one that violates fundamental moral norms such as the Golden Rule, rights and utilitarian considerations, and the guidelines in professional codes. The creative middle way solutions may not always satisfy the moral rigorist, who believes that moral norms must always be strictly applied, or the moral laxist, who believes that if moral norms cannot be applied rigorously, they should be abandoned altogether. Nevertheless, creative middle way solutions may often be the most satisfactory way of resolving boundary-crossing problems.

Lower levels of economic development often produce boundary-crossing problems, such as the ability to exploit workers who have few options. Certain conditions must be met for genuine exploitation to exist, and sometimes a moral problem may be far removed from a paradigm case of exploitation. Sometimes a situation that appears to be exploitation may not actually be one. Nevertheless, genuine exploitation can rarely, if ever, be justified. Bribery is also common in host countries with low levels of economic development, although it is by no means confined to such situations. The serious moral objections to bribery also indicate that it is rarely, if ever, justifiable. Extortion and grease payments are also especially common in lesser industrialized countries. They are less morally objectionable than bribery, and in some cases creative middle way solutions may be the best ways to handle the problems.

In some host countries, the basic social unit is the extended family or even some larger group, where members feel strong loyalties to one another. One aspect of this loyalty is the obligation to get other family members jobs. The moral and economic objections to nepotism suggest that a creative middle way might be appropriate between complete rejection of host country traditions and wholesale concession to nepotism.

In some host countries, business relationships are built on personal relationships, and these relationships are often cemented with gifts, many of which seem excessive by some host country standards. Restriction of gifts to sizes that may be larger than home country standards but not so large as to constitute bribery is a creative middle way between rejection of host country ways of doing business and actions that involve bribery.

The low level of scientific and technological sophistication suggests the need for paternalism. Weak paternalism, which actually preserves and protects a person's moral agency, can often be justified, but strong paternalism, which overrides moral agency for the sake of some substantial good, is more difficult to justify. It can only be justified in cases in which the good is considerable.

In some host countries, the tax rate is negotiated. This opens the way to bribery and inequitable distribution of taxes. Nevertheless, there may be some situations in which participating in this way of levying taxes without engaging in bribery and gaining an inequitably low tax rate is an acceptable creative middle way between having to leave the host country, on the one hand, and engaging in morally unjustifiable corruption, on the other hand.

NOTES

1. This is a modified version of an actual case presented in Lee A. Travis, *Power and Responsibility: Multinational Managers and Developing Country Concerns* (Notre Dame, IN: University of Notre Dame Press, 1997), pp. 315–338.
2. Ibid., p. 322.
3. Ibid., p. 322.
4. Ibid., p. 334.
5. See James F. Childress and John Macquarrie, eds., *The Westminster Dictionary of the Christian Church* (Philadelphia: Westminster Press, 1986), p. 499.

6. Ibid., p. 633.

7. See *The International Bill of Human Rights,* with forward by Jimmy Carter (Glen Ellen, CA: Entwhistle Books, 1981). No author.

8. James W. Nickel, *Making Sense of Human Rights: Philosophical Reflections on the Universal Declaration of Human Rights* (Berkeley: University of California Press, 1987), pp. 108–109.

9. Martha Nussbaum and Jonathan Glover, eds., *Women, Culture, and Development* (Oxford: Clarendon Press, 1995), pp. 83–85.

10. Robert E. Goodin, *Protecting the Vulnerable: A Reanalysis of Our Social Responsibilities* (Chicago: University of Chicago Press, 1985), pp. 195–196.

11. This scenario is a modification of one presented by Michael Philips titled "Bribery" in Patricia Werhane and Kendall D'Andrade, eds., *Profit and Responsibility* (New York: Edwin Mellon Press, 1985), pp. 197–220.

12. Thomas L. Carson, "Bribery, Extortion, and the 'Foreign Corrupt Practices Act,'" *Philosophy and Public Affairs,* 14, no. 1, 1985, pp. 66–90.

13. John T. Noonan, *Bribery* (New York: Macmillan, 1984).

14. Carson, "Bribery," p. 73.

15. Ibid., p. 79.

16. Ibid., p. 75.

17. For this case and related discussion, see Thomas Donaldson and Thomas W. Dunfee, "Toward a Unified Conception of Business Ethics: Integrative Social Contract Theory," *Academy of Management Review,* 19, no. 2, 1994, pp. 152–284.

18. See http://www.ti.com/corp/docs/company/citizen/ethics/market.shtml.

19. For a similar case and a similar conclusion, see Thomas Donaldson and Thomas W. Dunfee, *Ties that Bind: A Social Contracts Approach to Business Ethics* (Boston: Harvard Business School Press, 1999), pp. 198–207.

CASES

THE CASES LISTED HERE are presented for use in conjunction with materials in Chapters 1–10. They vary in length, complexity, and purpose. Some present factual events and circumstances. Others are fictional but realistic. Some present ethical problems for individual engineers. Others focus primarily on the corporate or institutional settings within which engineers work. Some, such as Case 44, "Where Are the Women?" focus on general problems within engineering as a profession. Others focus on large-scale issues such as global warming and the challenges and opportunities these issues pose for engineers, both individually and collectively. Some cases focus on wrongdoing and irresponsibility. Others illustrate exemplary engineering practice. A topical taxonomy of our cases appears next.

Many cases presented in previous editions of our book are not included here. However, most of them, and many others, are readily available on the Internet. Both the Online Ethics Center (www.onlineethics.org) and Texas A & M's Engineering Ethics website (www.ethics.tamu.edu) include Michael S. Pritchard, ed., *Engineering Ethics: A Case Study Approach,* a product of a National Science Foundation (NSF)-sponsored project. More than 30 cases and commentaries are presented. The Texas A & M website presents these cases in their original form, along with a taxonomy of the cases in accordance with their leading topical focus (e.g., safety and health, conflicts of interest, and honesty). (The cases are accessed under "1992 NSF Sponsored Engineering Ethics Cases.") Also included is an introductory essay by Pritchard. The Online Ethics Center presents the same cases with different individual titles, along with brief statements about each listed case. Cases and essays from two NSF-supported projects directed by Charles E. Harris and Michael J. Rabins are available at the Texas A & M website. These are also accessible at the Online Ethics Center (*Numerical and Design Problems* and *Engineering Ethics Cases from Texas A & M*). These appear under the heading "Professional Practice" and the subheading "Cases." The Online Ethics Center contains a wealth of other cases and essays that can be used in conjunction with our book. Of special interest is *Professional Ethics in Engineering Practice: Discussion Cases Based on NSPE BER Cases,* which provides access to cases and commentaries prepared by the National Society for Professional Engineer's Board of Ethical Review. These appear under the heading "Professional Practice" and the subheading "Cases" (*Discussion Cases from NSPE*).

LIST OF CASES

Acknowledging Limitations and Mistakes

Careers

Confidentiality

Conflicts of Interest

Dissent and Whistleblowing

CASE 1

Aberdeen Three

The Aberdeen Proving Ground is a U.S. Army facility where, among other things, chemical weapons are developed. The U.S. Army has used the facility to develop, test, store, and dispose of chemical weapons since World War II. Periodic inspections between 1983 and 1986 revealed serious problems with a part of the facility known as the Pilot Plant, including the following:

- Flammable and cancer-causing substances were left in the open.
- Chemicals that would become lethal if mixed were kept in the same room.
- Drums of toxic substances were leaking.

There were chemicals everywhere—misplaced, unlabeled, or poorly contained. When part of the roof collapsed, smashing several chemical drums stored below, no one cleaned up or moved the spilled substance and broken containers for weeks.[1]

When an external sulfuric acid tank leaked 200 gallons of acid into a nearby river, state and federal investigators were summoned to investigate. They discovered that the chemical retaining dikes were in a state of disrepair and that the system designed to contain and treat hazardous chemicals was corroded, resulting in chemicals leaking into the ground.[2]

On June 28, 1988, after 2 years of investigation, three chemical engineers—Carl Gepp, William Dee,

and Robert Lentz, now know as the "Aberdeen Three"—were criminally indicted for illegally handling, sorting, and disposing of hazardous wastes in violation of the Resource Conservation and Recovery Act (RCRA). Although the three engineers did not actually handle the chemicals, they were the managers with ultimate responsibility for the violations. Investigators for the Department of Justice concluded that no one above them was sufficiently aware of the problems at the Pilot Plant to be assigned responsibility for the violations. The three engineers were competent professionals who played important roles in the development of chemical weapons for the United States. William Dee, the developer of the binary chemical weapon, headed the chemical weapons development team. Robert Lentz was in charge of developing the processes that would be used to manufacture the weapons. Carl Gepp, manager of the Pilot Plant, reported to Dee and Lentz.

Six months after the indictment, the Department of Justice took the three defendants to court. Each defendant was charged with four counts of illegally storing and disposing of waste. William Dee was found guilty of one count, and Lentz and Gepp were found guilty on three counts each of violating the RCRA. Although each faced up to 15 years in prison and $750,000 in fines, they received sentences of 1,000 hours of community service and 3 years' probation. The judge justified the relatively light sentences on the grounds of the high standing of the defendants in the community and the fact that they had already incurred enormous court costs. Because the three engineers were criminally indicted, the U.S. Army could not assist them in their legal defense. This was the first criminal conviction of federal employees under RCRA.

CASE 2

Big Dig Collapse[3]

On July 10, 2006, a husband and wife were traveling through a connector tunnel in the Big Dig tunnel system in Boston. This system runs Interstate 93 beneath downtown Boston and extends the Massachusetts Turnpike to Logan Airport. As the car passed through, at least 26 tons of concrete collapsed onto it when a suspended concrete ceiling panel fell from above. The wife was killed instantly and the husband sustained minor injuries. The Massachusetts attorney general's office issued subpoenas next day to those involved in the Big Dig project. Soon, a federal investigation ensued.

The National Transportation Safety Board (NTSB) released its findings a year after the incident. The focus of the report was the anchor epoxy used to fasten the concrete panels and hardware to the tunnel ceiling. This product was marketed and distributed by Powers Fasteners, Inc., a company that specializes in the manufacturing and marketing of anchoring and fastening materials for concrete, masonry, and steel.

Investigators found that Powers distributed two kinds of epoxy: Standard Set and Fast Set. The latter type of epoxy, the one used in the collapsed ceiling tile, was susceptible to "creep," a process by which the epoxy deforms, allowing support anchors to pull free. The investigators concluded that this process allowed a ceiling tile to give way on July 10, 2006.

According to the NTSB report, Powers knew that Fast Set epoxy was susceptible to creep and useful for short-term load bearing only. Powers did not make this distinction clear in its marketing materials—the same materials distributed to tunnel project managers and engineers. Powers, the report continued, "failed to provide the Central Artery/Tunnel project with sufficiently complete, accurate, and detailed information about the suitability of the company's Fast Set epoxy for sustaining long-term tensile-loads." The report also noted that Powers failed to identify anchor displacement discovered in 1999 in portions of the Big Dig system as related to creep due to the use of Fast Set epoxy.

On the basis of the NTSB report, Powers was issued an involuntary manslaughter indictment by the Massachusetts attorney general's office just days after the release of the report. The indictment charged that "Powers had the necessary knowledge and the opportunity to prevent the fatal ceiling collapse but failed to do so."

The NTSB also targeted several other sources for blame in the incident (although no additional indictments were made). It concluded that construction contractors Gannett Fleming, Inc. and Bechtel/Parsons Brinkerhoff failed to account for the possibility of creep under long-term load conditions. The report indicated that these parties should have required that load tests be performed on adhesives before allowing their use and that the Massachusetts Turnpike Authority should have regularly inspected the portal tunnels. It asserted that if the Authority had conducted such inspections, the creep may have been detected early enough to prevent catastrophe.

The report provided recommendations to parties interested in the Big Dig incident. To the American Society of Civil Engineers, it advised the following:

> Use the circumstances of the July 10, 2006, accident in Boston, Massachusetts, to emphasize to your members through your publications, website, and conferences, as appropriate, the need to assess the creep characteristics of adhesive anchors before those anchors are used in sustained tensile-load applications.

To what extent must engineers educate themselves on the various materials being used and processes being employed in a project in order to ensure safety? If lack of knowledge played a part in causing the collapse, how might such understanding specifically help engineers to prevent an event like this in the future? How else might engineers work to avoid a similar catastrophe?

REFERENCES

1. National Transportation Safety Board, Public Meeting of July 10, 2007, "Highway Accident Report Ceiling Collapse in the Interstate 90 Connector Tunnel, Boston, Massachusetts," July 10, 2006. This document can be accessed online at www.ntsb.gov/Publictn/2007/HAR-07-02.htm.
2. The Commonwealth of Massachusetts Office of the Attorney General, "Powers Fasteners Indicted for Manslaughter in Connection with Big Dig Tunnel Ceiling Collapse." This document can be accessed online at www.mass.gov.

CASE 3

Bridges[4]

On August 1, 2007, the I-35W bridge over the Mississippi River in Minneapolis, Minnesota, collapsed during rush hour, resulting in 13 deaths and a multitude of injuries. The bridge was inspected annually dating from 1993 and every 2 years before that since its opening in 1967. The most recent inspection, conducted on May 2, 2007, cited only minor structural concerns related to welding details. At that time, the bridge received a rating of 4 on a scale from 0 to 9 (0 = shut down, 9 = perfect). The rating of 4, although signifying a bridge with components in poor condition, meant that the state was allowed to operate the bridge without any load restrictions.

A bridge rated 4 or less is considered to be "structurally deficient." According to the U.S. Department of Transportation, this label means that "there are elements of the bridge that need to be monitored and/or repaired. The fact that a bridge is 'deficient' does not imply that it is likely to collapse or that it is unsafe. It means it must be monitored, inspected, and maintained." In some cases, load restrictions are placed on structurally deficient bridges.

Although the cause of the I-35W collapse is still under investigation, the incident raises important questions about the state of U.S. bridges. In Minnesota, there are 1,907 bridges that are structurally deficient, which means they have also received a rating of 4 or lower on inspection. Bridges may also be considered "functionally obsolete," a label that the American Society of Civil Engineers (ASCE) Report Card for America's Infrastructure defines as a bridge that has "older design features and, while it is not unsafe for all vehicles, it cannot safely accommodate current traffic volumes, and vehicle sizes and weights." In 2003, 27.1 percent of bridges in the United States were deemed either structurally deficient or functionally obsolete.

The ASCE urges that "America must change its transportation behavior, increase transportation investment at all levels of government, and make use of the latest technology" to help alleviate the infrastructure problem involving the bridge system. In order for Americans to answer this charge, they must be aware of the problem. What role should engineers and engineering societies play in informing the public about the state of U.S. bridges? Should engineers lobby for congressional support and appropriate amounts of federal spending to be allocated to bridge repairs and reconstruction?

REFERENCES

1. ASCE, "Report Card for America's Infrastructure," 2005. This document can be accessed online at http://www.asce.org/reportcard/2005/index.cfm.
2. Minnesota Department of Transportation, "Interstate 35W Bridge Collapse," 2007. This document can be accessed online at http://www.dot.state.mn.us/i35wbridge/index.html.
3. U.S. Department of Transportation, Federal Highway Administration, "I-35 Bridge Collapse, Minneapolis, MN." This document can be accessed online at http://www.fhwa.dot.gov/pressroom/fsi35.htm.

CASE 4

Cadillac Chips[5]

Charged with installing computer chips that resulted in emitting excessive amounts of carbon dioxide from their Cadillacs, General Motors (GM) agreed in December 1995 to recall nearly 500,000 late-model Cadillacs and pay nearly $45 million in fines and recall costs. Lawyers for the Environmental Protection Agency (EPA) and the Justice Department contended that GM knew that the design change would result in pollution problems. Rejecting this claim, GM released a statement saying that the case was "a matter of interpretation" of complex regulations, but that it had "worked extremely hard to resolve the matter and avoid litigation."

According to EPA and Justice Department officials, the $11 million civil penalty was the third largest penalty in a pollution case, the second largest such penalty under the Clean Air Act, and the largest involving motor vehicle pollution. This was also the first case of a court ordering an automobile recall to reduce pollution rather than to improve safety or dependability.

Government officials said that in 1990 a new computer chip was designed for the engine controls of Cadillac Seville and Deville models. This was in response to car owners' complaints that these cars tended to stall when the climate control system was running. The chips injected additional fuel into the engine whenever this system was running. But this resulted in tailpipe emissions of carbon dioxide well in excess of the regulations.

Although the cars are usually driven with the climate control system running, tests used for certifying the meeting of emission standards were conducted when the system was not running. This was standard practice for emission tests throughout the automotive industry.

However, EPA officials argued that under the Clean Air Act, GM should have informed them that the Cadillac's design was changed in a way that would result in violating pollution standards under normal driving conditions. In 1970, the officials said, automobile manufacturers were directed not to get around testing rules by designing cars that technically pass the tests but that nevertheless cause avoidable pollution. GM's competitors, the officials contended, complied with that directive.

A GM spokesperson said that testing emissions with the climate control running was not required because "it was not in the rules, not in the regulations; it's not in the Clean Air Act." However, claiming that GM discovered the problem in 1991, Justice Department environmental lawyer Thomas P. Carroll objected to GM's continued inclusion of the chip in the 1992–1995 models: "They should have gone back and re-engineered it to improve the emissions."

In agreeing to recall the vehicles, GM said it now had a way of controlling the stalling problem without increasing pollution. This involves "new fueling calibrations," GM said, and it "should have no adverse effect on the driveability of the vehicles involved."

What responsibilities did GM engineers have in regard to either causing or resolving the problems with the Cadillac Seville and Deville models?

CASE 5

Cartex

Ben is assigned by his employer, Cartex, to work on an improvement to an ultrasonic range-finding device. While working on the improvement, he gets an idea for a modification of the equipment that might be applicable to military submarines. If this is successful, it could be worth a lot of money to his company. However, Ben is a pacifist and does not want to contribute in any way to the development of military hardware. So Ben neither develops the idea himself nor mentions it to anybody else in the company. Ben has signed an agreement that all inventions he produces on the job are the property of the company, but he does not believe the agreement applies to his situation because (1) his idea is not developed and (2) his superiors know of his antimilitary sentiments. Yet he wonders if he is ethically right in concealing his idea from his employers.

An interesting historical precedent: Leonardo Da Vinci recorded in his journal that he had discovered how to make a vessel that can move about underwater—a kind of submarine. However, he refused to share this idea with others on the grounds that he feared it would be used for immoral purposes. "I do not publish or divulge on account of the evil nature of men who would practice assassinations at the bottom of the seas, by breaking the ships in their lowest parts and sinking them together with the crews who are in them."[6]

CASE 6

Citicorp[7]

William LeMessurier was understandably proud of his structural design of the 1977 Citicorp building in downtown Manhattan. He had resolved a perplexing problem in a very innovative way. A church had property rights to a corner of the block on which the 59-story building was to be constructed. LeMessurier proposed constructing the building *over* the church, with four supporting columns located at the center of each side of the building rather than in the four corners. The first floor began the equivalent of nine stories above ground, thus allowing ample space for the church. LeMessurier used a diagonal bracing design that transferred weight to the columns, and he added a tuned mass damper with a 400-ton concrete block floating on oil bearings to reduce wind sway.

In June 1978, LeMessurier received a call from a student at a nearby university who said his professor claimed the Citicorp building's supporting columns should be on the corners instead of midway between them. LeMessurier replied that the professor did not understand the design problem, adding that the innovative design made it even more resistant to quartering, or diagonal, winds. However, since the New York City building codes required calculating the effects of only 90-degree winds, no one actually worked out calculations for quartering winds. Then he decided that it would be instructive for his own students to wrestle with the design problem.

This may have been prompted by not only the student's call but also a discovery LeMessurier had made just 1 month earlier. While consulting on a building project in Pittsburgh, he called his home office to find out what it would cost to weld the joints of diagonal girders similar to those in the Citicorp building. To his surprise, he learned that the original specification for full-penetration welds was not followed. Instead, the joints were bolted. However, since this still more than adequately satisfied the New York building code requirements, LeMessurier was not concerned.

However, as he began to work on calculations for his class, LeMessurier recalled his Pittsburgh discovery. He wondered what difference bolted joints might make to the building's ability to withstand quartering winds. To his dismay, LeMessurier determined that a 40 percent stress increase in some areas of the structure would result in a 160 percent increase in stress on some of the building's joints. This meant that the building was vulnerable to total collapse if certain areas were subjected to a "16-year storm" (i.e., the sort of storm that could strike Manhattan once every 16 years). Meanwhile, hurricane season was not far away.

LeMessurier realized that reporting what he had learned could place both his engineering reputation and the financial status of his firm at substantial risk. Nevertheless, he acted quickly and decisively. He drew up a plan for correcting the problem, estimated the cost and time needed for rectifying it, and immediately informed Citicorp owners of what he had learned. Citicorp's response was equally decisive. LeMessurier's proposed course of action was accepted and corrective steps were immediately undertaken. As the repairs neared completion in early September, a hurricane was reported moving up the coast in the direction of New York. Fortunately, it moved harmlessly out over the Atlantic Ocean, but not without first causing considerable anxiety among those working on the building, as well as those responsible for implementing plans to evacuate the area should matters take a turn for the worse.

Although correcting the problem cost several million dollars, all parties responded promptly and responsibly. Faced with the threat of increased liability insurance rates, LeMessurier's firm convinced its insurers that because of his responsible handling of the situation, a much more costly disaster may have been prevented. As a result, the rates were actually reduced.

Identify and discuss the ethical issues this case raises.

CASE 7

Disaster Relief[8]

Among the 24 recipients of the John D. and Catherine T. MacArthur Foundation Fellowships for 1995 was Frederick C. Cuny, a disaster relief specialist. The fellowship program is commonly referred to as a "genius program," but it is characterized by MacArthur executives as a program that rewards "hard-working experts who often push the boundaries of their fields in ways that others will follow."[9] The program, says Catherine Simpson, director of the awards program, is meant to serve as "a reminder of the importance of seeing as broadly as possible, of being willing to live outside of a comfort zone and of keeping your nerve endings open."[10]

Cuny's award was unusual in two respects. First, at the time the award was announced, his whereabouts were unknown, and it was feared that he had been executed in Chechnya. Second, he was a practicing engineer. Most MacArthur awards go to writers, artists, and university professors.

Ironically, although honored for his engineering achievements, Cuny never received a degree in engineering. Initially planning to graduate from the ROTC program at Texas A & M as a Marine pilot, he had to drop out of school in his second year due to poor grades. He transferred to Texas A & I, Kingsville, to continue his ROTC coursework, but his grades suffered there as well. Although he never became a Marine pilot, he worked effectively with Marine corps officers later in Iraq and Somalia.[11]

In Kingsville, Cuny worked on several community projects after he dropped out of school. He found his niche in life working in the barrios with poor Mexicans in Kingsville and formulated some common sense guidelines that served him well throughout his career. As he moved into disaster relief work, he understood immediately that the aid had to be designed for those who were in trouble in ways that would leave them in the position of being able to help themselves. He learned to focus on the main problem in any disaster to better understand how to plan the relief aid. Thus, if the problem was shelter, the people should be shown how to rebuild their destroyed homes in a better fashion than before. Similar approaches were adopted regarding famine, drought, disease, and warfare.

The first major engineering project Cuny worked on was the Dallas–Ft. Worth airport. However, attracted to humanitarian work, he undertook disaster relief work in Biafra in 1969. Two years later, at age 27, he founded the Intertect Relief and Reconstruction Corporation in Dallas. Intertect describes itself as

> a professional firm providing specialized services and technical assistance in all aspects of natural disaster and refugee emergency management—mitigation, preparedness, relief, recovery, reconstruction, resettlement—including program design and implementation, camp planning and administration, logistics, vulnerability analysis, training and professional development, technology transfer, assessment, evaluation, networking and information dissemination."[12]

Intertect also prides itself for its "multidisciplinary, flexible, innovative, and culturally-appropriate approach to problem-solving."[13] Obviously, such an enterprise requires the expertise of engineers. But it also must draw from social services, health and medical care professionals, sociology, anthropology, and other areas.

Fred Cuny was apparently comfortable working across disciplines. As an undergraduate he also studied African history. So, it is understandable that he would take a special interest in the course of the conflict between the Nigerian and Biafran governments in the late 1960s. In 1969, he announced to the Nigerian minister of the interior, "I'm from Texas. I'm here to study the war and try to suggest what can be done to get in humanitarian aid when it's over."[14] Rebuffed by the minister, Cuny then flew to Biafra and helped organize an airlift that provided short-term assistance to the starving Biafrans.

Cuny learned two important lessons from his Biafran work. First, food distribution in disaster relief often pulls people from their homes and working areas to distribution centers in towns and airports. Cuny commented, "The first thing I recognized was that we had to turn the system around and get people back into the countryside away from the airfield." Second, Cuny realized that public health is a major problem—one that can effectively be addressed only through careful planning. This requires engineering efforts to, for example, build better drains, roads, dwellings, and so on. At the same time, Cuny realized that relatively few engineers were in relief agencies: hence the founding of Intertect. Concerned to share his ideas with others, in 1983 Cuny published *Disasters and Development* (Oxford University Press), which provides a detailed set of guidelines for planning and providing disaster relief. A major theme of his book is that truly helpful relief requires careful study of local conditions in order to provide long-term assistance.

Despite its small size, since its founding in 1971, Intertect has been involved in relief projects in nearly 70 different countries during Cuny's career. His work came to the attention of wealthy Hungarian philanthropist George Soros, who provided him with funding to work on a number of major disaster relief projects.

An especially daring project was the restoration of water and heat to a besieged section of Sarajevo in 1993.[15] Modules for a water filtration system were specially designed to fit into a C-130 airplane that was flown from Zagreb (Croatia's capital) into Sarajevo. (Cuny commented that there were only 3 inches to spare on each side of the storage area.) In order to get the modules unnoticed through Serbian checkpoints, they had to be unloaded in less than 10 minutes.

Clearly, the preparation and delivery of the modules required careful planning and courage in execution. However, prior to that someone had to determine that such a system could be adapted to the circumstances in Sarajevo. When Cuny and his associates arrived in Sarajevo, for many the only source of water was from a polluted river. The river could be reached only by exposing oneself to sniper fire, which had already injured thousands and killed hundreds. Thus, residents risked their lives to bring back containers of water whose contaminated contents posed additional risks. Noting that Sarajevo had expanded downhill in recent years, and that the newer water system had to pump water uphill to Old Town Sarajevo, the Cuny team concluded that there must have been an earlier system for Old Town.[16] They located a network of old cisterns and channels still in good working order, thus providing them with a basis for designing and installing a new water filtration plant. This $2.5 million project was funded by the Soros Foundation, which also provided $2.7 million to restore heat for more than 20,000 citizens of Sarajevo.

Cuny told author Christopher Merrill, "We've got to say, 'If people are in harm's way, we've got to get them out of there. The first and most important thing is saving lives. Whatever it takes to save lives, you do it, and the hell with national sovereignty.'"[17] This philosophy lay behind his efforts to save 400,000 Kurds in northern Iraq after Operation Desert Storm, in addition to thousands of lives in Sarajevo; however, this may be what cost him his own life in Chechnya in 1995.

Perhaps Cuny's single most satisfying effort was in northern Iraq immediately following Operation Desert Storm. As soon as Iraq signed the peace treaty, Saddam Hussein directed his troops to attack the Shiites in the south and the Kurds in the north. The 400,000 Kurds fled into the mountains bordering Turkey, where the Turks prevented them from crossing the border. Winter was coming and food was scarce. President Bush created a no-fly zone over northern Iraq and directed the Marine Corps to rescue the Kurds in what was called "Operation Provide Comfort." The Marine general in charge hired Fred Cuny as a consultant, and Cuny quickly became, in effect, second in command of the operation.

When Operation Provide Comfort was regarded as no longer necessary, the Kurds held a farewell celebration at which the full Marine battalion marched before joyous crowds, with one civilian marching in the first row—Fred Cuny. Cuny had an enlargement of a photo of that moment hung over his desk in Dallas. The photo has the signature of the Marine general who led the parade.

Asked about his basic approach to disaster relief, Cuny commented: "In any large-scale disaster, if you can isolate a part that you can understand you will usually end up understanding the whole system."[18] In the case of Sarajevo, the main problems seemed to center around water and heat. So this is what Cuny and associates set out to address. In preparing for disaster relief work, Cuny was from the outset struck by the fact that medical professionals and materials are routinely flown to international disasters, but engineers and engineering equipment and supplies are not. So, his recurrent thought was, "Why don't you officials give first priority to, say, fixing the sewage system, instead of merely stanching the inevitable results of a breakdown in sanitary conditions?"[19]

It is unusual for engineers to receive the sort of public attention Fred Cuny did. We tend to take for granted the good work that engineers do. Insofar as engineers "make the news," more likely than not this is when an engineering disaster has occurred, a product is subjected to vigorous criticism, or an engineer has blown the whistle. Fred Cuny's stories are largely stories of successful humanitarian ventures.

Fred Cuny's untimely, violent death was tragic. In April 1995, while organizing a field hospital for victims in the conflict in Chechnya, Cuny, two Russian Red Cross doctors, and a Russian interpreter disappeared. After a prolonged search, it was concluded that all four were executed. Speculation is that Chechens may have been deliberately misinformed that the four were Russian spies. Cuny's article in the *New York Review of Books* titled "Killing Chechnya" was quite critical of the Russian treatment of Chechnya, and it gives some indication of why his views might well have antagonized Russians.[20] Already featured in the *New York Times*, the *New Yorker Magazine*, and the *New York Review of Books*, Cuny had attained sufficient national recognition that his disappearance received widespread attention and immediate response from President Clinton and government officials. Reports on the search for Cuny and colleagues regularly appeared in the press from early April until August 18, 1995, when his family finally announced that he was now assumed dead.

Many tributes have been made to the work of Fred Cuny. Pat Reed, a colleague at Intertect, was quoted soon after Cuny's disappearance: "He's one of the few visionaries in the emergency management field.

He really knows what he's doing. He's not just some cowboy."[21] At the Moscow press conference calling an end to the search, Cuny's son Chris said, "Let it be known to all nations and humanitarian organizations that Russia was responsible for the death of one of the world's great humanitarians."[22] William Shawcross fittingly concludes his article, "A Hero for Our Time," as follows:

> At the memorial meeting in Washington celebrating Fred's life it was clear that he had touched people in a remarkable way. He certainly touched me; I think he was a great man. The most enduring memorials to Fred are the hundreds of thousands of people he has helped—and the effect he has had, and will have, on the ways governments and other organizations try to relieve the suffering caused by disasters throughout the world.

AN AFTERWORD

It is certainly appropriate to make special note of extraordinary individuals such as Frederick C. Cuny for special praise. His life does seem heroic. However, we would do well to remember that even heroes have helpers. Cuny worked with others, both at Intertect and at the various other agencies with whom Intertect collaborated. There are unnamed engineers in Sarajevo with whom he worked. For example, his Sarajevo team was able to locate the old cisterns and channels through the assistance of local engineers (and historians).[23] Local engineers assisted in installing the water filtration system.

Furthermore, once the system was installed, the water had to be tested for purity. Here, a conflict developed between local engineers (as well as Cuny and specialists from the International Rescue Committee) and local water safety inspectors who demanded further testing. Convinced that they had adequately tested the water, the local engineers, Cuny, and the International Rescue Committee were understandably impatient. However, the cautious attitude of the water safety experts is understandable as well. Muhamed Zlatar, deputy head of Sarajevo's Institute for Water, commented, "The consequences of letting in polluted water could be catastrophic. They could be worse than the shelling. We could have 30,000 people come down with stomach diseases, and some of them could die."[24] Without presuming who might have been right, we might do well to remember Fran

Kelsey, the FDA official who, in 1962, refused to approve thalidomide until further testing was done. That is, in our rush to do good, caution should not be thrown to the winds.

Identify and discuss the ethical issues raised by the story of Frederick C. Cuny.

CASE 8

Electric Chair

Thanks in part to Theodore Bernstein, retired University of Wisconsin professor of electrical and computer engineering, apparently the electric chair is disappearing.[25] Once regarded as a more humane way of executing someone than beheading or hanging, the electric chair itself has a questionable history. For instance, the Death Penalty Information Center classifies 10 of the 149 electrocutions of the past 25 years as botched. Although, as Bernstein says, "You give enough shocks, you can kill anybody," it is not clear how much is enough—or too much.

Having spent three decades studying the effects of electricity on the human body, Bernstein has frequently testified in court and in hearings in an effort to help defendants avoid being placed in the electric chair. He comments,

> The substance of my testimony is pretty much always the same. I tell the court that most of the work on the electric chair was done with a seat-of-the-pants approach. The electrical design is poor. Every state has a different sequence of shocks. Many of the states use old equipment, and they don't test it very well. They'll have in the notebook or the protocols, "Check the equipment," or "Check the electrodes." What does that mean? They need to be more specific.[26]

The problem, says Bernstein, is that electrocution has always been controlled by people without background in biomedical engineering. This is also reflected in its beginnings in the late 19th century. Believing that the alternating current (AC) system of his competitor, George Westinghouse, was more dangerous than his own system of direct current (DC), Thomas Edison recommended the AC system for the

electric chair. Not wanting his company's reputation to be tied to the electric chair, Westinghouse provided funding to William Kemmler's attorneys in their effort to stop their client from becoming the first person executed in an electric chair. Edison testified that an electric chair that used alternating current would cause minimal suffering and instantaneous death. Although Kemmler's attorneys got Edison to admit that he knew little about the structure of the human body or conductivity in the brain, Edison's claims carried the day. According to Bernstein, Edison's "reputation made more of an impression than did his bioelectrical ignorance."[27]

Not only was Kemmler the first person executed in an electric chair but also he was the first person whose execution by electricity required more than one application of current, the second of which caused vapor and smoke to be emitted from Kemmler's body. Witnesses were dismayed by what they saw, with one physician commenting that using an electric chair "can in no way be regarded as a step in civilization."[28] According to Bernstein, a basic problem was that executioners knew very little about how electrocution causes death—and, he notes, executioners know little more even today.

Does electrocution "fry the brain"? Bernstein comments: "That's a lot of nonsense. The skull has a very high resistance, and current tends to flow around it." Actually, he says, electrocution usually causes cardiac arrest, and this may not be painless—and it may not be fatal on the first try.

Discuss the ethical issues surrounding Theodore Bernstein's chosen area of research and his role as a witness in the courtroom and legal hearings.

CASE 9

Fabricating Data[29]

INTRODUCTION

In recent years, the National Science Foundation (NSF), the National Institutes of Health (NIH), the Public Health Services (PHS), the Office of Scientific Integrity, and various scientific organizations such as the National Academy of Sciences have spent considerable time and effort in trying to agree on a definition of *scientific misconduct*. A good definition is needed in developing and implementing policies and regulations concerning appropriate conduct in research, particularly when federal funding is involved. This is an important area of concern because although serious scientific misconduct may be infrequent, the consequences of even a few instances can be widespread.

Those cases that reach the public's attention can cause considerable distrust among both scientists and the public, however infrequent their occurrence. Like lying in general, we may wonder which scientific reports are tainted by misconduct, even though we may be convinced that relatively few are tainted. Furthermore, scientists depend on each other's work in advancing their own. Building one's work on the incorrect or unsubstantiated data of others infects one's own research, and the chain of consequences can be quite lengthy as well as very serious. This is as true of honest or careless mistakes as it is of the intentional distortion of data, which is what *scientific misconduct* is usually restricted to. Finally, of course, the public depends on the reliable expertise of scientists in virtually every area of health, safety, and welfare.

Although exactly what the definition of scientific misconduct should include is a matter of controversy, all proposed definitions include the fabrication and falsification of data and plagiarism. As an instance of fraud, the fabrication of data is a particularly blatant form of misconduct. It lacks the subtlety of questions about interpreting data that pivot around whether the data have been "fudged" or "manipulated." Fabricating data is making it up, or *faking* it. Thus, it is a clear instance of a lie, a deliberate attempt to deceive others.

However, this does not mean that fabrications are easy to detect or handle effectively once they are detected; and this adds considerably to the mischief and harm they can cause. Two well-known cases illustrate this, both of which feature ambitious, and apparently successful, young researchers.

THE DARSEE CASE[30]

Dr. John Darsee was regarded as a brilliant student and medical researcher at the University of Notre Dame (1966–1970), Indiana University (1970–1974), Emory University (1974–1979), and Harvard University (1979–1981). He was regarded by faculty at all four institutions as a potential "all-star" with a great research future ahead of him. At Harvard, he reportedly often worked more than 90 hours a week as a research fellow in the Cardiac Research Laboratory headed by Dr. Eugene Braunwald. In less than 2 years at Harvard, he was first author of seven publications in very good scientific journals. His special area of research concerned the testing of heart drugs on dogs.

All of this came to a sudden halt in May 1981 when three colleagues in the Cardiac Research Laboratory observed Darsee labeling data recordings "24 seconds," "72 hours," "one week," and "two weeks." In reality, only minutes had transpired. Confronted by his mentor Braunwald, Darsee admitted the fabrication, but he insisted that this was the only time he had done this, and that he had been under intense pressure to complete the study quickly. Shocked, Braunwald and Darsee's immediate supervisor, Dr. Robert Kroner, spent the next several months checking other research conducted by Darsee in their lab. Darsee's research fellowships were terminated, and an offer of a faculty position was withdrawn. However, he was allowed to continue his research projects at Harvard for the next several months (during which time Braunwald and Kroner observed his work very closely).

Hopeful that this was an isolated incident, Braunwald and Kroner were shocked again in October. A comparison of results from four different laboratories in a National Heart, Lung, and Blood Institute Models Study revealed an implausibly low degree of invariability in data provided by Darsee. In short, his data looked "too good." Since these data had been submitted in April, there was strong suspicion that Darsee had been fabricating or falsifying data for some time. Subsequent investigations seemed to indicate questionable research practices dating back as far as his undergraduate days.

What were the consequences of John Darsee's misconduct? Darsee, we have seen, lost his research position at Harvard, and his offer of a faculty position was withdrawn. The NIH barred him from NIH funding or serving on NIH committees for 10 years. He left research and went into training as a critical care specialist. However, the cost to others was equally, if not more, severe. Harvard-affiliated Brigham and Women's Hospital became the first institution that NIH ever required to return funds ($122,371) because of research involving fraudulent data. Braunwald and colleagues had to spend several months investigating Darsee's research rather than simply continuing the work of the Cardiac Research Laboratory. Furthermore, they were severely criticized for carrying on their own investigation without informing NIH of their concerns until several months later. The morale and productivity of the laboratory were damaged. A cloud of suspicion hung over all the work with which Darsee was associated. Not only was Darsee's own research discredited but also, insofar as it formed an integral part of collaborative research, a cloud was thrown over published research bearing the names of authors whose work was linked with Darsee's.

The months of outside investigation also took others away from their main tasks and placed them under extreme pressure. Statistician David DeMets played a key role in the NIH investigation. Years later, he recalled the relief his team experienced when their work was completed:[31]

> For the author and the junior statistician, there was relief that the episode was finally over and we could get on with our careers, without the pressures of a highly visible misconduct investigation. It was clear early on that we had no room for error, that any mistakes would destroy the case for improbable data and severely damage our careers. Even without mistakes, being able to convince lay reviewers such as a jury using statistical arguments could still be defeating. Playing the role of the prosecuting statisticians was very demanding of our technical skills but also of

our own integrity and ethical standards. Nothing could have adequately prepared us for what we experienced.

Braunwald notes some positive things that have come from the Darsee case. In addition to alerting scientists to the need for providing closer supervision of trainees and taking authorship responsibilities more seriously, the Darsee incident contributed to the development of guidelines and standards concerning research misconduct by PHS, NIH, NSF, medical associations and institutes, and universities and medical schools. However, he cautions that no protective system is able to prevent all research misconduct. In fact, he doubts that current provisions could have prevented Darsee's misconduct, although they might have resulted in earlier detection. Furthermore, he warns that good science does not thrive in an atmosphere of heavy "policing" of one another's work:[32]

> The most creative minds will not thrive in such an environment and the most promising young people might actually be deterred from embarking on a scientific career in an atmosphere of suspicion. Second only to absolute truth, science requires an atmosphere of openness, trust, and collegiality.

Given this, it seems that William F. May is right in urging the need for a closer examination of character and virtue in professional life.[33] He says that an important test of character and virtue is what we do when no one is watching. The Darsee case and Braunwald's reflections seem to confirm this.

Many who are caught having engaged in scientific misconduct plead that they were under extreme pressure, needing to complete their research in order to meet the expectations of their lab supervisor, to meet a grant deadline, to get an article published, or to survive in the increasingly competitive world of scientific research. Although the immediate stakes are different, students sometimes echo related concerns: "I knew how the experiment should have turned out, and I needed to support the right answer"; "I needed to get a good grade"; "I didn't have time to do it right; there's so much pressure." Often these thoughts are accompanied by another—namely that this is only a classroom exercise and that, of course, one will not fabricate data when one becomes a scientist and these pressures are

CASE 9 • Fabricating Data **245**

absent. What the Darsee case illustrates is that it is naive to assume such pressures will vanish. Therefore, the time to begin dealing with the ethical challenges they pose is now, not later (when the stakes may be even higher).

THE BRUENING CASE[34]

In December 1983, Dr. Robert Sprague wrote an 8-page letter, with 44 pages of appendices, to the National Institute of Mental Health documenting the fraudulent research of Dr. Stephen Breuning.[35] Breuning fabricated data concerning the effects psychotropic medication has on mentally retarded patients. Despite Breuning's admission of fabricating data only 3 months after Sprague sent his letter, the case was not finally resolved until July 1989. During that 5½-year interval, Sprague was a target of investigation (in fact, he was the first target of investigation), he had his own research endeavors severely curtailed, he was subjected to threats of lawsuits, and he had to testify before a U.S. House of Representatives committee. Most painful of all, Sprague's wife died in 1986 after a lengthy bout with diabetes. In fact, his wife's serious illness was one of the major factors prompting his "whistleblowing" to NIH. Realizing how dependent his diabetic wife was on reliable research and medication, Sprague was particularly sensitive to the dependency that the mentally retarded, clearly a vulnerable population, have on the trustworthiness of not only their caregivers but also those who use them in experimental drug research.

Writing 9 years after the closing of the Bruening case, Sprague obviously has vivid memories of the painful experiences he endured and of the potential harms to participants in Bruening's studies. However, he closes the account of his own experiences by reminding us of other victims of Bruening's misconduct—namely psychologists and other researchers who collaborated with Bruening without being aware that he had fabricated data.

Dr. Alan Poling, one of those psychologists, writes about the consequences of Bruening's misconduct for his collaborators in research. Strikingly, Poling points out that between 1979 and 1983, Bruening was a contributor to 34 percent of all published research on the psychopharmacology of mentally retarded people. For those not involved in the research, initial doubts may, however unfairly, be cast on all these publications. For those involved in the research, efforts need to be made in each case to determine to what extent, if any, the validity of the research was affected by Bruening's role in the study. Even though Bruening was the only researcher to fabricate data, his role could contaminate an entire study. In fact, however, not all of Bruening's research did involve fabrication. Yet, convincing others of this is a time-consuming, demanding task. Finally, those who cited Bruening's publications in their own work may also suffer "guilt by association." As Poling points out, this is especially unfair in those instances in which Bruening collaborations with others involved no fraud at all.

THE ISSUES

The Darsee and Bruening cases raise a host of ethical questions about the nature and consequences of scientific fraud:

- What kinds of reasons are offered for fabricating data?
- Which, if any, of those reasons are *good* reasons—that is, reasons that might *justify* fabricating data?
- Who is likely to be harmed by fabricating data? Does actual harm have to occur in order for fabrication to be ethically wrong?
- What responsibilities does a scientist or engineer have for checking the trustworthiness of the work of other scientists or engineers?
- What should a scientist or engineer do if he or she has reason to believe that another scientist or engineer has fabricated data?
- Why is honesty in research important to the scientific and engineering communities?
- Why is honesty in research important for the public?
- What might be done to diminish the likelihood that research fraud occurs?

READINGS

For readings on scientific integrity, including sections on the fabrication of data and a definition of scientific misconduct, see Nicholas Steneck, *ORI Introduction to Responsible Conduct in Research* (Washington, DC: Office of Research Integrity, 2004); *Integrity and Misconduct in Research* (Washington, DC: U.S.

Department of Health and Human Services, 1995); *On Being a Scientist,* 2nd ed. (Washington, DC: National Academy Press, 1995); and *Honor in Science* (Research Triangle Park, NC: Sigma Xi, The Scientific Research Society, 1991).

CASE 10

Gilbane Gold

The fictional case study presented in the popular videotape *Gilbane Gold* focuses on David Jackson, a young engineer in the environmental affairs department of ZCORP, located in the city of Gilbane.[36] The firm, which manufactures computer parts, discharges lead and arsenic into the sanitary sewer of the city. The city has a lucrative business in processing the sludge into fertilizer, which is used by farmers in the area.

To protect its valuable product, Gilbane Gold, from contamination by toxic discharges from the new high-tech industries, the city has imposed highly restrictive regulations on the amount of arsenic and lead that can be discharged into the sanitary sewer system. However, recent tests indicate that ZCORP may be violating the standards. David believes that ZCORP must invest more money in pollution-control equipment, but management believes the costs will be prohibitive.

David faces a conflict situation that can be characterized by the convergence of four important moral claims. First, David has an obligation as a good employee to promote the interests of his company. He should not take actions that unnecessarily cost the company money or damage its reputation. Second, David has an obligation—based on his personal integrity, his professional integrity as an engineer, and his special role as environmental engineer—to be honest with the city in reporting data on the discharge of the heavy metals. Third,

David has an obligation as an engineer to protect the health of the public. Fourth, David has a right, if not an obligation, to protect and promote his own career.

The problem David faces is this: How can he do justice to all of these claims? If they are all morally legitimate, he should try to honor all of them, and yet they appear to conflict in the situation. David's first option should be to attempt to find a creative middle way solution, despite the fact that the claims appear to be incompatible in the situation. What are some of the creative middle way possibilities?[37]

One possibility would be to find a cheap technical way to eliminate the heavy metals. Unfortunately, the video does not directly address this possibility. It begins in the midst of a crisis at ZCORP and focuses almost exclusively on the question of whether David Jackson should blow the whistle on his reluctant company. For a detailed exploration of some creative middle way alternatives, see Michael Pritchard and Mark Holtzapple, "Responsible Engineering: *Gilbane Gold* Revisited," *Science and Engineering,* 3, no. 2, April 1997, pp. 217–231.

Another avenue to explore in *Gilbane Gold* is the attitudes toward responsibility exhibited by the various characters in the story. Prominent, for example, are David Jackson, Phil Port, Diane Collins, Tom Richards, Frank Seeders, and Winslow Massin. Look at the transcript (available at www.niee.org/pd.cfm?pt=Murdough). What important similarities and differences do you find?

CASE 11

Green Power?[38]

The growing consensus among scientists that carbon emissions are contributing to global warming is also beginning to have a significant impact on local energy policies and projects. For example, Fort Collins, Colorado, has a Climate Wise energy program to go with its official motto, "Where renewal is a way of life." Local reduction of carbon emissions is one of the city's global aims.

At the same time, local communities such as Fort Collins have continued, if not growing, energy needs. AVA Solar and Powertech Uranium are proposing ways of helping to meet these needs. Working with Colorado State University, AVA has developed a manufacturing process to make electricity-producing solar panels. Solar energy has popular appeal and is typically given high marks in regard to "green" technology. Local critics, however, have some worries about the AVA project. The process uses cadmium, which raises concerns about cancer. AVA's director of strategic planning, Russ Kanjorski, acknowledges that the use of cadmium will call for careful environmental monitoring, particularly in the discharge of water, and that monitoring practices are still in the developmental stage.

Powertech Uranium proposes drilling for uranium, which can be used to create nuclear power. Nuclear power promises to reduce carbon emissions, but it lacks solar power's popularity. Although Governor Bill Ritter, Jr., is strongly committed to what he calls "the new energy economy," this does not favor uranium mining. In fact, there are long-term, unresolved scientific and technological worries about extracting, processing, and disposing of uranium.

Complicating matters is that both projects seem to have great economic potential for the companies and the local economy. As Kirk Johnson states, "There is no doubt that new money is chasing new energy."

Meanwhile, Johnson observes, local environmentalists such as Dan Bihn are genuinely puzzled. Bihn is an electrical engineer and environmental consultant on the Fort Collins Electric Utilities Board. Johnson quotes Bihn as saying "I think nuclear needs to be on the table, and we need to work through this thing and we can't just emotionally react to it." What is Bihn's emotional reaction to Powertech's proposal? "Deep down inside," he told Johnson, "my emotional reaction is that we should never do this."

Lane Douglas, a spokesperson for Powertech and its Colorado land and project manager, urges that its company's proposal be judged on facts, not prejudice. "The science will either be good science or it won't," Douglas says. "We're just saying give us a fair hearing."

Local citizens such as Ariana Friedlander are striving to be consistent in evaluating the proposals. Skeptical about uranium mining, she adds, "But we shouldn't be giving the other guys a pass because they're sexy right now."

Discuss the ethical issues raised by the Fort Collins circumstances. What responsibilities do engineers have in regard to issues like these? When Dan Bihn says we shouldn't *just* emotionally react to these issues, do you think he is saying that he should *ignore* his own emotional reaction? (Why do you suppose he characterizes this as "deep down inside"?) What do you think Lane Douglas has in mind by appealing to "good science" in resolving the issues about uranium mining? Do you think "good science" alone can provide the answers?

CASE 12

Greenhouse Gas Emissions[39]

On November 15, 2007, the Ninth Circuit Court of Appeals in San Francisco rejected the Bush administration's fuel economy standards for light trucks and sport utility vehicles. The three-judge panel objected that the regulations fail to take sufficiently into account the economic impact that tailpipe emissions can be expected to have on climate change. The judges also questioned why the standards were so much easier on light trucks than passenger cars. (The standards hold that by 2010 light trucks are to average 23.5 mpg, whereas passenger cars are to average 27.5 mpg.)

Although it is expected that an appeal will be made to the U.S. Supreme Court, this ruling is one of several recent federal court rulings that urge regulators to consider the risk of climate change in setting standards for carbon dioxide and other heat-trapping gas emissions from industry.

Patrick A. Parenteau, Vermont Law School environmental law professor, is quoted as saying, "What this says to me is that the courts are catching up with climate change and the law is catching up with climate change. Climate change has ushered in a whole new era of judicial review."[40]

One of the judges, Betty B. Fletcher, invoked the National Environmental Policy Act in calling for cumulative impacts analyses explicitly taking into account the environmental impact of greenhouse gas emissions. Acknowledging that cost–benefit analysis may appropriately indicate realistic limits for fuel economy standards, she insisted that "it cannot put a thumb on the scale by undervaluing the benefits and overvaluing the costs of more stringent standards."

Finally, Judge Fletcher wrote, "What was a reasonable balancing of competing statutory priorities 20 years ago may not be a reasonable balancing of those priorities today."

Given recent court trends, what implications are there for the responsibilities (and opportunities) of engineers working in the affected areas?

CASE 13

"Groupthink" and the Challenger *Disaster*

The video *Groupthink* presents Irving Janis's theory of "groupthink" in the form of a case study of the 1986 *Challenger* disaster (discussed in Chapters 7 and 8). As we indicate in Chapter 2, Janis characterizes "groupthink" as a set of tendencies of cohesive groups to achieve consensus at the expense of critical thinking.

View the video and then discuss the extent to which you agree with the video's suggestion that groupthink could have been a significant factor leading up to the *Challenger* disaster. (This video is available from CRM Films, McGraw-Hill Films, 1221 Avenue of the Americas, New York, NY 10020. 1-800-421-0833.)

CASE 14

Halting a Dangerous Project

In the mid 1980s, Sam was Alpha Electronics' project leader on a new contract to produce manufactured weaponry devices for companies doing business with NATO government agencies.[41] The devices were advanced technology land mines with electronic controls that could be triggered with capacitor circuits to go off only at specified times, rather than years later when children might be playing in old minefields. NATO provided all the technical specifications and Alpha Electronics fulfilled the contract without problems. However, Sam was concerned that one new end user of this device could negate the safety aspects of the trigger and make the land mines more dangerous than any others on the market.

After the NATO contract was completed, Sam was dismayed to learn that Alpha Electronics had signed another contract with an Eastern European firm that had a reputation of stealing patented devices and also of doing business with terrorist organizations. Sam halted the production of the devices. He then sought advice from some of his colleagues and contacted the U.S. State Department's Office of Munitions Controls. In retrospect, he wishes he had also

contacted the Department of Commerce's Bureau of Export Administration, as well as the Defense Department. He ruefully acknowledges that the issue would have been brought to a close much more quickly.

The contract that Sam unilaterally voided by his action was for nearly $2 million over 15 years. Sam noted that no further hiring or equipment would have been needed, so the contract promised to be highly profitable. There was a $15,000 penalty for breaking the contract.

On the basis of global corporate citizenship, it was clear that Alpha Electronics could legally produce the devices for the NATO countries but not for the Eastern European company. The Cold War was in full swing at that time.

On the basis of local corporate citizenship, it was clear that Alpha Electronics had to consider the expected impact on local communities. In particular, there was no guarantee regarding to whom the Eastern European company would be selling the devices and how they would end up being used.

Sam took matters into his own hands without any foreknowledge of how his decision would be viewed by his company's upper management, board of directors, or fellow workers, many of whom were also company stockholders. Happily, Sam was never punished for his unilateral action of halting production. He recently retired from Alpha Electronics as a corporate-level vice president. He was especially gratified by the number of Alpha employees who were veterans of World War II, the Korean War, and the Vietnam War who thanked him for his action.

Sam strongly believed his action was the right thing to do, both for his company and for the public welfare. What ideas typically covered in an engineering ethics course might support that conviction?

CASE 15

Highway Safety Improvements[42]

David Weber, age 23, is a civil engineer in charge of safety improvements for District 7 (an eight-county area within a midwestern state). Near the end of the fiscal year, the district engineer informs David that delivery of a new snow plow has been delayed, and as a consequence the district has $50,000 in uncommitted funds. He asks David to suggest a safety project (or projects) that can be put under contract within the current fiscal year.

After a careful consideration of potential projects, David narrows his choice to two possible safety improvements. Site A is the intersection of Main and Oak Streets in the major city within the district. Site B is the intersection of Grape and Fir Roads in a rural area.

Pertinent data for the two intersections are as follows:

	Site A	Site B
Main road traffic (vehicles/day)	20,000	5,000
Minor road traffic (vehicles/day)	4,000	1,000
Fatalities per year (3-year average)	2	1
Injuries per year (3-year average)	6	2
PD* (3-year average)	40	12
Proposed improvement	New signals	New signals
Improvement cost	$50,000	$50,000

*PD refers to property damage-only accidents.

A highway engineering textbook includes a table of average reductions in accidents resulting from the installation of the types of signal improvements David proposes. The tables are based on studies of intersections in urban and rural areas throughout the United States during the past 20 years.

	Urban	Rural
Percent reduction in fatalities	50	50
Percent reduction in injuries	50	60
Percent reduction in PD	25	−25*

*Property damage-only accidents are expected to increase because of the increase in rear-end accidents due to the stopping of high-speed traffic in rural areas.

David recognizes that these reduction factors represent averages from intersections with a wide range of physical characteristics (number of approach lanes, angle of intersection, etc.), in all climates, with various mixes of trucks and passenger vehicles, various approach speeds, various driving habits, and so on. However, he has no special data about sites A and B that suggest relying on these tables is likely to misrepresent the circumstances at these sites.

Finally, here is additional information that David knows:

1. In 1975, the National Safety Council (NSC) and the National Highway Traffic Safety Administration (NHTSA) both published dollar scales for comparing accident outcomes, as shown below:

	NSC	NHTSA
Fatality	$52,000	$235,000
Injury	3,000	11,200
PD	440	500

A neighboring state uses the following weighting scheme:

Fatality	9.5 PD
Injury	3.5 PD

2. Individuals within the two groups pay roughly the same transportation taxes (licenses, gasoline taxes, etc.).

Which of the two site improvements do you think David should recommend? What is your rationale for this recommendation?

CASE 16

Hurricane Katrina

As we have noted in the text, until approximately 1970 nearly all engineering codes of ethics held that the engineer's first duty is fidelity to his or her employer and clients. However, soon after 1970, most codes insisted that "Engineers shall hold paramount the safety, health, and welfare of the public." Whatever may have precipitated this change in the early 1970s, recent events—ranging from the collapse of Manhattan's Twin Towers on September 11, 2001, to the collapse of a major bridge in Minneapolis/St. Paul on August 1, 2007—make apparent the vital importance of this principle. The devastation wreaked by Hurricane Katrina along the Gulf of Mexico coastline states of Louisiana, Mississippi, and Alabama in late August 2005 is also a dramatic case in point.

Hardest hit was Louisiana, which endured the loss of more than 1,000 lives, thousands of homes, damage to residential and nonresidential property of more than $20 billion, and damage to public infrastructure estimated at nearly $7 billion. Most severely damaged was the city of New Orleans, much of which had to be evacuated and which suffered the loss of more than 100,000 jobs. The city is still reeling, apparently having permanently lost much of its population and only slowly recovering previously habitable areas.

At the request of the U.S. Army Corp of Engineers (USACE), the ASCE formed the Hurricane Katrina External Review Panel to review the comprehensive work of USACE's Interagency Performance Evaluation Task Force. The resulting ASCE report, *The New Orleans Hurricane Protection System: What Went Wrong and Why,* is a detailed and eloquent statement of the ethical responsibilities of engineers to protect public safety, health, and welfare.[43]

The ASCE report documents engineering failures, organizational and policy failures, and lessons learned for the future. Chapter 7 of the report ("Direct Causes of the Catastrophe") begins as follows:[44]

> What is unique about the devastation that befell the New Orleans area from Hurricane Katrina—compared to other natural disasters—is that much of the destruction was the result of engineering and engineering-related policy failures.

From an engineering standpoint, the panel asserts, there was an underestimation of soil strength that rendered the levees more vulnerable than they should have been, a failure to satisfy standard factors of safety in the original designs of the levees and pumps, and a failure to determine and communicate clearly to the public the level of hurricane risk to which the city and its residents were exposed. The panel concludes,[45]

> With the benefit of hindsight, we now see that questionable engineering decisions and management choices, and inadequate interfaces within and between organizations, all contributed to the problem.

This might suggest that blame-responsibility is in order. However, the panel chose not to pursue this line, pointing out instead the difficulty of assigning blame:[46]

> No one person or decision is to blame. The engineering failures were complex, and involved numerous decisions by many people within many organizations over a long period of time.

Rather than attempt to assign blame, the panel used the hindsight it acquired to make recommendations about the future. The report identifies a set of critical actions the panel regards as necessary. These actions fall under one of four needed shifts in thought and approach:[47]

- Improve the understanding of risk and firmly commit to safety.
- Repair the hurricane protection system.
- Reorganize the management of the hurricane protection system.
- Insist on engineering quality.

The first recommended action is that safety be kept at the forefront of public priorities, preparing for the possibility of future hurricanes rather than allowing experts and citizens alike to fall into a complacency that can come from the relative unlikelihood of a repeat performance in the near future.

The second and third recommendations concern making clear and quantifiable risk estimates and communicating them to the public in ways that enable nonexperts to have a real voice in determining the acceptability or unacceptability of those risks.

The next set of recommendations concern replacing the haphazard, uncoordinated hurricane protection "system" with a truly organized, coherent system. This, the panel believes, calls for "good leadership, management, and someone in charge."[48] It is the panel's recommendation that a high-level licensed engineer, or a panel of highly qualified, licensed engineers, be appointed with full authority to oversee the system:[49]

> The authority's overarching responsibility will be to keep hurricane-related safety at the forefront of public priorities. The authority will provide leadership, strategic vision, definition of roles and responsibilities, formalized avenues of communication, prioritization of funding, and coordination of critical construction, maintenance, and operations.

The panel's seventh recommendation is to improve interagency coordination. The historical record thus far, the panel maintains, is disorganization and poor mechanisms for interagency communication:[50]

> Those responsible for maintenance of the hurricane protection system must collaborate with system designers and constructors to upgrade their inspection, repair, and operations to ensure that the system is hurricane-ready and flood-ready.

Recommendations 8 and 9 relate to the upgrading and review of design procedures. The panel points out

that "ASCE has a long-standing policy that recommends independent external peer review of public works projects where performance is critical to public safety, health, and welfare."[51] This is especially so where reliability under emergency conditions is critical, as it clearly was when Hurricane Katrina struck. The effective operation of such an external review process, the panel concludes, could have resulted in a significant reduction in the amount of (but by no means all) destruction in the case of Hurricane Katrina.

The panel's final recommendation is essentially a reminder of our limitations and a consequent ethical imperative to "place safety first":[52]

> Although the conditions leading up to the New Orleans catastrophe are unique, the fundamental constraints placed on engineers for any project are not. Every project has funding and/or schedule limitations. Every project must integrate into the natural and manmade environment. Every major project has political ramifications.
>
> In the face of pressure to save money or to make up time, engineers must remain strong and hold true to the requirements of the profession's canon of ethics, never compromising the safety of the public.

The panel concludes with an appeal to a broader application of the first Fundamental Canon of ASCE's Code of Ethics. Not only must the commitment to protect public safety, health, and welfare be the guiding principle for New Orleans' hurricane protection system but also "it must be applied with equal rigor to every aspect of an engineer's work—in New Orleans, in America, and throughout the world."[53]

Reading the panel's report in its entirety would be a valuable exercise in thinking through what ASCE's first Fundamental Canon requires not only regarding the Hurricane Katrina disaster but also regarding other basic responsibilities to the public that are inherent in engineering practice.

A related reading is "Leadership, Service Learning, and Executive Management in Engineering: The Rowan University Hurricane Katrina Recovery Team," by a team of engineering students and faculty advisors at Rowan University.[54] In their abstract, the authors identify three objectives for the Hurricane Katrina Recovery Team Project:

> The main objective is to help distressed communities in the Gulf Coast Region. Second, this project seeks to

not only address broader social issues but also leave a tangible contribution or impact in the area while asking the following questions: What do we as professional engineers have as a responsibility to the communities we serve, and what do we leave in the community to make it a better, more equitable place to live? The last objective is the management team's successful assessment of the experience, including several logistical challenges. To this end, this article seeks to help other student-led projects by relaying our service learning experience in a coherent, user-friendly manner that serves as a model experience.

CORPORATE RESPONSES

Supportive corporate responses to the Katrina hurricane were swift. By mid-September 2005, more than $312 million worth of aid had been donated by major corporations, much of it by those with no plants or businesses in the afflicted areas.[55] Engineers have played a prominent role in these relief efforts, as they did after the 9/11 Twin Towers attack and

the Asian tsunami disaster. Hafner and Deutsch comment,[56]

> With two disasters behind them, some companies are applying lessons they have learned to their hurricane-related philanthropy. GE is a case in point. During the tsunami, the company put together a team of 50 project engineers—experts in portable water purification, energy, health care, and medical equipment.
>
> After Hurricane Katrina, GE executives took their cues from Jeffrey R. Immelt, GE's chief executive, and reactivated the same tsunami team for New Orleans. "Jeff told us, 'Don't let anything stand in the way of getting aid where it's needed,'" said Robert Corcoran, vice president for corporate citizenship.

Discuss how, with corporate backing, engineers who subscribe to Fred Cuny's ideas about effective disaster relief in his *Disasters and Development* (Oxford University Press, 1983) might approach the engineering challenges of Katrina.

CASE 17

Hyatt Regency Walkway Disaster

Approximately 4 years after its occurrence, the tragic 1981 Kansas City Hyatt Regency walkway collapse was in the news again. A November 16, 1985, *New York Times* article reported the decision of Judge James B. Deutsch, an administrative law judge for Missouri's administrative hearing commission. Judge Deutsch found two of the hotels structural engineers guilty of gross negligence, misconduct, and unprofessional conduct.

The ASCE may have influenced this court ruling. Just before the decision was made, ASCE announced a policy of holding structural engineers responsible for structural safety in their designs. This policy reflected the recommendations of an ASCE committee that convened in 1983 to examine the disaster.

The project manager, Judge Deutsch is quoted as saying, displayed "a conscious indifference to his professional duties as the Hyatt project engineer who was primarily responsible for the preparation of design drawings and review of shop drawings for that project." The judge also cited the chief engineer's failure

to closely monitor the project manager's work as "a conscious indifference to his professional duties as an engineer of record."

This court case shows that engineers can be held responsible not only for their own conduct but also for the conduct of others under their supervision. It also holds that engineers have special *professional* responsibilities.

Discuss the extent to which you think engineering societies should play the sort of role ASCE apparently did in this case. To what extent do you think practicing engineers should support (e.g., by becoming members) professional engineering societies' attempts to articulate and interpret the ethical responsibilities of engineers?

The Truesteel Affair is a fictionalized version of circumstances similar to those surrounding the Hyatt Regency walkway collapse. View this video and discuss the ethical issues it raises. (This film is available from Fanlight Productions, 47 Halifax St., Boston, MA 02130. 1-617-524-0980.)

CASE 18

Hydrolevel[57]

"A conflict of interest is like dirt in a sensitive gauge," one that can not only soil one person's career but also taint an entire profession.[58] Thus, as professionals, engineers must be ever alert to signs of conflict of interest. The case of the *American Society of Mechanical Engineers (ASME) v. Hydrolevel Corporation* shows how easily individuals, companies, and professional societies can find themselves embroiled in expensive legal battles that tarnish the reputation of the engineering profession as a whole.

In 1971, Eugene Mitchell, vice president for sales at McDonnell and Miller, Inc., located in Chicago, was concerned about his company's continued dominance in the market for heating boiler low-water fuel cutoff valves that ensure that boilers cannot be fired without sufficient water in them because deficient water could cause an explosion.

Hydrolevel Corporation entered the low-water cutoff valve market with an electronic low-water fuel supply cutoff that included a time delay on some of its models. Hydrolevel's valve had won important approval for use from Brooklyn Gas Company, one of the largest installers of heating boilers. Some Hydrolevel units added the time-delay devices so the normal turbulence of the water level at the electronic probe would not cause inappropriate and repeated fuel supply turn-on and turn-off. Mitchell believed that McDonnell and Miller's sales could be protected if he could secure an interpretation stating that the Hydrolevel time delay on the cutoff violated the ASME B-PV code. He referred to this section of the ASME code: "Each automatically fired steam or vapor system boiler shall have an automatic low-water fuel cutoff, so located as to automatically cut off the fuel supply when the surface of the water falls to the lowest visible part of the water-gauge glass."[59] Thus, Mitchell asked for an ASME interpretation of the mechanism for operation of the Hydrolevel device as it pertained to the previously mentioned section of the code. He did not, however, specifically mention the Hydrolevel device in his request.

Mitchell discussed his idea several times with John James, McDonnell and Miller's vice president for research. In addition to his role at McDonnell

and Miller, James was on the ASME subcommittee responsible for heating boilers and had played a leading role in writing the part of the boiler code that Mitchell was asking about.

James recommended that he and Mitchell approach the chairman of the ASME Heating Boiler Subcommittee, T. R. Hardin. Hardin was also vice president of the Hartford Steam Boiler Inspection and Insurance Company. When Hardin arrived in Chicago in early April on other business, the three men went to dinner at the Drake Hotel. During dinner, Hardin agreed with Mitchell and James that their interpretation of the code was correct.

Soon after the meeting with Hardin, James sent ASME a draft letter of inquiry and sent Hardin a copy. Hardin made some suggestions, and James incorporated Hardin's suggestions in a final draft letter. James's finalized draft letter of inquiry was then addressed to W. Bradford Hoyt, secretary of the B-PV Boiler and Pressure Vessel Committee.

Hoyt received thousands of similar inquiries every year. Since Hoyt could not answer James's inquiry with a routine, prefabricated response, he directed the letter to the appropriate subcommittee chairman, T. R. Hardin. Hardin drafted a response without consulting the whole subcommittee, a task he had authorization for if the response was treated as an "unofficial communication."

Hardin's response, dated April 29, 1971, stated that a low-water fuel cutoff must operate immediately. Although this response did not say that Hydrolevel's time-delayed cutoff was dangerous, McDonnell and Miller's salesmen used Hardin's conclusion to argue against using the Hydrolevel product. This was done at Mitchell's direction.

In early 1972, Hydrolevel learned of the ASME letter through one of its former customers who had a copy of the letter. Hydrolevel then requested an official copy of the letter from ASME. On March 23, 1972, Hydrolevel requested an ASME review and ruling correction.

ASME's Heating and Boiler Subcommittee had a full meeting to discuss Hydrolevel's request, and it confirmed part of the original Hardin interpretation.

James, who had replaced Hardin as chairman of the subcommittee, refrained from participating in the discussion but subsequently helped draft a critical part of the subcommittee's response to Hydrolevel. The ASME response was dated June 9, 1972.

In 1975, Hydrolevel filed suit against McDonnell and Miller, Inc., ASME, and the Hartford Steam Boiler Inspection and Insurance Company, charging them with conspiracy to restrain trade under the Sherman Antitrust Act.

Hydrolevel reached an out-of-court settlement with McDonnell and Miller and Hartford for $750,000 and $75,000, respectively. ASME took the case to trial. ASME officials believed that, as a society, ASME had done nothing wrong and should not be liable for the misguided actions of individual volunteer members acting on their own behalf. After all, ASME gained nothing from such practices. ASME officials also believed that a pretrial settlement would set a dangerous precedent that would encourage other nuisance suits.

Despite ASME arguments, however, the jury decided against ASME, awarding Hydrolevel $3.3 million in damages. The trial judge deducted $800,000 in prior settlements and tripled the remainder in accordance with the Clayton Act. This resulted in a decision of $7,500,000 for Hydrolevel.

On May 17, 1982, ASME's liability was upheld by the second circuit. The Supreme Court, in a controversial 6-3 vote, found ASME guilty of antitrust violations. The majority opinion, delivered by Justice Blackmun, read as follows:

> ASME wields great power in the nation's economy. Its codes and standards influence the policies of numerous states and cities, and has been said about "so-called voluntary standards" generally, its interpretation of guidelines "may result in economic prosperity or economic failure, for a number of businesses of all sizes throughout the country," as well as entire segments of an industry....ASME can be said to be "in reality an extragovernmental agency, which prescribes rules for the regulation and restraint of interstate commerce." When it cloaks its subcommittee officials with the authority of is reputation, ASME permits those agents to affect the destinies of businesses and thus gives them power to frustrate competition in the marketplace.[60]

The issue of damages was retried in a trial lasting approximately 1 month. In June, the jury returned a verdict of $1.1 million, which was tripled to $3.3 million. Parties involved were claiming attorney's fees in excess of $4 million, and a final settlement of $4,750,000 was decreed.

Following the decision, ASME revised it's procedures as follows:

> In the wake of the Hydrolevel ruling, the Society has changed the way it handles codes and standards interpretations, beefed up its enforcement and conflict-of-interest rules, and adopted new "sunset" review procedures for its working bodies.
>
> The most striking changes affect the Society's handling of codes and standards interpretations. All such interpretations must now be reviewed by at least five persons before release; before, the review of two people was necessary. Interpretations are available to the public, with replies to nonstandard inquiries published each month in the Codes and Standards section of ME or other ASME publications. Previously, such responses were kept between the inquirer and the involved committee or subcommittee. Lastly, ASME incorporates printed disclaimers on the letterhead used for code interpretations spelling out their limitations: that they are subject to change should additional information become available and that individuals have the right to appeal interpretations they consider unfair.
>
> Regarding conflict-of-interest, ASME now requires all staff and volunteer committee members to sign statements pledging their adherence to a comprehensive and well-defined set of guidelines regarding potential conflicts. Additionally, the Society now provides all staff and volunteers with copies of the engineering code of ethics along with a publication outlining the legal implications of standards activities.
>
> Finally, the Society now requires each of its councils, committees, and subcommittees to conduct a "sunset" review of their operations every 2 years. The criteria include whether their activities have served the public interest and whether they have acted cost-effectively, in accordance with Society procedures.[61]

Conflict-of-interest cases quickly become complicated, as the following questions illustrate:

- How could McDonnell and Miller have avoided the appearance of a conflict of interest? This applies to both Mitchell and James.
- What was T. R. Hardin's responsibility as chair of the B-PV Code Heating Boiler Subcommittee? How could he have handled things differently to protect the interests of ASME?

- What can engineering societies do to protect their interests once a conflict of interest is revealed?
- Was the final judgment against ASME fair? Why or why not?

- Have ASME's revised conflict-of-interest procedures addressed the problems fully? Why or why not?

CASE 19

Incident at Morales

Incident at Morales is a multistage video case study developed by the National Institute for Engineering Ethics (NIEE). It involves a variety of ethical issues faced by the consulting engineer of a company that is in a hurry to build a plant so that it can develop a new chemical product that it hopes will give it an edge on the competition. Issues include environmental, financial, and safety problems in an international

setting. Interspersed between episodes are commentaries by several engineers and ethicists involved in the production of the video. Information about ordering the video is available from the NIEE or the Murdough Center for Engineering Ethics (www.niee.org/pd.cfm?pt=Murdough). The full transcript of the video and a complete study guide are available online from the Murdough Center.

CASE 20

Innocent Comment?

Jack Strong is seated between Tom Evans and Judy Hanson at a dinner meeting of a local industrial engineering society. Jack and Judy have an extended discussion of a variety of concerns, many of which are related to their common engineering interests. At the conclusion of the dinner, Jack turns to Tom, smiles, and says, "I'm sorry not to have talked with you more tonight, Tom, but Judy's better looking than you."

Judy is taken aback by Jack's comment. A recent graduate from a school in which more than 20 percent of her classmates were women, she had been led to believe that finally the stereotypical view that women are not as well suited for engineering as men was finally going away. However, her first job has

raised some doubts about this. She was hired into a division in which she is the only woman engineer. Now, even after nearly 1 year on the job, she has to struggle to get others to take her ideas seriously. She wants to be recognized first and foremost as a good engineer. So, she had enjoyed "talking shop" with Jack. But she was stunned by his remark to Tom, however innocently it might have been intended. Suddenly, she saw the conversation in a very different light. Once again, she sensed that she was not being taken seriously enough as an engineer.

How should Judy respond to Jack's remark? Should she say anything? Assuming Tom understands her perspective, what, if anything, should he say or do?

CASE 21

Late Confession

In 1968, Norm Lewis was a 51-year-old doctoral candidate in history at the University of Washington.[62] While taking his final exam in the program, he excused himself to go to the bathroom, where he

looked at his notes. For the next 32 years, Lewis told no one. At age 83, he decided to confess, and he wrote to the president of the university admitting that he had cheated and that he had regretted it ever since.

Commenting on the case, Jeanne Wilson, president of the Center for Academic Integrity remarked, "I think there is an important lesson here for students about the costs of cheating. He has felt guilty all these years, and has felt burdened by this secret, believing that he never really earned the degree he was awarded." Wilson's position is that the University of Washington should not take action against Lewis, given his confession, his age, and the fact that, after all, he did complete his coursework and a dissertation.

But, she added, "On the other hand, I think an institution might feel compelled to revoke the degree if we were talking about a medical or law degree or license, or some other professional field such as engineering or education, and the individual were younger and still employed on the basis of that degree or license."

Discuss the ethical issues this case raises, both for Dr. Lewis and for University of Washington officials. Evaluate Jeanne Wilson's analysis, especially as it might apply to engineers.

CASE 22

Love Canal[63]

INTRODUCTION

Degradation of the environment resulting from human activity is certainly not a phenomenon of recent origin. As early as the 15th century, long before the beginning of the industrial revolution, London was already being plagued by noxious air pollution resulting from the burning of coal and wood. However, the extent of the effect of environmental pollution was greatly increased following the end of World War II by the exponential expansion of industrial activity in developed nations, employing vast quantities of fossil fuels and synthetic chemicals. Today's environmental concerns are regional, national, and global, as well as local.

The ongoing educational, social, and political movement, which has raised the consciousness of people in the United States and throughout the world about environmental concerns, began in the early 1960s. Its initiation is often attributed to the popular response to *Silent Spring,* the eloquent book by marine biologist Rachel Carson about the dire effects of the overuse of pesticides and other chemical poisons, which was published in 1962. The ensuing environmental movement has spawned numerous local, regional, national, and international organizations— many rather militant—that have used numerous tactics to press their demands for the preservation of clean air, pure water, and unspoiled land. In response to these demands, legislative bodies have enacted all manner of regulations and numerous agencies have been charged with the task of environmental protection.

This increase in environmental activity has been accompanied by much controversy. Entrepreneurs, property owners, industrial workers, politicians, scientists, and people in all other walks of life differ with regard to the relative value they accord to the benefits and costs associated with restrictions on freedom of action designed to protect the environment. A wide variety of ethics and values issues arise in the course of attempting to balance such demands as property rights and the entrepreneurial freedom to pursue profits against the ecological need to curtail those rights and restrict that freedom.

One of the most contentious environmental issues has been how to respond to the discovery of many thousands of hazardous toxic dumps that have resulted from decades of virtually unrestricted disposal of toxic industrial waste. This issue was first widely publicized as a result of the health emergency declared by the New York State Department of Health in 1978 in response to shocking revelations about the problems caused by improper waste disposal in the now infamous Love Canal dump site. The actions and reactions of the corporation that disposed of the waste in question, public officials, residents, the media, and scientists involved in the Love Canal controversy serve as excellent illustrations of many of the ethics issues associated with efforts to protect the public from environmental pollution.

BACKGROUND

During the late 19th century, numerous canals were built by entrepreneurs to unify waterways into efficient shipping systems. One such canal was begun in 1894 by venture capitalist William Love in the Niagara Falls area of New York State. Within a few years, an economic depression undermined Love's financial plans and the partially completed project was abandoned.

Dubbed "Love Canal" by the local residents, it was used as a swimming hole and an ice rink. In 1942, faced with the need for a place to dispose of toxic waste from the manufacture of chlorinated hydrocarbons and caustics, the Hooker Electrochemical Corporation (currently Hooker Chemical and Plastics, a subsidiary of Occidental Petroleum Corporation) leased the canal as a waste dump. In 1947, Hooker bought the canal and the surrounding land. Between 1942 and 1950, more than 21,000 tons of chemicals, including such potent toxins as benzene, the pesticide lindane, polychlorinated dioxins, PCBs, and phosphorous, were deposited in the canal, which Hooker had lined with cement. Having exhausted the canal's potential as a waste dump, Hooker then installed an impermeable cap that was supposed to prevent water from entering and promoting seepage of the toxins, and the former canal disappeared from view beneath a layer of fill.

In the early 1950s, the local school board was confronted with the need to build a new school to accommodate an increasing population of children. The board knew that Hooker was anxious to get rid of the Love Canal property and began making inquiries. Hooker has claimed that it resisted and warned the board of education that the buried chemicals made the site inappropriate for school construction. The property sale was consummated for $1.00 in 1953, but the company asserts that it gave in because the board would otherwise have taken the land by eminent domain. Whether Hooker was as reluctant as it says it was and as assertive in cautioning the board about the hazards is impossible to determine. Existing minutes of the meetings in question do not fully support Hooker's version of the proceedings, and none of the board members are still alive. What is clear is that the deed that was negotiated contains a clause exempting Hooker from any "claim, suit, or action" due to future human exposure to the buried chemicals.

An elementary school was built in the middle of the property and the surrounding land was sold by the school board to developers who built 98 homes along the former canal banks and approximately 1,000 additional houses in the Love Canal neighborhood. The construction of the school, houses, and associated utilities resulted in the breaching of parts of the canal's cap and its cement walls.

THE CASE

The first known case of exposure to the buried toxins occurred in 1958 when three children suffered chemical burns from waste that had resurfaced at the former canal site. Both Hooker Chemical and city officials were officially informed, but neither the Niagara Falls Health Department nor any other public agency took any action in response to that event or to numerous other complaints during the next 20 years. Hooker's records reveal that it investigated the initial incident and several other reports and quickly became convinced that the very large reservoir of toxins was not likely to be contained. Hooker did nothing to convey this knowledge to the Love Canal homeowners, who had never been informed about the nature of the potential hazard. In testimony two decades later, Hooker acknowledged that its failure to issue a warning was due to concern that this might be interpreted as liability for possible harm despite the clause in their property sales deed.

By 1978, occupants of the homes in the area had begun to organize what was to become the Love Canal Homeowners Association (LCHA), under the highly competent and aggressive leadership of Lois Gibbs. Investigative newspaper reporter Michael Brown helped publicize the plight of the many deeply concerned local residents who had encountered evidence of toxins resurfacing in or around their property. Chemicals had been observed in the form of viscous fluids seeping into both yards and basements, pervasive odors in homes, and a stench emanating from storm sewer openings.

Love Canal soon became the first hazardous waste site to be featured in TV news reports and to get front-page headline billing in newspapers and magazines in New York State and nationally. Embarrassed by the past failure of officials to respond to the clear indications of a serious problem, both the New York State Department of Health (NYSDH) and the EPA quickly became involved. Tests soon revealed a wide variety of noxious chemicals in the air in Love Canal homes and an excess frequency of miscarriages among women living in homes adjacent to the former canal site. A public health emergency was declared on August 2, 1978, by the New York State Commissioner of Health. A few days later, Governor Hugh Carey announced that New York State would purchase the 239 homes nearest to the canal and assist the displaced families in relocating. These abandoned homes were fenced in and work was soon begun on a plan to construct an elaborate drainage system including trenches, wells, and pumping stations to prevent further outward migration of the toxins.

These initial actions, which quickly followed the emergence of Love Canal as a national "cause célèbre," ultimately cost the state and federal governments in excess of $42 million. Public officials quickly recognized that a continued preemptive response to potential health problems at Love Canal was likely to exceed available emergency funds in the state's coffers. Furthermore, it was known that thousands of other toxic waste sites existed throughout the country that might pose similar threats to numerous other communities. Thus, it is not surprising that the concerns and demands of the owners of the 850 homes outside the inner evacuated circle were not to be satisfied by either state or federal officials in a similar manner.

The NYSDH did conduct a survey study of the residents in the remaining homes, which led to an announcement in early fall that the rest of the neighborhood was safe, posing no increased health risk. As subsequently revealed, this assurance had been based on only one health issue examined by the survey. The department had concluded that the miscarriage rate in the homes beyond the fence did not exceed normal rates—a conclusion based on a methodology that was subsequently seriously questioned. The many other possible health effects of chemical exposure had not entered into the NYSDH evaluation.

Citing the fact that chemical seepage was evident beyond the evacuated area and that families living there appeared to be experiencing unusual health problems, members of the LCHA rejected the department's assurances. They demanded more definitive studies, and when they did not get a satisfactory response from either the NYSDH or the EPA, they sought scientific aid from outside the government's environmental health establishment.

Beverly Paigen, a cancer research scientist who worked for the NYSHD Roswell Park Memorial Institute in nearby Buffalo, agreed to volunteer her services in an unofficial capacity. Her professional interests

included the variation among individuals in their responses to chemical toxins and she anticipated that in addition to helping the Love Canal residents, her involvement might also result in identifying appropriate subjects for her research work. Dr. Paigen designed a survey aimed at investigating several potential effects of exposure to chemicals. She used a different set of assumptions about the mechanism and likely path of the flow of the dissolved toxins that seeped out of the canal. Based on her model, Dr. Paigen found that miscarriages were significantly higher among women living in homes most likely to be in the path of the chemical plume. She also found much higher than normal rates of birth defects and evidence of serious nervous system toxicity as well as elevated incidences of asthma and urological problems for residents of these homes.

In early November 1978, Dr. Paigen presented the results of her "unofficial" research to her NYSDH superiors. After a delay of 3 months, the new New York State Commissioner of Health publicly announced that after reevaluating its own data it also found excess miscarriages and birth defects in homes in previously "wet" regions of the Love Canal neighborhood and promised additional studies of Dr. Paigen's other findings. However, the action taken in response to these results puzzled and dismayed both the residents and Dr. Paigen. Families with children younger than 2 years of age or with women who could prove they were pregnant were to be relocated at state expense but only until the youngest child reached the age of 2 years. Women who were trying to become pregnant, or those who thought they were in the early stages of pregnancy when the fetus is most sensitive to toxins but who could not yet prove they were pregnant with tests available at that time, were denied permission to join the group that was evacuated.

During the next 1½ years, the frustration and the militancy of the LCHA members increased as the additional studies promised by the commissioner failed to materialize. On the federal-level EPA lawyers had become convinced by media reports and public appeals from Love Canal residents claiming a variety of toxin-related illnesses that hundreds of additional families should be moved away. They sought a court order from the Department of Justice requiring Hooker Chemical to pay for the relocations. When the Justice Department responded by demanding evidence that the inhabitants who remained in the Love Canal neighborhood were at risk, the EPA commissioned a quick "pilot" study to determine whether residents had suffered chromosome damage that could be attributed to chemical exposure. This study, which was to subsequently receive much criticism from the scientific community both because of its specific design and because, at the time, chromosome studies were notoriously difficult to interpret, did provide the type of evidence the EPA was seeking. On the basis of finding "rare chromosomal aberrations" in 11 of 36 subjects tested, the scientist who performed the study concluded that inhabitants of the area were at increased risk for a variety of adverse health outcomes.

On May 19, 1980, when two EPA representatives went to the LCHA office in one of the evacuated homes to announce the results of the chromosome study, they were greeted by irate homeowners who proceeded to lock them in the office for 5 hours until FBI agents arrived and demanded their release. This tactic, which received the anticipated media coverage, had the desired effect. With the intervention of high-ranking officials in the Executive Branch, and undoubtedly with the support of President Carter, funds were made available for the relocation of several hundred additional Love Canal families.

A conclusion that can clearly be drawn from this and many subsequent environmental controversies is that politics, public pressure, and economic considerations all take precedence over scientific evidence in determining the outcome. Another aspect of the Love Canal case that is characteristic of such events is that the victims, although hostile to Hooker Chemical, directed most of their rage at an indecisive, aloof, often secretive and inconsistent public health establishment.

Lawsuits against Occidental Petroleum Corporation, which bought Hooker Chemical in 1968, were initiated by both the state of New York and the U.S. Justice Department to cover costs of the cleanup and the relocation programs and also by more than 2,000 people who claimed to have been personally injured by the buried chemicals. In 1994, Occidental agreed to pay $94 million to New York in an out-of-court settlement, and the following year the federal case was settled for $129 million. Individual victims have thus far won in excess of $20 million from the corporation.

In early 1994, it was announced that the cleanup of the condemned homes in Love Canal had been completed and it was safe to move back to the area. The real estate company offering the inexpensive refurbished homes for sale had chosen to rename the area "Sunrise City."

READINGS AND RESOURCES

A wealth of written and audiovisual material is available on Love Canal and other environmental controversies. Searching the electronic catalogue of any public or academic library or using an Internet search engine should prove very fruitful.

For a colorful discussion of the early events in the Love Canal case by the investigative reporter who initiated the media coverage of the issue, and for a personal version of the events by the woman who organized the LCHA and went on to become a national leader of citizen's toxic waste organizing, see

Michael Brown, *Laying Waste* (New York: Pantheon, 1979).

Lois Gibbs, *Love Canal: My Story,* as told to Murray Levine (Albany: State University of New York Press, 1981).

For a thought-provoking article that focuses on the political and ethical dimensions of the case by the scientist who volunteered her services to the Love Canal residents, see

Beverly Paigen, "Controversy at Love Canal," *The Hastings Center Report,* June 1982, pp. 29–37.

For a report written by the public health, transportation, and environmental agencies of New York State, see

New York State Department of Health, Office of Public Health, "Love Canal, a Special Report to the Governor and Legislature," with assistance of New York State Department of Transportation and New York State Department of Environmental Conservation (Albany, NY: New York State Department of Health, Office of Public Health, 1981).

For two additional perspectives on the controversy, see

Adeline Levine, *Love Canal: Science, Politics
and People* (Lexington, MA: Lexington
Books, 1982).

L. Gardner Shaw, *Citizen Participation in Gov-
ernment Decision Making: The Toxic Waste
Threat at Love Canal, Niagara Falls, New
York* (Albany: State University of New York,
Nelson A. Rockefeller Institute of Govern-
ment, 1983).

For articles published in science news journals, see

Barbara J. Culliton, "Continuing Confusion over
Love canal," *Science,* 209, August 19,
1980, pp. 1002–1003.

"Uncertain Science Pushes Love Canal Solutions
to Political, Legal Arenas," *Chemical &
Engineering News,* August 11, 1980,
pp. 22–29.

For comments on the plan to rehabilitate, rename,
and repopulate the Love Canal neighborhood, see

Rachel's Hazardous Waste News, 133, June 13,
1989.

For a highly informative collection of essays,
comments, and analysis on a wide variety of issues
in environmental ethics, see

D. Van Deveer and C. Pierce, *Environmental
Ethics and Policy Book* (Belmont, CA:
Wadsworth, 1994).

THE ISSUES

The following are significant questions of ethics and
values raised by this case:

• Beverly Paigen, the research scientist who volunteered her services to the Love Canal residents, commented in reference to her differences with her
superiors in the NYSDH, "I thought our differences
could be resolved in the traditional scientific manner
by examining protocols, experimental design, and statistical analysis. But I was to learn that actual facts
made little difference in resolving our disagreements—
the Love Canal controversy was predominantly
political in nature, and it raised a series of questions

that had more to do with values than science." Consider the differences in the values that might be of
greatest importance to a Love Canal resident, the
New York State Commissioner of Health, a scientist
doing research sanctioned by either the New York
State Department of Environmental Conservation or
the EPA, an independent scientist (like Dr. Paigen)
who was doing volunteer research for the residents,
and a typical citizen of the state of New York. In
what respects might these value differences lead
them to conflicting decisions about what should
have been done in response to the Love Canal disaster
and how to do it?

• Is it reasonable to demand that the ethical duty of
public officials is to respond to an environmental problem by objectively examining the scientific facts and
the potential hazards to local residents, independent
of economic and political considerations?

• One of the charges raised against the NYSDH and
the health commissioner was that the public health
establishment would not divulge the details of the
studies that led to its decisions, held many closed
meetings, and even refused to reveal the names of
members who served on consultation panels it established. Do you think that there might be an ethical justification for such public agencies to refuse public
access to such information? If so, does this seem to
apply to the Love Canal situation?

• Another accusation was that state employees
sympathetic to the Love Canal residents were har-
assed and punished. For example: Dr. Paigen's ability
to raise funds for her research work was curtailed by
the Roswell Park Memorial Institute, causing the professional staff to charge the administration with scientific censorship; her mail arrived opened and taped
shut; her office was searched; and when she was subjected to a state income tax audit, she discovered
newspaper clippings about her Love Canal activities
in the auditor's file. In addition, when William Friedman, who had been the Department of Environmental
Conservation's regional director, pressed state officials to take a less conservative approach to protecting the health of Love Canal residents, he was
promptly demoted to staff engineer. This type of reaction by the political power structure seems morally in-
defensible, but it is by no means unique to the Love
Canal case.

• Another values issue is the extent of evidence needed to justify action to protect public health. In order for the scientific community to accept as fact research showing that a specific health effect is caused by a particular agent, the statistical analysis of the data must indicate with more than 95 percent certainty that the observed effect could not occur by chance. This high but clearly arbitrary standard has been adopted to protect the integrity of the body of accepted scientific facts. But should public health officials demand, as they often do, the same standard before taking action? For example, if evidence shows that there is an 80 percent chance that exposure to some chemical in the environment may cause a serious adverse health effect, should health officials refuse to inform the public of the risk or take action to prevent exposure until further studies—which may take months or even years—raise the certainty of the causal relationship to 95 percent?

• It is common in environmental controversies for those who believe they are at risk to become distrustful of public officials in charge of investigating their concerns. This was certainly the case in the Love Canal controversy. It is unusual for a citizens group to be able to obtain the volunteer services of an independent expert with qualifications like those of Dr. Paigen and they are not likely to have the financial resources necessary to hire their own consultant. Furthermore, although Dr. Paigen was able to provide valuable scientific services, she was unable to gain access to and assess much of the evidence that the public officials used as the basis for their decisions. Dr. Paigen and others have suggested that the ethical solution to this problem is to provide public funds to groups such as the LCHA with which they can hire their own experts and which they can use to hire a qualified advocate who will be given access to all public data and a voice in the decision-making process.

• The Hooker Chemical Company did not violate any then-existing specific environmental regulations by disposing of toxic waste in Love Canal or by selling the land to the school board. However, the courts have found Hooker financially liable for the harm that was the ultimate result of their disposal practices. This decision was largely based on the judgment that Hooker possessed the scientific expertise to be able to anticipate that dumping waste chemicals in the canal was likely to result in a public health threat. It was also argued that Hooker acted irresponsibly by not informing the public of the risks it discovered in 1958. Should corporations be required to use their knowledge to avoid activities that may cause public harm?

• In recent years, the issues of environmental justice and equity have been raised within the environmental movement. Minority populations, and poor people in general, have produced persuasive data showing that they are far more likely to be exposed to environmental pollution from factories or waste disposal facilities than more affluent white people. In the Love Canal case, the initial neighborhood population was neither poor nor did it have a high percentage of minority members. Of course, those who chose to live there were not aware of the pollution risk. It is likely, however, that the inexpensive houses now being offered to induce people to move back into the area after remediation is supposed to have made it safe will attract primarily the poor. One proposal that has been put forth in response to demand for environmental justice is to provide some form of reward to those who live in neighborhoods where exposure to environmental toxins is significantly higher than average. Would this be an ethical practice? What other steps might be taken to promote environmental equity in an ethical manner?

• In our society, environmental risks are generally evaluated in economic terms. However, the assignment of economic value to human health, a pristine forest, or a smog-free vista is surely not an objective exercise. What other means might be used to evaluate environmental risks and benefits?

• We generally assign value to things in anthropogenic terms. We consider how humans will be affected by an activity that will cause pollution or degrade an ecosystem. Some environmental ethicists have proposed that we should adopt a biocentric perspective in which living things and natural objects are assigned intrinsic value independent of human concerns. How do you respond to the assertion that nature does not exist solely for the purpose of being exploited by humans?

Although there is no explicit mention of engineers in this case study, it is not difficult to imagine that engineers, too, were involved in the events resulting in the creation of the Love Canal hazard, as well as in the cleanup. Discuss the types of responsibilities that engineers have in regard to the prevention of hazards such as this from occurring in the future. What, if any, public roles might they play in helping the public understand what is at stake and how the issues should be addressed?

CASE 23

Member Support by IEEE

In the mid-1970s, the New York City Police Department operated an online computerized police car dispatching system called SPRINT. Upon receiving a telephoned request for police assistance, a dispatcher would enter an address into a computer and the computer would respond within seconds by displaying the location of the nearest patrol car. By reducing the response time for emergency calls, the SPRINT system probably saved lives.

In 1977, another system, PROMIS, was being considered by New York City prosecutors using the same host computer as that for SPRINT. The PROMIS system would provide names and addresses of witnesses, hearing dates, the probation statuses of defendants, and other information that would assist prosecutors or arresting officers who wanted to check the current status of apprehended perpetrators. This project was being managed by the Criminal Justice Coordinating Council, or Circle Project, a committee of high-level city officials that included the deputy mayor for criminal justice, the police commissioner, and Manhattan District Attorney Robert Morgenthau as chairman.

The committee employed a computer specialist as project director, who in turn hired Virginia Edgerton, an experienced system analyst, as senior information scientist to work under his supervision. Soon after being employed, Edgerton expressed concern to the project director about the possible effect on SPRINT's response time from loading the computer with an additional task, but he instructed her to drop the

matter. Edgerton then sought advice from her professional society, the Institute of Electrical and Electronics Engineers (IEEE).

After an electrical engineering professor at Columbia University agreed that her concerns merited further study, she sent a memorandum to the project director requesting a study of the overload problem. He rejected the memorandum out of hand, and Edgerton soon thereafter sent copies of the memorandum with a cover letter to the members of the Circle Project's committee. Immediately following this, Edgerton was discharged by the project director on the grounds that she had, by communicating directly with the committee members, violated his orders. He also stated that the issues she had raised were already under continuing discussion with the police department's computer staff, although he gave no documentation to support this claim.

The case was then investigated by the Working Group on Ethics and Employment Practices of the Committee on the Social Implications of Technology (CSIT) of the IEEE, and subsequently by the newly formed IEEE Member Conduct Committee. Both groups agreed that Edgerton's actions were fully justified. In 1979, she received the second IEEE–CSIT Award for Outstanding Service in the Public Interest. After her discharge, Edgerton formed a small company selling data-processing services.[64]

Discuss the supporting role played by IEEE in this case. Does this provide electrical and electronic engineers an ethical basis for joining or supporting IEEE?

CASE 24

Moral Development[65]

The introduction of ethics into engineering education raises an important question about moral education: Shouldn't a student's introduction to ethics occur much earlier than the college level? The answer to this question is "yes, it should," and in fact, whether formally or informally, it does—in the home, in religious upbringing, on the playground, and in the schools. However, as children move into adulthood, their moral background needs to be adapted to new and more complex settings, such as the engineering workplace. This means that young engineers still have much to learn about ethics. Still, the importance of one's moral upbringing for addressing the ethical challenges facing professionals should not be underestimated.

Children's introduction to ethics, or morality, occurs rather early. They argue with siblings and playmates about what is fair or unfair. The praise and blame they receive from parents, teachers, and others encourage them to believe that they are capable of some degree of responsible behavior. They are both recipients and dispensers of resentment, indignation, and other morally reactive attitudes. There is also strong evidence that children, even as young as age 4 years, seem to have an intuitive understanding of the difference between what is merely conventional (e.g., wearing certain clothes to school) and what is morally important (e.g., not throwing paint in another child's face).[66] Therefore, despite their limited experience, children typically have a fair degree of moral sophistication by the time they enter school.

What comes next is a gradual enlargement and refinement of basic moral concepts—a process that, nevertheless, preserves many of the central features of those concepts. All of us can probably recall examples from our childhood of clear instances of fairness, unfairness, honesty, dishonesty, courage, and cowardice that have retained their grip on us as paradigms, or clear-cut illustrations, of basic moral ideas. Philosopher Gareth Matthews states,[67]

> A young child is able to latch onto the moral kind, bravery, or lying, by grasping central paradigms of that kind, paradigms that even the most mature and sophisticated moral agents still count as paradigmatic. Moral

development is...enlarging the stock of paradigms for each moral kind; developing better and better definitions of whatever it is these paradigms exemplify; appreciating better the relation between straightforward instances of the kind and close relatives; and learning to adjudicate competing claims from different moral kinds (classically the sometimes competing claims of justice and compassion, but many other conflicts are possible). This makes it clear that, although a child's moral start may be early and impressive, there is much conflict and confusion that needs to be sorted through. It means that there is a continual need for moral reflection, and this does not stop with adulthood, which merely adds new dimensions.

Nevertheless, some may think that morality is more a matter of subjective feelings than careful reflection. However, research by developmental psychologists such as Jean Piaget, Lawrence Kohlberg, Carol Gilligan, James Rest, and many others provides strong evidence that important as feelings are, moral reasoning is a fundamental part of morality as well.[68] Piaget and Kohlberg, in particular, performed pioneering work showing that there are significant parallels between the cognitive development of children and their moral development. Many of the details of their accounts have been hotly disputed, but a salient feature that survives is that moral *judgment* involves more than just feelings. Moral judgments (e.g., "Smith acted wrongly in fabricating the lab data") are amenable to being either supported or criticized by *good reasons*.

Kohlberg's account of moral development has attracted a very large following among educators, as well as an increasing number of critics. He characterizes development in terms of an invariable sequence of six stages.[69] The first two stages are highly self-interested and self-centered. Stage 1 is dominated by the fear of punishment and the promise of reward. Stage 2 is based on reciprocal agreements ("You scratch my back, and I'll scratch yours"). The next two stages are what Kohlberg calls conventional morality. Stage 3 rests on the approval and disapproval of friends and peers. Stage 4 appeals to "law and order" as necessary for social cohesion and order. Only the last two stages embrace what Kohlberg calls critical, or postconventional, morality. In these two stages, one acts on self-chosen principles that can be used to evaluate the appropriateness of responses in the first four stages. Kohlberg has been criticized for holding that moral development proceeds in a rigidly sequential manner (no stage can be skipped, and there is no regression to earlier stages); for assuming that later stages are more adequate morally than earlier ones; for being male biased in overemphasizing the separateness of individuals, justice, rights, duties, and abstract principles at the expense of equally important notions of interdependence, care, and responsibility; for claiming that moral development follows basically the same patterns in all societies; for underestimating the moral abilities of younger children; and for underestimating the extent to which adults employ critical moral reasoning. We do not attempt to address these issues here.[70] Nevertheless, whatever its limitations, Kohlberg's theory makes some important contributions to our understanding of moral education. By describing many common types of moral reasoning, it invites us to be more reflective about how we and those around us typically do arrive at our moral judgments. It invites us to raise critical questions about how we *should* arrive at those judgments. It encourages us to be

more autonomous, or critical, in our moral thinking rather than simply letting others set our moral values for us and allowing ourselves to accept without any questions the conventions that currently prevail. It brings vividly to mind our self-interested and egocentric tendencies and urges us to employ more perceptive and consistent habits of moral thinking. Finally, it emphasizes the importance of giving reasons in support of our judgments.

For a provocative presentation of Kohlberg's theory of moral development, see the video *Moral Development* (CRM Educational Films, McGraw-Hill Films, 1221 Avenue of the Americas, New York, NY. 1-800-421-0833). This video simulates the famous Milgram experiments on obedience, in which volunteers are led to believe that they are administering shocks to other volunteers in an experiment on learning and punishment. Kohlberg's theory is used to characterize the different kinds of responses of volunteers to instructions to administer shocks. Viewers can use this video as a stimulus for reflecting on their own and others' responses to moral challenges. Engineers can also ask the question of whether there are any ethical problems in assisting someone to develop the types of equipment needed to conduct experiments like Kohlberg's.

CASE 25

Oil Spill?[71]

Peter has been working with the Bigness Oil Company's local affiliate for several years, and he has established a strong, trusting relationship with Jesse, manager of the local facility. The facility, on Peter's recommendations, has followed all of the environmental regulations to the letter, and it has a solid reputation with the state regulatory agency. The local facility receives various petrochemical products via pipelines and tank trucks, and it blends them for resale to the private sector.

Jesse has been so pleased with Peter's work that he has recommended that Peter be retained as the corporate consulting engineer. This would be a significant advancement for Peter and his consulting firm, cementing Peter's steady and impressive rise in the firm. There is talk of a vice presidency in a few years.

One day, over coffee, Jesse tells Peter a story about a mysterious loss in one of the raw petrochemicals

he receives by pipeline. Sometime during the 1950s, when operations were more lax, a loss of one of the process chemicals was discovered when the books were audited. There were apparently 10,000 gallons of the chemical missing. After running pressure tests on the pipelines, the plant manager found that one of the pipes had corroded and had been leaking the chemical into the ground. After stopping the leak, the company sank observation and sampling wells and found that the product was sitting in a vertical plume, slowly diffusing into a deep aquifer. Because there was no surface or groundwater pollution off the plant property, the plant manager decided to do nothing. Jesse thought that somewhere under the plant there still sits this plume, although the last tests from the sampling wells showed that the concentration of the chemical in the groundwater within 400 feet of the surface was essentially zero. The wells were capped, and the story never appeared in the press.

Peter is taken aback by this apparently innocent revelation. He recognizes that state law requires him to report all spills, but what about spills that occurred years ago, where the effects of the spill seem to have dissipated? He frowns and says to Jesse, "We have to report this spill to the state, you know."

Jesse is incredulous. "But there *is* no spill. If the state made us look for it, we probably could not find it; and even if we did, it makes no sense whatever to pump it out or contain it in any way."

"But the law says that we have to report...," replies Peter.

"Hey, look. I told you this in confidence. Your own engineering code of ethics requires client confidentiality. And what would be the good of going to the state? There is nothing to be done. The only thing that would happen is that the company would get into trouble and have to spend useless dollars to correct a situation that cannot be corrected and does not need remediation."

"But...."

"Peter, let me be frank. If you go to the state with this, you will not be doing anyone any good—not the company, not the environment, and certainly not your own career. I cannot have a consulting engineer who does not value client loyalty."

What are the ethical issues in this case? What factual and conceptual questions need to be addressed? How do you think Peter should deal with this situation?

CASE 26

Peter Palchinsky: Ghost of the Executed Engineer[72]

Peter Palchinsky grew up in Russia in the late 19th century. He was paid a small stipend by the tsarist government to attend St. Petersburg School of Mines. He supplemented this small income by working summers in factories, railroads, and coal mines. This impressed on him the importance of paying close attention to the living conditions of workers.

After graduating in 1901, Palchinsky was assigned by the government to an investigative team studying methods of increasing coal production in the Ukraine's Don River basin to support Russia's growing industrialization. He visited the living quarters of the miners and found barracks with no space between bunks and cracks in the walls so wide that snow blew over the workers as they slept. Underpaid, the workers also suffered from poor health and low morale. His report on these conditions marked the start of his pioneering work in the developing field of industrial engineering.

However, because of this report, Palchinsky was sentenced to 8 years of house arrest in Irkutsk, Siberia, charged with working with anarchists to overthrow the tsarist government. Nevertheless, he continued to be used by tsarist officials as a consultant because his recommendations led to increased production whenever they were followed. After 3 years of house arrest, Palchinsky and his wife escaped to western Europe, where he continued his work on increasing the productivity of workers and published multivolume studies on facility planning for the governments of Holland, Italy, and France. He was recognized in 1913, at the age of 38, as one of the leading and most productive engineers in Europe. Through the efforts of his wife, he was pardoned so that he could return to Russia.

For the next 3 years, Palchinsky served as a consultant to the tsarist government while establishing several engineering organizations. After the overthrow of the tsars in February 1917, he worked for the Russian provisional government. Following the Bolshevik Revolution in October 1917, Palchinsky and other officers of the provisional government were imprisoned. A number of these officials were executed, but Lenin was persuaded to use Palchinsky's skills for the good of the Bolshevik government. This began a decade of Palchinsky consultancies interrupted by stays in Siberian gulags for his outspoken views that conflicted with Soviet doctrine regarding engineering projects.

Palchinsky was especially critical of Stalin's massive engineering projects, complaining about careless disregard of both engineering and humanitarian issues. Stalin's projects included the world's largest hotel, university, steel mill, power plant, and canal. In the latter project alone, it is estimated that more than 5,000 slave laborers lost their lives and were buried in the foundations of the canal.

Palchinsky's planning studies for what was to be the world's largest dam and supplier of electricity in Dneprostroi opposed the government's final plan. All of his engineering and humanitarian warnings were ignored, and the dam never met its objectives. Palchinsky was next asked to do a planning study for a complex of blast furnaces and steel finishing mills in Magnitogorsk, designed to be the largest such facility in the world. Again, he called attention to many government engineering and humanitarian shortcomings. These warnings were ignored, and Palchinsky was sent back to Siberia. Slave labor was used to build the steel mill, which never came close to meeting its objectives.

In 1929, on Stalin's orders, Palchinsky was secretly taken from his prison and shot. In secret files uncovered as the result of the *glasnost* policy in Russia in the early 1990s, Palchinsky wrote that no government regime could survive the Bolshevik's inhumanity. He predicted that the Russian government would fall before the end of the 20th century (which it did). During the 1920s, the number of engineers decreased from approximately 10,000 to 7,000, with most simply disappearing. Peter Palchinsky sacrificed

his life during this time fighting for the engineering and humanitarian concerns in which he believed.

Loren Graham's *Ghost of the Executed Engineer* portrays Palchinsky as a visionary and prophetic engineer. The "ghost" of Palchinsky, Graham suggests, can be seen in the Soviet Union's continued technological mistakes in the 60 years following his death, culminating in the 1986 Chernobyl nuclear disaster and the dissolution of the Soviet Union in 1991.

Ironically, although praising Palchinsky for his integrity, forthrightness, and vision, Graham concludes his book with a mixed verdict:[73]

> It is quite probably that Palchinsky's execution resulted from his refusal, even under torture, to confess to crimes he did not commit. Palchinsky always prided himself on being a rational engineer. One can question whether his final act was rational, but one cannot question its bravery.

Discuss whether it can be rational to be willing to die rather than confess to crimes to which one has not committed. (Those familiar with Plato's *Crito* might compare Palchinsky's situation with that of Socrates, who also gave up his life rather than compromise his integrity.) How much personal sacrifice should one be willing to make to maintain one's professional integrity?

CASE 27

Pinto[74]

In the late 1960s, Ford designed a subcompact, the Pinto, that weighed less than 2,000 pounds and sold for less than $2,000. Anxious to compete with foreign-made subcompacts, Ford brought the car into production in slightly more than 2 years (compared with the usual 3½ years). Given this shorter time frame, styling preceded much of the engineering, thus restricting engineering design more than usual. As a result, it was decided that the best place for the gas tank was between the rear axle and the bumper. The differential housing had exposed bolt heads that could puncture the gas tank if the tank were driven forward against them upon rear impact.

In court, the crash tests were described as follows:[75]

> These prototypes as well as two production Pintos were crash tested by Ford to determine, among other

things, the integrity of the fuel system in rear-end accidents....Prototypes struck from the rear with a moving barrier at 21-miles-per-hour caused the fuel tank to be driven forward and to be punctured, causing fuel leakage....A production Pinto crash tested at 21-miles-per-hour into a fixed barrier caused the fuel tank to be torn from the gas tank and the tank to be punctured by a bolt head on the differential housing. In at least one test, spilled fuel entered the driver's compartment.

Ford also tested rear impact when rubber bladders were installed in the tank, as well as when the tank was located above rather than behind the rear axle. Both passed the 20-mile-per-hour rear impact tests.

Although the federal government was pressing to stiffen regulations on gas tank designs, Ford contented that the Pinto met all applicable federal safety standards at the time. J. C. Echold, director of automotive safety for Ford, issued a study titled "Fatalities Associated with Crash Induced Fuel Leakage and Fires."[76] This study claimed that the costs of improving the design ($11 per vehicle) outweighed its social benefits. A memorandum attached to the report described the costs and benefits as follows:

Benefits

Savings	180 burn deaths, 180 serious burn injuries, 2,100 burned vehicles
Unit cost	$200,000 per death, $67,000 per injury, $700 per vehicle
Total benefits	180 × $200,000 plus
	180 × $67,000 plus
	2100 × $700 = $49.15 million

Costs

Sales	11 million cars, 1.5 million light trucks
Unit cost	$11 per car, $11 per truck
Total costs	11,000,000 × $11 plus
	1,500,000 × $11 = $137 million

The estimate of the number of deaths, injuries, and damage to vehicles was based on statistical studies. The $200,000 for the loss of a human life was based on an NHTSA study, which estimated social costs of a death as follows:[77]

Component	1971 Costs
Future productivity losses	
Direct	$132,000
Indirect	41,300
Medical costs	
Hospital	700
Other	425
Property damage	1,500
Insurance administration	4,700
Legal and court	3,000
Employer losses	1,000
Victim's pain and suffering	10,000
Funeral	900
Assets (lost consumption)	5,000
Miscellaneous accident cost	200
Total per fatality	$200,725

Discuss the appropriateness of using data such as these in Ford's decision regarding whether or not to make a safety improvement in its engineering design. If you believe this is not appropriate, what would you suggest as an alternative? What responsibilities do you think engineers have in situations like this?

CASE 28

Profits and Professors

A *Wall Street Journal* article reports:

High-tech launches from universities frequently can't get off the ground without a steady supply of students, who are often the most talented and the most willing to toil around the clock. But intense schedules on the job can keep students from doing their best academic work. And when both student and teacher share a huge financial incentive to make a company a success, some professors might be tempted to look the other way when studies slip or homework gets in the way.[78]

In some instances, the article claims, students seriously consider leaving school before completing their degrees in order devote themselves more fully to work that is financially very attractive.

In 1999, Akamai won the MIT Sloan eCommerce Award for Rookie of the Year, an award to the startup company that seems most likely to dominate its field. The article comments,

No company has been more closely tied to MIT. The firm has its roots in a research project directed by Mr. Leighton [Computer Systems Engineering professor at MIT] about 3 years ago. Daniel Lewin, one of Mr. Leighton's graduate students, came up with a key idea for how to apply algorithms, or numerical instructions for computers, to Internet congestion problems.[79]

Soon, Mr. Leighton and Mr. Lewin teamed up to form Akamai, hiring 15 undergraduates to help code the algorithms.

They tried to separate their MIT and Akamai responsibilities. Mr. Leighton advised Mr. Lewin to get a second professor to co-sign his master's thesis "because he worried about the appearance of conflict in his supervising Mr. Lewin's academic work while also pursuing a business venture with him." It turns out that the co-signer was someone involved in Mr. Lewin's original research project, who sometime after the completion of Mr. Lewin's thesis became a part-time research scientist at Akamai.

Akamai continues to rely heavily on MIT students as employees. However, it does not hire students full-time before they have completed their undergraduate degree. Still, the opportunities seem very attractive. According to the article, Luke Matkins took a summer job with Akamai in the summer after his sophomore year. By age 21, prior to completing his degree, he was making $75,000 a year and was given 60,000 shares of stock estimated to be worth more than $1 million.

Mr. Matkins grades suffered because his work left him too little time to complete all of his homework assignments. However, he apparently has no regrets: "Mr. Matkins says the prospect of being a millionaire by his senior year is 'very cool.' He loves MIT, but in many ways, he says, Akamai has become his real university. 'There are different ways to learn stuff,' he says. 'I've learned more at Akamai than I would in a classroom.'"[80]

The article notes that Mr. Lewin's doctoral dissertation will be based on his work at Akamai, although he'll probably need permission from the Akamai board of directors to use some of the material. The article concludes, "He will also probably need approval from Akamai's chief scientist, Mr. Leighton, who, it turns out, is his PhD adviser."[81]

Identify and discuss the ethical issues that the previous account raises.

CASE 29

Pulverizer

Fred is a mechanical engineer who works for Super Mulcher Corporation. It manufactures the Model 1 Pulverizer, a 10-hp chipper/shredder that grinds yard waste into small particles that can be composted and blended into the soil. The device is particularly popular with homeowners who are interested in reducing the amount of garden waste deposited in landfills.

The chipper/shredder has a powerful engine and a rapidly rotating blade that can easily injure operators if they are not careful. During the 5 years the Model 1 Pulverizer has been sold, there have been 300 reported accidents with operators. The most common accident occurs when the discharge chute gets plugged with shredded yard waste, prompting the operator to reach into the chute to unplug it. When operators reach in too far, the rotating blades can cut off or badly injure their fingers.

Charlie Burns, president of Super Mulcher, calls a meeting of the engineers and legal staff to discuss ways to reduce legal liability associated with the sale of the Model 1 Pulverizer. The legal staff suggest several ways of reducing legal liability:

• Put bright yellow warning signs on the Model 1 Pulverizer that say, "Danger! Rapidly rotating blades. Keep hands out when machine is running!"

• Include the following warning in the owner's manual: "Operators must keep hands away from the rotating blades when machine is in operation."

• State in the owner's manual that safe operation of the Model 1 Pulverizer requires a debris collection bag placed over the discharge chute. State that operators are not to remove the debris collection bag while the Model 1 Pulverizing is running. If the discharge chute plugs, the owner is instructed to turn off the Model 1 Pulverizer, remove the debris collection bag, replace the debris collection bag, and restart the engine.

From operating the Model 1 Pulverizer, Fred knows the discharge chute has a tendency to plug. Because the machine is difficult to restart, there is a great temptation to run the unit without the debris collection bag—and to unplug the discharge chute while the unit is still running.

For each of the following scenarios, discuss the various ways Fred attempts to resolve the problem:

Scenario 1: Fred suggests to his engineering colleagues that the Model 1 Pulverizer should be redesigned so it does not plug. His colleagues reply that the company probably cannot afford the expense of reengineering the Model 1, and they conclude that the legal staff's recommendations should be sufficient. Dissatisfied, in his spare time Fred redesigns the Model 1 Pulverizer and solves the plugging problem in an affordable way.

Scenario 2: Fred says nothing to his colleagues about the impracticality of requiring the machine to be

run with the debris collection bag. He accepts the legal staff's advice and adds the warning signs' and owner's manual instructions. No changes are made in the design of the Model 1 Pulverizer.

Scenario 3: Fred suggests to his engineering colleagues that they try to convince management that the Model 1 Pulverizer should be redesigned so that it does not plug. They agree and prepare a redesign plan that will cost $50,000 to implement. Then they take their plan to management.

CASE 30

Reformed Hacker?

According to John Markoff's "Odyssey of a Hacker: From Outlaw to Consultant," John T. Draper is attempting to become a "white-hat" hacker as a way of repaying society for previous wrongdoing.[82] In the early 1970s, Draper became known as "Cap'n Crunch" after discovering how to use a toy whistle in the Cap'n Crunch cereal box to access the telephone network in order to get free telephone calls. While serving time in jail for his misdeeds, he came up with the early design for EasyWriter, IBM's first word-processing program for its first PC in 1981. However, says Markoff, in subsequent years Draper used his skills to hack into computer networks, became a millionaire, lost jobs, and experienced homelessness.

Now, however, Draper has been enlisted to help operate an Internet security software and consulting firm that specializes in protecting the online property

of corporations. Draper says, "I'm not a bad guy." However, realizing there are bound to be doubters, he adds, "But I'm being treated like a fox trying to guard the hen house." SRI International's computer security expert Peter Neumann summarizes the concern:

> Whether black hats can become white hats is not a black-and-white question. In general, there are quite a few black hats who have gone straight and become very effective. But the simplistic idea that hiring overtly black-hat folks will increase your security is clearly a myth.

Discuss the ethical issues this case raises. What might reasonably convince doubters that Draper has, indeed, reformed? Are customers of the consulting firm entitled to know about Draper's history and his role at the firm?

CASE 31

Resigning from a Project

In 1985, computer scientist David Parnas resigned from an advisory panel of the Strategic Defense Initiative Organization (SDIO).[83] He had concluded that SDI was both dangerous and a waste of money. His concern was that he saw no way that any software program could adequately meet the requirements of a

good SDI system.[84] His rationale for resigning rested on three ethical premises.[85] First, he must accept responsibility for his own actions rather than rely on others to decide for him. Second, he must not ignore or turn away from ethical issues. In Parnas's case, this means asking whether what he is doing is of any benefit to society. Finally, he "must make sure that I am solving the real problem, not simply providing short-term satisfaction to my supervisor."

However, Parnas did more than resign from the panel. He also undertook public opposition to SDI. This was triggered by the failure of SDIO and his fellow panelists to engage in scientific discussion of the technical problems he cited. Instead, Parnas says, he received responses such as "The government has decided; we cannot change it." "The money will be spent; all you can do is make good use of it." "The system will be built; you cannot change that." and "Your resignation will not stop the program."[86] To this, Parnas replied,

> It is true, my decision not to toss trash on the ground will not eliminate litter. However, if we are to eliminate litter, I must decide not to toss trash on the ground. We all make a difference.

As for his part, Parnas regarded himself as having a responsibility to help the public understand why he was convinced that the SDI program could not succeed, thus enabling them to decide for themselves.[87]

Parnas's concerns did not stop with SDI. He also expressed concerns about research in colleges and universities:[88]

> Traditionally, universities provide tenure and academic freedom so that faculty members can speak out on issues such as these. Many have done just that. Unfortunately, at U.S. universities there are institutional pressures in favor of accepting research funds from any source. A researcher's ability to attract funds is taken as a measure of his ability.

Identify and discuss the ethical issues raised by David Parnas. Are there other ethical issues that should be discussed?

CASE 32

Responsible Charge[89]

Ed Turner graduated from Santa Monica College (a 2-year school) with an associate degree in 1961. He worked for 8 years for the City of Los Angeles in its engineering department and took the professional Engineer in Training exam in California. As a result, he received a Civil Engineering/Professional Engineering license in the state of Idaho. To get his license, he had to work under the direction of already licensed supervisors and be strongly recommended for licensure by all of them. Because he did not have a BS degree in engineering from an accredited school, his experience had to be exemplary.

In the late 1960s, Turner moved to the city of Idaho Falls and went to work for the Department of Public Works. As a licensed professional engineer in 1980, he had sign-off authority for all engineering work done in the city. His problems with the city started when he refused to approve some engineering designs for public works projects. One such project omitted the sidewalk, requiring students to walk in street traffic on their way to school. The public works director and mayor responded to his refusal by demoting him and moving him out of his office to a new and smaller work area. They appointed an

unlicensed nonengineer as city engineering administrator to replace him and sign off on all engineering work. This was in violation of Idaho state law.

Turner stayed on that new job as long as he could to keep an eye on engineering work in the city and because he needed an income to support his family. Finally, he was dismissed, and he and his wife had to sort potatoes and do custodial work in order to survive and to finance a court appeal.

The Idaho Job Service Department approved his request for unemployment insurance coverage, but the city of Idaho Falls succeeded in getting that ruling reversed. The Idaho Industrial Commission eventually overturned the city's ruling, and Turner ultimately received his unemployment insurance.

Turner and the American Engineering Alliance (AEA) of New York managed to obtain the support of 22 states in his case against Idaho Falls for wrongful discharge and for not having responsible charge of engineering work. The Idaho State Board of Professional Engineers and the National Society of Professional Engineers (NSPE) also supported him, as did the ASME, the ASCE, the AEA, as well as several other important professional societies. Ed's wife,

Debra, played a significant role throughout the 4-year litigation. In addition to keeping the court files in order, she was on the witness stand and was cross-examined by the city's lawyers.

Many individuals cognizant of the issues involved, including one of the authors of this text, volunteered their services to Turner on a pro bono basis and submitted depositions. However, the depositions were not admitted by the Idaho Falls city court that was hearing the case, and the case was thrown out of the court because the papers submitted to the Idaho Falls judge were late and on the wrong forms.

Fortunately, the story does have a happy ending. On the advice of many, and with a new lawyer, Ed's former lawyer was sued for malpractice at a court in another city. In order for a malpractice suit to be successful, the jury must first vote that the original case was winnable, and then it must separately determine that there was malpractice involved. Turner won both those decisions, with the court admonishing the government of Idaho Falls that it had violated state law. Although the settlement was large, after legal fees and taxes were paid, it was clear that Turner was not, in his words, "made whole." But he resumed practicing as a licensed professional civil engineer and happy that he was able to contribute to his profession and to public safety. It is noteworthy that in response to the devastation caused by Hurricane Katrina in 2005, Ed and his wife Debra spent months doing volunteer work in Alabama to provide aid to its victims.

CASE 33

Scientists and Responsible Citizenry

As a young man, Harrison Brown (1917–1986) played a prominent role in the Manhattan Project at the University of Chicago and Oak Ridge. In 1943, he became assistant director of chemistry for the Oak Ridge Plutonium Project. During the very few years it took to develop the atomic bomb, Brown and many of his fellow research scientists had serious and deep discussions of their responsibilities as scientists. After the bomb was used in 1945, Brown immediately wrote a book, *Must Destruction Be Our Destiny* (Simon & Schuster, 1946), in which he articulated his concerns and those of his colleagues. An ardent advocate for the establishment of an international body that could peaceably control the spread and possible use of atomic weapons, in the space of 3 months in 1946 he gave more than 100 speeches throughout the country presenting the basic arguments of his book.

It is noteworthy that on the jacket of this book, Albert Einstein is quoted as saying the following:

> One feels that this book is written by a man who is used to responsible work. It gives a clear, honest, and vivid description of the atom bomb as a weapon of war, objective and without any exaggeration. It gives a clear discussion, free of rhetoric, of the special international problems and the possibilities for their solution. Everyone who reads this book carefully will be enabled—and one hopes stimulated—to contribute to a sensible solution of the present dangerous situation.

It is also noteworthy that the subtitle of *Must Destruction Be Our Destiny* is *A Scientist Speaks as a Citizen*. This subtitle reflects the modesty, yet firmness of conviction, with which Brown undertook his effort to communicate his concerns to the public. He was very sensitive to the claim that scientists should restrict themselves to questions of science. Without crediting scientists with special expertise regarding the social or political implications of science and technology, he responded by pointing out that scientists working on the atomic bomb had the advantage of knowing about the potential uses and consequences of this weapon some time before the general public did, and they had given this much careful thought. Convinced that the "man in the street" needs to be well informed before presenting social and political opinions about matters of great importance, Brown held that scientists have a responsibility to acquire and communicate needed information to lay audiences so that they are able to exercise better judgment.

As for himself, Brown said in his preface, "I have written as a man in the street, as an ordinary citizen, possessing primarily the fundamental desires to live freely, comfortably, and unafraid." Implicit here is the notion that *this* ordinary citizen also possessed information needed by all other ordinary citizens— information that, he was convinced, would enable them to join hands with those scientists who "have had the advantage of months and years to become acquainted with the problems and to think of them as would any reasonably literate and sensitive persons." He added, "As scientists we have indicated the problems—as citizens we have sought the answers."

Of course, Harrison Brown the scientist and Harrison Brown the ordinary citizen were one and the same person. He also chose to pursue a career at the California Institute of Technology, holding joint appointments in the geology and humanities divisions. In other words, he deliberately chose an interdisciplinary path in higher education. This is further reflected in his joining the Emergency Committee of Atomic Scientists as Vice Chair (with Albert Einstein serving as Chair) in 1947, his role as editor-in-chief of *The Bulletin of Atomic Scientists,* his service as foreign secretary of the National Academy of Sciences (1962–1974), and his service as science advisor to the presidential campaigns of Adlai Stevenson and Robert Kennedy.

Apparently, Harrison Brown's commitments as citizen–scientist did not interfere with his commitments to "pure science." He continued his scientific studies on meteorites, along with work in mass spectroscopy, thermal diffusion, fluorine and plutonium chemistry, geochemistry, and planetary structure. In 1947, at age 30, he became the youngest scientist ever to receive the annual award for making "the most notable contribution to science," based on his report, "Elements in Meteorites and the Earth's Origins." In 1952, he received the American Chemical Society's Award in Pure Chemistry.

In his second book, *The Challenge of Man's Future* (Viking Press, 1954), and in subsequent writings throughout the next three decades, Harrison Brown argued that technological advancement, population growth, the desire for increased living standards throughout the world, and limited food, mineral, and energy resources call for urgent consideration by scientists and ordinary citizens alike. Convinced that we have the power, intelligence, and imagination to deal with the challenges posed by these developments, he insisted, however, that this "necessitates an

understanding of the relationships between man, his natural environment, and his technology."

The comments of three Nobel Prize winners were quoted on the jacket of this second book. One of them, Albert Einstein, said,

> We may well be grateful to Harrison Brown for this book on the condition of mankind as it appears to an erudite, clear-sighted, critically appraising scientist.... The latest phase of technical–scientific progress, with its fantastic increase of population, has created a situation fraught with problems of hitherto unknown dimensions.... This objective book has high value.

Harrison Brown died in 1986. Twenty years later, Harvard University's John Holdren, Teresa and John Heinz Professor of Environmental Policy and Director of the Program on Science, Technology, and Public Policy in the John F. Kennedy School of Government, recalled reading *The Challenge of Man's Future* years before as a high school student. In a speech titled, "Science, Technology, and the State of the Word: Some Reflections after September 11," he said that prior to reading that book and C. P. Snow's *The Two Cultures,* his ambition was to become the chief design engineer at Boeing. Moved by these books, he decided that, instead, he wanted to "work on the great problems of the human condition that sit at the intersection of disciplines, the intersection of the natural sciences and the social sciences where science, technology, and the public policy come together" (www.spusa.org/pubs/speeches/holdrenspeech.html).

At the outset of his speech, Holdren said that he would be sharing his reflections in the way he thought Harrison Brown would if he were still alive—focusing on what we can now (and should have been able to earlier) clearly understand about the relationships among science, technology, and the state of the world prior to September 11, 2001. Most important, he indicated that he would be talking "in terms of what socially responsible scientists and technologists should be striving to contribute to these issues, not just the issues in the aftermath of September 11th but the still wider ones at this immensely important intersection of science and technology and the human condition."

CASE 34

Sealed Beam Headlights

It is important to realize that engineering success typically requires the collaborative efforts of engineers rather than simply the efforts of one individual. An early safety problem in the automotive industry was the unreliability of headlights due to the fact that they were inadequately protected from moisture and the resulting rusting. In the late 1930s, a group of General Electric engineers worked together to develop the sealed beam headlight, which promised to reduce sharply the number of fatalities caused by night driving.[90] To accomplish this, it was necessary to involve engineers in collaborative research, design, production, economic analysis, and governmental regulation. Although the need for headlight improvement was widely acknowledged, there was also widespread skepticism about its technical and economic feasibility. By 1937, the GE team provided the technical feasibility of the sealed beam headlight. However, the remaining task was to persuade car builders and designers to cooperate with each other in support of the innovation, as well as to convince regulators of its merits.

Given this skepticism, there is little reason to suppose that the GE engineers were simply doing what they were told—namely to develop a more adequate headlamp. Apparently, the consensus was that this could not be done, so the engineers had to overcome considerable resistance. This was no ordinary task, as evidenced by the remarks of another engineer of that era:

> The reaching of the consensus embodied in the specifications of the sealed beam headlamp is an achievement which commands the admiration of all who have any knowledge of the difficulties that were overcome. It is an achievement not only in illuminating engineering, but even more in safety engineering, in human engineering, in the art of cooperation.[91]

The difficulties faced by this group of engineers should remind us that enthusiasm for desirable ends needs to be tempered with realism. Other demands and constraints may discourage undertaking such projects. Nevertheless, looking for opportunities to accomplish such ends, as well as taking advantage of these opportunities when they arise, is desirable. Discuss the abilities and qualities of character that contribute to the success of projects such as the sealed beam headlight. Can you think of other examples of collaborative engineering success?

CASE 35

Service Learning[92]

Current Accreditation Board for Engineering and Technology (ABET) requirements for accredited engineering programs in the United States include helping students acquire "an understanding of the ethical characteristics of the engineering profession and practice."[93] ABET 2000 more specifically requires engineering programs to demonstrate that their graduates also understand the impact of engineering in a global and social context, along with a knowledge of current issues related to engineering. The recent surge of interest in service learning in engineering education presents students with creative, hands-on possibilities to meet these ABET expectations.

Service learning involves combining community service and academic study in ways that invite reflection on what one learns in the process. Given ABET 2000's requirement that students be involved in a "major design experience" that includes ethical factors in addition to economic, environmental, social, and political factors, the idea of service learning in engineering may be especially promising. But this idea is important for another reason. Much of the engineering ethics literature dwells on the negative—wrongdoing, its prevention, and appropriate sanctioning of misconduct. These will always be fundamental concerns. However, there is more to engineering ethics. There is the more positive side that focuses on doing one's work responsibly and well—whether in the workplace or in community service.

Given the common association of engineering ethics with wrongdoing and its prevention, it might be asked whether community service should be regarded as a part of engineering ethics at all. However, it is not uncommon for other professions to include pro bono service as an important feature of their professional ethics. This is based in large part on the recognition that professions provide services that may be needed by anyone but which not everyone can afford or gain easy access to. Medical and legal services readily come to mind. But this is no less true of engineering.

Is this acknowledged in engineering codes of ethics? It is in at least two—those of the NSPE and the ASCE. Emphasizing the crucial impact that engineering has on the public, the Preamble of NSPE's Code of Ethics for Engineers states that engineering "requires adherence to the highest principles of ethical conduct on behalf of the public, clients, employers, and the profession." Following this, the code lists as its first Fundamental Canon that engineers are to hold paramount the safety, health, and welfare of the public in the performance of their professional duties. Under section III. Professional Obligations, provision 2 reads, "Engineers shall at all times strive to serve the public interest." Subsection a under this obligation reads, "Engineers shall seek opportunities to be of constructive service in civic affairs and work for the advancement of the safety, health, and well-being of their community."

Noteworthy here is the assertion that engineers are to seek opportunities to be of service to the community. Furthermore, there is no qualifier, "in the performance of their professional duties." This suggests that engineers' obligations in regard to public well-being are not restricted to their responsibilities within their place of employment.

The first Fundamental Canon of ASCE's code reads, "Engineers shall hold paramount the safety, health, and welfare of the public and shall strive to comply with the principles of sustainable development in the performance of their professional duties." Subsection e, directly under this, reads, "Engineers should seek opportunities to be of constructive service in civic affairs and work for the advancement of the safety, health, and well-being of their communities, and the protection of the environment through the practice of sustainable development." Subsection f

reads, "Engineers should be committed to improving the environment by adherence to the principles of sustainable development so as to enhance the quality of life of the general public."

Although the NSPE and ASCE provisions are rather broadly stated, they do provide a rationale for concluding that, at least from the perspective of two major professional engineering societies, community service is an important feature of engineering ethics.

Many worry that students today are part of a "me-generation." At the same time, however, there has been a marked increase in student interest in volunteer work. Until fairly recently, there has not been a strong correlation between students' academic pursuits and the types of volunteer work they undertake. Noting this lack of correlation, organizations such as Campus Compact have made concerted efforts to encourage the development of academic programs that explicitly encourage students to seek volunteer work related to their course of academic study and to reflect quite self-consciously on the connections.[94]

Academic areas such as teacher education and the health care professions immediately suggest themselves as candidates for service learning programs. Students preparing to become teachers can offer tutorial or mentoring services to the schools, students in nursing programs can volunteer their services to nursing homes or other health care facilities, and so on. But engineering students, even early on in their programs, can volunteer tutorial services to the schools, particularly in areas of computer science, math, science, and technology that are relevant to engineering. For example, while at the University of South Alabama, Edmund Tsang's Introduction to Mechanical Engineering course included a service learning project.[95] Engineering student teams worked with the Mobile school system and its Southeastern Consortium for Minorities in Engineering program. Students in this class designed equipment for teachers and middle school students that illustrated basic principles of motion, energy, and force and mathematical modeling.

To illustrate the potential value of service learning projects for both students and those who benefit from their projects, it is helpful to discuss an example in some detail. This was a project undertaken some years ago by a group of electrical engineering students at Texas A & M in Tom Talley's senior design course.[96] This course was intended to help prepare students for the challenges in project design and management that they will confront in industry. In this case, the students were also introduced to community service.

Team members were undecided about what project to undertake until Tom Talley shared with them a letter he had received from the Brazos Valley Rehabilitation Center. The letter identified a need for an Auditory Visual Tracker (AVIT) to help in evaluating and training visual skills in very young children with disabilities. Most students, Talley said, end up only building a working prototype. However, in this case, he pointed out, "The students took on the project knowing that it was larger and potentially more expensive for them to produce than might be expected of a typical project."

"We like that it was a project that was going to be genuinely used," said team member Robert D. Siller, "It wasn't going to just end up in a closet. It's actually helping someone." Myron Moodie added, "When we presented the AVIT to the center, we got to see some of the kids use it. It was worth it watching the way the children like it." However, completion of the project was anything but easy. One complication was that the team was interdisciplinary. It included a student from management, which meant that the team was introduced to the project management environment, giving the endeavor a more industry-like flavor than was typical of projects in Talley's design class. To further complicate matters, the management student was seriously injured in a car accident during the semester; but she was able to continue in the project. By the end of the semester, the project was not quite completed. However, the students were so committed to providing a usable AVIT for the rehabilitation center that they stayed on after the semester.

What seems obvious from student comments is that they found the service aspect of their experience very rewarding. Whether this encouraged them to continue to seek out community service opportunities once they were fully employed engineers can be, of course, only a matter for speculation. Another matter for speculation is that this experience speaks positively about the kinds of engineers these students could be expected to become in their places of employment. Tom Talley, at least, was quite optimistic. He said, "They clearly went above and beyond—that's Aggie spirit. Someone is going to get some fine young engineers." This comment can be taken to include what

can be expected from these students both as engineers in the workplace and as civic-minded contributors to the public good.

This particular kind of project—one taken to completion and one involving direct interaction with those being helped—can enhance students' understanding and appreciation of responsibilities they have both on the job and in community service. In this case, the project went well beyond designing a prototype; everything worked out well. However, this required very careful attention to the specific needs of the center's staff and the children who were in need of assistance. This is a very important lesson in responsible engineering, whether volunteer or work related.

From a service learning perspective, two limitations of this example should be noted. First, although the students apparently did reflect on the significance of the service aspects of their experience, this was not a specific objective of the project. Service learning is distinguished by it's deliberate combining of service and study: "One of the characteristics of service learning that distinguishes it from volunteerism is its balance between the act of community service by participants and reflection on that act, in order both to provide better service and to enhance the participants' own learning."[97] This project was not simply an instance of volunteerism; it was a class project. However, it was a project primarily in engineering design and, from the perspective of the class, only incidentally did it involve community service. Nevertheless, this is the sort of project that could be undertaken with the full service learning objectives in mind; many of those objectives were, in fact, fulfilled even though this was not part of the official class agenda.

Second, a point related to the first, the AVIT project stood virtually alone. There may have been other projects that lent themselves to service learning objectives that were undertaken by students in Tom Talley's design class or in other design classes at Texas A & M. But service learning in engineering as a planned, coordinated activity requires a much more sustained effort. A second example illustrates this point.

An early service learning program in engineering, the student-initiated Case Engineering Support Group (CESG) at Case Western Reserve University was founded in 1990 as a nonprofit engineering service organization composed of engineering students who "design and build custom equipment to assist the disabled in therapy or normal daily activities."[98] According to the CESG brochure, the equipment is given to individuals at therapy centers at no cost. CESG has received donations of equipment from industry, financial support from the National Science Foundation and the Case Alumni Association, legal services from Case's Law School Clinic, and cooperation and support from the medical and health care community in Cleveland.

In CESG's first year, 18 students completed 6 projects. During the 1995–1996 academic year, 120 students completed 60 projects, as well as follow-up work on previous projects. At that time, CESG supported four major programs:[99]

• Custom Product Development Program: working with faculty members designing, manufacturing, and providing at no cost to individuals adaptive devices and equipment to help them gain a higher level of independent living skills; working with physicians and physical, occupational, and speech therapists in adapting, modifying, and providing devices and equipment.

• Technology Lender Program: repairing and adapting donated computer equipment and designing specialized software for those with special communication, vocational, or educational needs.

• Toy Modification Program: providing specially adapted toys to families of children with disabilities and to hospitals, and presenting related workshops to junior and senior high school students to stimulate interest in engineering as a career.

• Smart Wheelchair Project: working with the Cleveland Clinic Foundation's Seating/Wheeled Mobility Clinic, Invacare Corporation, and engineers at the NASA Lewis Research Center to design, modify, and improve the 'smart wheelchair,' which is fit with special sensors and artificial intelligence routines.

Recent years have seen the rapid growth of service learning programs in engineering. The *International Journal for Service Learning in Engineering* was launched in 2006. This periodical provides detailed accounts of service learning projects written by faculty and students. Learn and Serve America's National Service-Learning Clearinghouse provides a comprehensive list of web resources on service learning in engineering, as well as a list of print resources (www.servicelearning.org). Three web references warrant special mention here:

Engineers Without Borders (www.ewb-usa.org). Established in 2000, this is a national, nonprofit organization that offers help developing areas throughout the world with their engineering needs. It has the goal of "involving and training a new kind of internationally responsible engineering student." This website lists all the EWB-USA registered student chapters, along with their websites. EWB-USA also has a *Wikipedia* entry (http://en.wikipedia .org). It is identified as a member of the "Engineers Without Borders" international network. EWB-USA's projects typically involve the design and construction of water, sanitation, energy, and shelter systems in projects initiated by and completed with the host communities. According to the *Wikipedia* entry, "These projects are initiated by, and completed with, contributions from the host community, which is trained to operate the systems without external assistance. In this way, EWB-USA ensures that its projects are appropriate and self-sustaining."

Engineering Projects in Community Service (EPICS) National Program (http://epicsnational.ecn.purdue .edu). EPICS is described as integrating "highly mentored, long-term, large-scale, team-based, multidisciplinary design projects into the undergraduate engineering curriculum.... Teams work closely with a not-for-profit organization in the community to define, design, build, test, deploy, and support projects that significantly improve the organization's ability to serve the community."

Service-Learning in Engineering: A Resource Guidebook (www.compact.org/publications). Developed by William Oaks and published by Campus Compact, this guidebook introduces the idea of service learning in engineering and provides models from the EPICS program, course descriptions and syllabi, and evaluation tools. It can be downloaded from the Campus Compact website.

CASE 36

Shortcut?

Bruce Carson's civil engineering firm has a contract with the state to specify the route of a new road connecting two major cities. Bruce determines that the shortest workable path will save 20 minutes from what would otherwise be a 2-hour trip, but it would require the state to destroy a farm house that has been in the Jones family for 150 years. Bruce visits the Jones family to get some idea of what it would cost the state to purchase their home and the land immediately surrounding it.

Not surprisingly, the prospect of losing the home their family has maintained for the past 150 years is very upsetting to the family. "What's 20 minutes compared to 150 years of family tradition?" objects Robert Jones, who has lived in the farmhouse the entire 63 years of his life. The family insists that no amount of money would tempt them to sell their home to the state, or to anyone else for that matter.

Bruce knows that one option would be for the state to exercise "eminent domain" and condemn the farmhouse. Should he recommend this to the state? Why or why not?

CASE 37

"Smoking System"[100]

Philip Morris Companies reported testing a microelectronic cigarette holder that eliminates all smoke except that exhaled by the smoker. Battery powered, it is expected to cost approximately $50. The result of years of research, it cost approximately $200 million to develop.

Tentatively called the Accord, the device uses cigarettes that are 62 millimeters long (compared with the standard 85 millimeters). Users will have to remember to recharge the Accord's battery (a 30-minute process, but extra batteries can be purchased). A cigarette is inserted into the 4-inch long, 1½-inch wide device. A microchip senses when the cigarette is puffed and transmits powers to eight heating blades. A display shows the remaining battery charge and indicates how many puffs are left in the eight-puff cigarette. The device also contains a catalytic converter that burns off residues.

Supporters of this product say it will be welcomed by smokers who currently refrain from smoking in their homes or cars for the sake of nonsmoking family members, guests, and passengers. Although smokers will inhale the same amount of tar and nicotine as from conventional "ultralight" cigarettes, 90 percent of second-hand smoke will be eliminated. Furthermore, the same smoking restriction rules in public places will apply to the device.

Critics claim that the Accord will simply reinforce addiction to cigarettes. Richard A. Daynard, chair of the Tobacco Products Liability Project at Boston's Northeastern University School of Law, an anti-tobacco organization, asks, "Who would use an expensive and cumbersome thing like this if they weren't hooked? There is something grim and desperate about it. This is hardly the Marlboro Man, getting on his horse and checking the battery." He also expresses concern that children might be encouraged to smoke since the Accord would enable them to hide smoking from their parents. However, Philip Morris replies that the Accord has a locking device for parents.

Consider the following questions:

• Imagine that it is several years ago and you have just received your engineering degree. You are in search of your first job. You are invited to interview with a research division of Philip Morris that is about to begin research to develop the Accord. Would you have any reservations about accepting such a position? Discuss.

• If you have some reservations, would the fact that this job pays $10,000 more per year than any other offer you have convince you to take the Philip Morris offer?

• Assuming you took the job, what kinds of ethical concerns might you have about how the device should be designed? For example, would you agree that it should have a locking device?

CASE 38

Software for a Library[101]

A small library seeks a software system to catalogue its collection and keep records of materials checked out of the library. Currently, the records of who has checked out what, when materials are due, and the like are kept in a file drawer behind the check-out desk. These records are confidential. Patrons are assured that these records are not accessible to anyone other than library personnel. But, of course, drawers can be opened when no one is looking. What assurance is there that the software systems under consideration will provide as much, if not greater, security? Assuming that no one in the library is a software

specialist, the library has no alternative but to place its trust in someone who presumably has the requisite expertise. How concerned should that expert be (again, bearing in mind that even the best system is not completely sleuthproof)? Furthermore, what assurance has the library that it is not being oversold or undersold in general? To what extent should software specialists be concerned with determining precisely what the various needs of the library are—and to try to meet those needs rather than offer more than is necessary in order to secure greater profit or less than is needed in order to come in with a lower bid?

CASE 39

Sustainability

Scientists, engineers, and the government are publicly expressing urgent concern about the need to address the challenges of sustainable scientific and technological development. Global warming, for example, raises concern about glacial meltdown and consequent rising ocean levels threatening coastal cities. A related concern is the lowering of levels of freshwater in the American West as a result of lowered levels of accumulated mountain snow. In Joe Gertner's "The Future Is Drying Up," Nobel laureate Steven Chu, director of the Lawrence Berkeley National Laboratory, is cited as saying that even optimistic projections for the second half of the 21st century indicate a 30 to 70 percent drop in the snowpack level of the Sierra Nevada, provider of most of northern California's water.[102] Gertner goes on to discuss other likely freshwater problems that will have to be faced by Western states as a result of both global warming and the consumption needs and demands of an increasing population. He also outlines some of the efforts of engineers to address these problems aggressively now rather than wait until it is too late to prevent disaster.[103]

We noted in Chapter 9 that most engineering society codes of ethics do not make direct statements about the environmental responsibilities of engineers. However, in 2007 the NSPE joined the ranks of

engineering societies that do. Under section III. Professional Obligations, provision 2 reads, "Engineers shall at all times strive to serve the public interest." Under this heading, there is a new entry, d: "Engineers are encouraged to adhere to the principles of sustainable development in order to protect the environment for future generations." Footnote 1 addresses the conceptual question of what is meant by "sustainable development": "'Sustainable development' is the challenge of meeting human needs for natural resources, industrial products, energy, food, transportation, shelter, and effective waste management while conserving and protecting environmental quality and the natural resource base essential for future development."

Although this definition of sustainable development leaves many fundamental conceptual and value questions in need of further analysis (e.g., What are human needs? What is meant by "environmental quality"?), it provides a general framework for inquiry. It also identifies a variety of fundamental areas of concern (e.g., food, transportation, and waste management). Of course, responsibilities in these areas do not fall only on engineers. Government officials, economists, business leaders, and the general citizenry need to be involved as well. Thus, a basic question relates to how those who need to work together might best do so and what role engineers might play. We offer three illustrations for discussion. The first is an early effort to involve students from different disciplines in a project that supports sustainable development. The second is the recent proliferation of centers and institutes for sustainability on college campuses throughout the country. The third is service learning opportunities in support of sustainable design and development.

RENEWABLE ENERGY[104]

Dwayne Breger, a civil and environmental engineer at Lafayette College, invited junior and senior engineering, biology, and environmental science students to apply to be on an interdisciplinary team to design a project that would make use of farmland owned by Lafayette College in a way that supports the college mission. Twelve students were selected for the project: two each from civil and environmental engineering, mechanical engineering, chemical engineering, and Bachelor of Arts in engineering, plus three biology majors and one in geology and environmental geosciences. These students had minors in such areas as economics and business, environmental science, chemistry, government, and law. The result of the project was a promising design for a biomass farm that could provide an alternative, renewable resource for the campus steam plant.[105]

Professor Breger regards projects such as this as providing important opportunities for students to involve themselves in work that contributes to restructuring our energy use toward sustainable resources. ABET's *Engineering Criteria 2000* for evaluating engineering programs includes the requirement that engineering programs demonstrate that their graduates have "an understanding of professional and ethical responsibility," "the broad education necessary to understand the impact of engineering solutions in a global and societal context," and "a knowledge of contemporary issues." Criterion 4 requires that students have "a major design experience" that includes consideration of the impact on design of such factors as economics, sustainability, manufacturability, ethics, health, safety, and social and political issues.[106] Discuss how the Lafayette College project might satisfy criterion 4, especially the ethical considerations.

ACADEMIC CENTERS FOR SUSTAINABILITY

Historically, joint research in colleges and universities is done within separate disciplines rather than in collaboration with other disciplines. Thus, biologists collaborate with other biologists, chemists with other chemists, economists with other economists, and political scientists with other political scientists. The recent emergence of centers and institutes for sustainability represents a significant and important break from that tradition.

In September 2007, the Rochester Institute of Technology initiated the Golisano Institute for Sustainability.[107] Noting that it is customary for new programs to be run by just one discipline, Nabil Nasr, the institute director, comments, "But the problem of sustainability cuts across economics, social elements, engineering, everything. It simply cannot be solved by one discipline, or even by coupling two disciplines."[108]

Dow Chemical has recently given the University of California at Berkeley $10 million to establish a sustainability center. Dow's Neil Hawkins says, "Berkeley has one of the strongest chemical engineering schools in the world, but it will be the M.B.A.'s who understand areas like microfinance solutions to drinking water problems."[109] The center is in Berkeley's Center for Responsible Business, directed by Kellie A. McElhaney. Commercialization of research undertaken by students and professors is expected. However, McElhaney notes, "Commercialization takes forever if the chemical engineers and the business types do not coordinate. So think how much easier it will be for chemistry graduates to work inside a company if they already know how to interact with the business side."[110]

Discuss how considerations of ethics might enter into the collaborative efforts of centers and institutes for sustainability.

SERVICE LEARNING OPPORTUNITIES

The first two issues of the recently launched *International Journal for Service Learning* feature three articles promoting the notion that service learning projects can provide hands-on opportunities to undertake sustainable design and development. In "Service Learning in Engineering and Science for Sustainable

Development," Clarion University of Pennsylvania physicist Joshua M. Pearce urges that undergraduates should have opportunities to become involved in projects that apply appropriate technologies for sustainable development.[111] Especially concerned with alleviating poverty in the developing world, Pearce argues,

> The need for development is as great as it has ever been, but future development cannot simply follow past models of economic activity, which tended to waste resources and produce prodigious pollution. The entire world is now paying to clean up the mess and enormous quantities of valuable resources have been lost for future generations because of the Western model of development. For the future, the entire world population needs ways to achieve economic, social, and environmental objectives *simultaneously*.

He cites successful projects in Haiti and Guatemala that make use of readily available materials in the locales in which they have been undertaken.

In "Learning Sustainable Design through Service," Stanford University PhD students Karim Al-Khafaji and Margaret Catherine Morse present a service learning model based on the Stanford chapter of Engineers for a Sustainable World to teach sustainable design.[112] They illustrate this model in discussing a Stanford project in the Andaman Islands that focused on rebuilding after the December 26, 2004, earthquake and tsunami. Behind such projects is a student-led course, "Design for a Sustainable World," that seeks to

- Develop students' iterative design skills, project management and partnership-building abilities, sustainability awareness, cultural sensitivity, empathy, and desire to use technical skills to promote peace and human development.
- Help developing communities ensure individuals' human rights via sustainable, culturally appropriate, technology-based solutions.
- Increase Stanford University's stewardship of global sustainability.[113]

In "Sustainable Building Materials in French Polynesia," John Erik Anderson, Helena Meryman, and Kimberly Porsche, graduate students at the University of California at Berkeley's Department of Civil and Environmental Engineering, provide a detailed, technical description of a service learning project designed to assist French Polynesians in developing a system for the local manufacturing of sustainable building materials.[114]

CASE 40

Testing Water... and Ethics

The video *Testing Water...and Ethics* is a fictional portrayal of a young engineer facing his first professional dilemma. He attempts to solve the problem by treating it as analogous to a design problem in engineering. He also employs the method of seeking a creative middle way. This video is available from the Institute for Professional Practice, 13 Lanning Road, Verona, NJ 07044-2511 (phone, 1-888-477-2723; e-mail, Bridge2PE@aol.com).

CASE 41

Training Firefighters[115]

Donald J. Giffels, civil engineer and president of a large engineering consulting firm, was puzzled by the design of a government facility to train firefighters dealing with fire crashes of airplanes. His firm was under contract to do the civil engineering work for installing equipment at the facility. Because it contaminates the soil, jet fuel had recently been replaced by liquid propane for simulating crash fires. However, Giffels was concerned about a lack of design specificity in a number of areas crucial to safety (e.g., sprinkler systems, safeguards against flashbacks, fuel quantity, and fuel controls). Furthermore, no design analysis was submitted. Giffels concluded that none existed. However, none of this fell within the direct responsibility of Giffels's firm, whose contract was simply to do the civil engineering work required for installation.

Nevertheless, Giffels concluded that his firm could not simply let this go. He contacted the designers and asked them how they could justify putting their professional seal of approval on the design. They replied, "We don't need to. We're the government." Giffels agreed, but he persisted (to the point, he suspects, of making a pest of himself). Noting that it is easy to be a minimalist (e.g., stay within the law), Giffels worried that one might nevertheless fail to fulfill a responsibility to society. He contacted another engineering firm that had installed a similar design at 10 sites. It, too, he said, had been concerned about safety when looking at the designs. It contacted a mechanical engineering firm, asking it to do a design study. This request was turned down because of liability fears. So, the civil engineering firm asked the government agency to write a letter absolving it of

any responsibility in case of mishaps due to the inadequate design.

While not contesting the legality of this firm's way of dealing with the problem, Giffels insisted that this was not the correct way to proceed. His company refused to proceed with the installation until the safety issues were adequately addressed. The government agency agreed to bring in three other firms to deal with the concerns. Giffels firm's contract was modified to provide assurances that the safety issues would be addressed. Giffels stresses the importance of being able to communicate effectively about these matters—a communication responsibility. Good communication, he says, is essential to getting others on board.

Although successful in his efforts to ensure safety, Giffels says that this is not a story that would receive press notice. However, *not* resisting, he insists, might well have resulted in press coverage—such as from the deaths of firefighters going through their simulations.

Discuss the ethical challenges facing Giffels and his strategy in dealing with them.

CASE 42

TV Antenna[116]

Several years ago, a TV station in Houston decided to strengthen its signal by erecting a new, taller (1,000-foot) transmission antenna in Missouri City, Texas. The station contracted with a TV antenna design firm to design the tower. The resulting design employed twenty 50-foot segments that would have to be lifted into place sequentially by a jib crane that moved up with the tower. Each segment required a lifting lug to permit that segment to be hoisted off the flatbed delivery truck and then lifted into place by the crane. The actual construction of the tower was done by a separate rigging firm that specialized in such tasks.

When the rigging company received the 20th and last tower segment, it faced a new problem. Although the lifting lug was satisfactory for lifting the segment horizontally off the delivery truck, it would not enable the segment to be lifted vertically. The jib crane cable interfered with the antenna baskets at the top of the segment. The riggers asked permission

from the design company to temporarily remove the antenna baskets and were refused. Officials at the design firm said that the last time they gave permission to make similar changes, they had to pay tens of thousands of dollars to repair the antenna baskets (which had been damaged on removal) and to remount and realign them correctly.

The riggers devised a solution that was seriously flawed. They bolted an extension arm to the tower section and calculated the size of the required bolts based on a mistaken model. A sophomore-level engineering student who had taken a course in statics could have detected the flaw, but the riggers had no engineers on their staff. The riggers, knowing they lacked engineering expertise, asked the antenna design company engineers to review their proposed solution. The engineers again refused, having been ordered by company management not only not to look at the drawings but also not to visit the construction site during the lifting of the last segment. Management of the design firm feared that they would be held liable if there were an accident. The designers also failed to suggest to the riggers that they should hire an engineering consultant to examine their lifting plans.

When the riggers attempted to lift the top section of the tower with the microwave baskets, the tower fell, killing seven men. The TV company was taping the lift of the last segment for future TV promotions, and the videotape shows the riggers falling to their death.

Consider how you would react to watching that tape if you were the design engineer who refused to look at the lifting plans or if you were the company executive who ordered the design engineer not to examine the plans.

To take an analogy, consider a physician who examines a patient and finds something suspicious in an area outside her specialty. When asking advice from a specialist, the physician is rebuffed on the grounds that the specialist might incur a liability. Furthermore, the specialist does not suggest that the patient should see a specialist.

What conceptions of responsibility seemed most prevalent in this case? Can you suggest other conceptions that might have helped avoid this tragedy?

CASE 43

Unlicensed Engineer[117]

Charles Landers, former Anchorage assemblyman and unlicensed engineer for Constructing Engineers, was found guilty of forging partner Henry Wilson's signature and using his professional seal on at least 40 documents. The falsification of the documents was done without Wilson's knowledge, who was away from his office when they were signed. Constructing Engineers designs and tests septic systems. The signed and sealed documents certified to the Anchorage city health department that local septic systems met city wastewater disposal regulations. Circuit Judge Michael Wolverton banned Landers for 1 year from practicing as an engineer's, architect's, or land surveyor's assistant. The judge also sentenced him to 20 days in jail, 160 hours of community service, $4,000 in fines, and 1 year of probation. Finally, Landers was ordered to inform property owners about the problems with the documents, explain how he would rectify the problem, and pay for a professional engineer to review, sign, and seal the documents.

Assistant Attorney General Dan Cooper had requested the maximum penalty: a 4-year suspended sentence and $40,000 in fines. Cooper argued that "the 40 repeated incidents make his offense the most serious within the misuse of an engineer's seal." This may have been the first time a case like this was litigated in Alaska. The Attorney General's office took on the case after seeking advice from several professional engineers in the Anchorage area.

According to Cooper, Landers said he signed and sealed the documents because "his clients needed something done right away." (The documents were needed before proceeding with property transactions.) Lander's attorney, Bill Oberly, argued that his client should be sentenced as a least offender since public health and safety were not really jeopardized—subsequent review of the documents by a professional engineer found no violations of standards (other than forgery and the misuse of the seal). The documents were resubmitted without needing changes.

However, Judge Wolverton contended that Lander's actions constituted a serious breach of public trust. The public, he said, relies on the word of those, like professional engineers, who are entrusted with special responsibilities: "Our system would break down completely if the word of individuals could not be relied upon."

The judge also cited a letter from Richard Armstrong, chairman of the Architects, Engineers, and Land Surveyors Board of Registration for Alaska's Department of Commerce and Economic Development. Armstrong said,

> Some of the reasons for requiring professional engineers to seal their work are to protect the public from unqualified practitioners; to assure some minimum level of competency in the profession; to make practicing architects, engineers, and land surveyors responsible for their work; and to promote a level of ethics in the profession. The discovery of this case will cast a shadow of doubt on other engineering designed by properly licensed individuals.

Identify and discuss the ethically important elements in this case. How relevant is it that subsequent review showed that none of the falsified documents needed to be changed? (Although Judge Wolverton did not impose the maximum penalty, he did not treat Landers as a least offender.)

CASE 44

Where Are the Women?[118]

Although women have become more prevalent in engineering schools during the past few decades, they still make up only approximately 20 percent of engineering school undergraduates in the United States. Even this percentage is somewhat misleading. Women are more prevalent in some engineering fields than others. For example, more than 30 percent of the undergraduates in chemical engineering departments are women, but only 13 percent of the undergraduates in mechanical engineering and electrical engineering are women.[119] Eighteen percent of all engineering PhDs are awarded to women. There are even fewer women faculty in engineering schools. The higher the faculty rank, the fewer women there are. At the top rank of full professor, less than 5 percent are women.[120] This means that engineering students in the United States are taught and mentored almost exclusively by males, that there are few women faculty serving as role models for female students, and that engineering more generally remains dominated by men.

As interesting comparisons, women receive 57 percent of all baccalaureate degrees in the United States and 55 percent of all social science PhDs, women make up at least 50 percent of the students in medical and law schools, and 28 percent of full professors in the social sciences are women.[121] Therefore, what is happening in engineering schools? No doubt, there are a number of contributing factors to the fact that there are so few women in engineering. But many common beliefs about women and academic advancement in engineering prove to be without merit when the evidence is examined.

Belief Evidence

Belief	Evidence
1. Women are not as good in mathematics as men.	Female performance in high school mathematics now matches that of males.
2. It is only a matter of time before the issue of "underrepresentation" on faculties is resolved; it is a function of how many women are qualified to enter these positions.	Women's representation decreases with each step up the tenure track and academic leadership hierarchy, even in fields that have had a large proportion of women doctorates for 30 years.
3. Women are not as competitive as men. Women do not want jobs in academe.	Similar proportions of men and women with science and engineering doctorates plan to enter postdoctoral study or academic employment.
4. Women and minorities are recipients of favoritism through affirmative action programs.	Affirmative action is meant to broaden searches to include more women and minority group members but not to select candidates on the basis of race or sex, which is illegal.
5. Academe is a meritocracy.	Although scientists like to believe that they "choose the best" based on objective criteria, decisions are influenced by factors—including biases about race, sex, geographic location of a university, and age—that have nothing to do with the quality of the person or work being evaluated.
6. Changing the rules means that standards of excellence will be deleteriously affected.	Throughout a scientific career, advancement depends on judgments of one's performance by more senior scientists and engineers. This process does not optimally select and

(Continued)

Belief Evidence (Continued)

advance the best scientists and engineers because of implicit bias and disproportionate weighting of qualities that are stereotypically male. Reducing these sources of bias will foster excellence in science and engineering fields.

7. Women faculty are less productive than men.	The publication productivity of women science and engineering faculty has increased during the past 30 years and is now comparable to that of men. The critical factor affecting publication productivity is access to institutional resources; marriage, children, and elder care responsibilities have minimal effects.
8. Women are more interested in family than in careers.	Many women scientists and engineers persist in their pursuit of academic careers despite severe conflicts between their roles as parents and as scientists and engineers. These efforts, however, are often not recognized as representing the high level of dedication to their careers they represent.
9. Women take more time off due to childbearing, so they are a bad investment.	On average, women take more time off during their early careers to meet caregiving responsibilities, which fall disproportionately to women. However, by middle age, a man is likely to take more sick leave than a woman.
10. The system as currently configured has worked well in producing great science; why change it?	The global competitive balance has changed in ways that undermine America's traditional science and engineering advantages. Career impediments based on gender, racial, or ethnic bias deprive the nation of talented and accomplished researchers.[122]

Recently, a number of academic researchers have attempted to separate the myths from the facts about why so few women hold senior-level and leadership engineering positions. One plausible explanation is that slight disparities accumulate over time to disadvantage women and advantage men. Subconscious expectations tied to gender (gender schemas) are an important source of these disparities. We expect, for example, men to be the primary earners and women to be the primary providers of child care. A full range of studies on the influence of gender schemas in assessments of professional competence shows quite convincingly that over time, gender schemas contribute significantly to female engineering faculty being consistently underrated and male engineering faculty being consistently overrated.[123] Gender schemas are held unconsciously by both men and women and subtly influence perceptions and judgments made about one another.[124] Experimental data show, for example, that letters of reference for professional women tend to be shorter and to contain twice

as many doubt-raisers (e.g., "she has a somewhat challenging personality"), more grindstone adjectives (e.g., "hardworking" or "conscientious"), and fewer standout adjectives (e.g., "brilliant") as letters for men.[125] Other studies show that women tend to feel less entitled to high salaries and less confident in their mathematical abilities even when their actual performance levels equal those of male peers. Men are expected to be strong and assertive (leaders) and women to be nurturing listeners. As a result, women holding positions of leadership often must work harder to demonstrate actual leadership.

Because most of the faculty and administrators at engineering schools, both male and female, genuinely wish to advance and promote more women, focusing on gender schemas is especially relevant to advancing women in engineering fields. Virginia Valian, a researcher on gender schemas, makes this point. She writes, "The moral of the data on gender schemas is that good intentions are not enough; they will not guarantee the impartial and fair evaluation that we all hold as an ideal."[126] As engineering schools attempt to recruit and advance more women, it is important to assess the ways in which and the degree to which harmful gender schemas serve as barriers to women's advancement. At some institutions, such as the University of Michigan, such efforts have involved conducting gender schema workshops, forming focus groups, conducting interviews, and collecting survey data to assess the prevalence of gender schemas contributing to underrating women faculty in science, technology, engineering, and mathematics fields.[127]

One hypothesis is that once the harmful implicit schemas are made explicit, we can begin to address them at individual, departmental, and institutional levels and, at the very least, decrease their harmful impact.

Identify and discuss some of the subtle expectations both men and women have about gender. How do these gender schemas influence the advancement and promotion of women in engineering? Can you think of any examples from your own experience of men being advantaged and women being disadvantaged as a result of gender schemas?

CASE 45

XYZ Hose Co.[128]

Farmers use anhydrous ammonia to fertilize their fields. The anhydrous ammonia reacts violently with water, so care must be exercised in disbursing it. Farmers' cooperatives rent anhydrous ammonia in pressurized tanks equipped with wheels so the tanks can be pulled by tractors. The farmers also rent or purchase hoses that connect the tanks to perforated hollow blades that can be knifed through the soil to spread the ammonia. Leaks from the hose are potentially catastrophic.

For years, the industry standard hose was made of steel-meshed reinforced rubber, which was similar in construction to steel-reinforced automobile tires. Two separate trade associations had established these industry-wide standards.

Approximately 15 years ago, a new, heavy-duty plastic became available that could replace the steel in the hoses. The plastic-reinforced hoses were less expensive, lighter, and easier to process than the steel-braided rubber. The new hose met the industry standards. One company, the XYZ Hose Company, began marketing the plastic-reinforced hose to farmers. Officials of XYZ knew, as a result of tests by a consultant at a nearby state agricultural college, that the plastic did not react immediately to the anhydrous ammonia;

however, over the years the plastic did degrade and lose some of its mechanical properties. Accordingly, they put warnings on all the hoses they manufactured, indicating that they should be replaced periodically.

After the product had been on the market a few years, several accidents occurred in which the XYZ hoses ruptured during use and blinded and severely injured the farmers using them. Litigation followed, and XYZ argued in its defense that the farmers had misused the hoses and not heeded the replacement warnings. This defense was unsuccessful, and XYZ made substantial out-of-court settlements.

XYZ has since dropped this product line and placed advertisements in farmers' trade journals and producers' cooperatives newsletters asking farmers to turn in their XYZ hoses for full refunds. The advertisements state that the hoses are "obsolete," not that they are unsafe.

Identify and discuss the ethical issues this case raises, paying special attention to relevant, key ideas presented in this chapter. What are the relevant facts? What factual, conceptual, and application issues are there? What methods for resolving these issues might be used?

NOTES

1. Steven Weisskoph, "The Aberdeen Mess," *Washington Post Magazine*, January 15, 1989.
2. *The Aberdeen Three*, a case prepared under National Science Foundation grant number DIR-9012252. The principal investigators were Michael J. Rabins, Charles E. Harris, Jr., Charles Samson, and Raymond W. Flumerfelt. The complete case is available at the Texas A & M Engineering Ethics website (http://ethics.tamu.edu).
3. Case study prepared by Ryan Pflum, MA philosophy student at Western Michigan University.

4. Case study prepared by Ryan Pflum.

5. This account is based on John H. Cushman, Jr., "G.M. Agrees to Cadillac Recall in Federal Pollution Complaint," *New York Times,* December 1, 1995, pp. A1 and A12.

6. Leonardo Da Vinci, *The Notebooks of Leonardo Da Vinci,* vol. I, Edward MacCurdy, ed. (New York: George Braziller, 1939), p. 850. Cited in Mike Martin and Roland Schinzinger, *Ethics in Engineering,* 3rd ed. (New York: McGraw-Hill, 1996), p. 246.

7. This account is based on Joe Morgenstern, "The Fifty-Nine Story Crisis," *The New Yorker Magazine,* May 29, 1995, 49–53. For more on William LeMessurier and the Citicorp building, see the Online Ethics Center for Engineering (http://www.onlineethics.org/CMS/profpractice/exemp.index.aspx).

8. Much of what follows is based on Michael S. Pritchard, "Professional Responsibility: Focusing on the Exemplary," *Science and Engineering Ethics,* 4, 1998, pp. 230–233. In addition to sources cited here, there is an excellent PBS *Frontline* documentary on Cuny, "The Lost American." This is available at PBS Video, P.O. box 791, Alexandria, VA 22313-0791. There is a wealth of additional information on Cuny online at http://www.pbs.org/wgbh/pages/frontline/shows/cuny/bio/chron.html. Also, Cuny is featured as a moral leader on the Online Ethics Center for Engineering.

9. Karen W. Arenson, "Missing Relief Expert Gets MacArthur Grant," *New York Times,* June 13, 1995, p. A12.

10. Ibid.

11. Scott Anderson's gripping account of Frederick Cuny's life portrays him as a person with many foibles and shortcomings who still managed to save the lives of thousands threatened by man-made and natural disasters. See Scott Anderson, *The Man Who Tried to Save the World: The Dangerous Life and Mysterious Disappearance of Fred Cuny* (New York: Doubleday, 1999).

12. From Intertect's corporate brochure.

13. Ibid.

14. Quoted in William Shawcross, "A Hero of Our Time," *New York Review of Books,* Nov. 30, 1995, p. 35. The next paragraph is based on Shawcross's article.

15. The following is based on Chuck Sudetic, "Small Miracle in a Siege: Safe Water for Sarajevo," *New York Times,* January 10, 1994, pp. A1 and A7.

16. This account is based on "The Talk of the Town," *The New Yorker,* 69, no. 39, Nov. 22, 1993, pp. 45–46.

17. Anderson, *The Man Who Tried to Save the World,* p. 120.

18. Ibid.

19. Ibid. This expresses a thought attributed to Cuny.

20. Frederick C. Cuny, "Killing Chechnya," *The New York Review of Books,* April 6, 1995, pp. 15–17.

21. Marilyn Greene, "Texas Disaster Relief 'Visionary' Vanishes on Chechnya Mission," *USA Today,* May 10, 1995, p. A10.

22. Shawcross, "A Hero of Our Time," p. 39.

23. "Talk of the Town," p. 46.

24. Sudetic, "Small Miracle in a Siege," p. A7.

25. John Allen, "The Switch," *On Wisconsin,* Fall 2001, pp. 38–43.

26. Ibid., p. 42.

27. Ibid., p. 41.

28. Ibid.

29. These case studies were written by Theodore D. Goldfarb and Michael S. Pritchard for their *Ethics in the Science Classroom* (http://www.onlineethics.org). This text is a product of two National Science Foundation grants on teaching ethics in the science classroom: SBR-9601284 and SBR-932055.

30. Sources for information on the Darsee case include Sharen Begley, with Phyllis Malamud and Mary Hager, "A Case of Fraud at Harvard," *Newsweek,* February 4, 1982, pp. 89–92; Richard Knox, "The Harvard Fraud Case: Where Does the Problem Lie?" *Journal of the American Medical Association,* 249, no. 14, April 3, 1983, pp. 1797–1807; Walter W. Stewart, "The Integrity of the Scientific Literature," *Nature,* 325, January 15, 1987, pp. 207–214; Eugene Braunwald, "Analysing Scientific Fraud," *Nature,* 325, January 15, 1987, pp. 215–216; and Eugene Brunwald, "Cardiology: The John Darsee Experience," in David J. Miller and Michel Hersen, eds., *Research Fraud in the Behavioral and Biomedical Sciences* (New York: Wiley, 1992), pp. 55–79.

31. David DeMets, "Statistics and Ethics in Medical Research," *Science and Engineering Ethics,* 5, no. 1, January 1999, p. 111. At the 1994 Teaching Research Ethics for Faculty Workshop at Indiana University's Poynter Center, DeMets recounted in great detail the severe challenges

he and his team of statisticians faced in carrying out their investigation.

32. Eugene Braunwald, "Cardiology: The John Darsee Experience," in David J. Miller and Michel Hersen, eds., *Research Fraud in the Behavioral and Biomedical Sciences* (New York: Wiley, 1992), pp. 55–79.

33. William F. May, "Professional Virtue and Self-Regulation," in Joan Callahan, ed., *Ethical Issues in Professional Life* (New York: Oxford University Press, 1988), p. 408.

34. For readings on the Bruening case, see Robert L. Sprague, "The Voice of Experience," *Science and Engineering Ethics*, 4, no. 1, 1998, p. 33; and Alan Poling, "The Consequences of Fraud," in Miller and Hersen, pp. 140–157.

35. Robert L. Sprague, "The Voice of Experience," *Science and Engineering Ethics*, Vol. 4, 1, 1998, p. 33.

36. This video was produced by the National Society for Professional Ethics (Alexandria, VA) in 1989. Information about obtaining it can be found at the Murdough Center for Engineering Ethics website, http://www.niee.org/pd.cfm?pt= Murdough. This website also contains the entire transcript for this video.

37. One would be to find a cheap technical way to eliminate the heavy metals. Unfortunately, the video does not directly address this possibility. It begins in the midst of a crisis at ZCORP and focuses almost exclusively on the question of whether David Jackson should blow the whistle on his reluctant company. For a detailed exploration of some creative middle way alternatives, see Michael Pritchard and Mark Holtzapple, "Responsible Engineering: *Gilbane Gold* Revisited," *Science and Engineering*, 3, no. 2, April 1997, pp. 217–231.

38. This case is based on Kirk Johnson, "A Deeply Green City Confronts Its Energy Needs and Nuclear Worries," *New York Times*, November 19, 2007 (http://www.nytimes.com/2007/11/19/us/19collins.html?th=&emc=).

39. This case is based on Felicity Barringer and Micheline Maynard, "Court Rejects Fuel Standards on Trucks," *New York Times*, Nov. 16, 2007 (http://www.nytimes.com/2007/11/16/business/16fuel.html?th&emc=th).

40. Ibid.

41. We first learned of this true case (with names changed) from Sam's daughter, who was an honor student in two of the authors' engineering ethics classes and a member of a team of students from that class that competed in the College Ethics Bowl competition held at Loyola/Marymount College in Los Angeles. She suggested that the team present a case based on her father's experience. The team won the competition with its discussion of the case described here (which has been reviewed by "Sam" for accuracy).

42. This is an adaptation of a case developed by James Taylor, Civil Engineering, Notre Dame University.

43. ASCE Hurricane Katrina External Review Panel, *The New Orleans Hurricane Protection System: What Went Wrong and Why* (Reston, VA: American Society for Civil Engineers, 2007). Available at http://www.asce.org/static/hurricane/erp.cfm.

44. Ibid., p. 47.

45. Ibid., p. 61.

46. Ibid.

47. Ibid., p.73.

48. Ibid., p. 79.

49. Ibid.

50. Ibid.

51. Ibid., p. 81.

52. Ibid., p. 82.

53. Ibid., p. 82.

54. Jacqueline Finger, Joseph Lopez, III, Christopher Barallus, Matthew Parisi, Fred Rohs, John Schmalzel, Amrinder Kaur, DeMond S. Miller, and Kimberly Rose, "Leadership, Service Learning, and Executive Management in Engineering: The Rowan University Hurricane Katrina Recovery Team," *International Journal for Service Learning in Engineering*, 2, no. 2, Fall 2007.

55. Katie Hafner and Claudia H. Deutsch, "When Good Will Is Also Good Business," *New York Times*, September 14, 2005 (http://nytimes.com).

56. Ibid.

57. This account is drawn from R. W. Flumerfelt, C. E. Harris, M. J. Rabins, and C. H. Samson, eds., *Introducing Ethics Case Studies into Required Undergraduate Engineering Courses*, NSF Grant no. DIR-9012252, November 1992. The full version is available at the Texas A & M Engineering Ethics website (http://ethics.tamu.edu).

58. Paula Wells, Hardy Jones, and Michael Davis, *Conflicts of Interest in Engineering*, Module Series in Applied Ethics, Center for the Study of Ethics in the Professions, Illinois Institute of Technology (Dubuque, IA: Kendall/Hunt, 1986), p. 20.

288 CASES

59. American Society of Mechanical Engineers, Boiler and Pressure Vessel Code, section IV, paragraph HG-605a.

60. Charles W. Beardsley, "The Hydrolevel Case—A Retrospective," *Mechanical Engineering,* June 1984, p. 66.

61. Ibid., p. 73.

62. This case is based on an article that appeared in *The Seattle Times,* July 24, 2000.

63. This case study was written by Theodore D. Goldfarb and appeared in Theodore D. Goldfarb and Michael S. Pritchard, *Ethics in the Science Classroom* (http://www.onlineethics.org). It is a product of two National Science Foundation grants on teaching ethics in the science classroom: SBR-9601284 and SBR-932055.

64. This case is based on Stephen H. Unger's account in *Controlling Technology: Ethics and the Responsible Engineer* (New York: Holt, Rinehart & Winston, 1994), pp. 27–30.

65. Much of this account is adapted from Theodore D. Goldfarb and Michael S. Pritchard for their *Ethics in the Science Classroom* (http://www.onlineethics.org). This text is a product of two National Science Foundation grants on teaching ethics in the science classroom: SBR-9601284 and SBR-932055.

66. See Richard A. Shweder, Elliot Turiel, and Nancy C. Much, "The Moral Intuitions of the Child," in John H. Flavell and Lee Ross, eds., *Social Cognitive Development: Frontiers and Possible Futures* (Cambridge, UK: Cambridge University Press, 1981), p. 288.

67. Gareth Matthews, "Concept Formation and Moral Development," in James Russell, ed., *Philosophical Perspectives on Developmental Psychology* (Oxford: Basil Blackwell, 1987), p. 185.

68. For balanced, accessible discussions of recent findings in moral development, see William Damon, *The Moral Child* (New York: Free Press, 1988); and Daniel K. Lapsley, *Moral Psychology* (Boulder, CO: Westview Press, 1996).

69. See, for example, Lawrence Kohlberg, *The Philosophy of Moral Development: Essays on Moral Development,* Vol. 1 (San Francisco: Harper & Row, 1981).

70. Michael Pritchard has written extensively on many of them elsewhere. See his *On Becoming Responsible* (Lawrence: University Press of Kansas, 1991); *Reasonable Children* (Lawrence: University Press of Kansas, 1996); and "Kohlbergian Contributions to Educational Programs

for the Moral Development of Professionals," *Educational Psychology Review,* 11, no. 4, 1999, pp. 397–411. James Rest, Muriel Bebeau, Darcia Narvaez, and Stephen Thoma have developed what they call a neo-Kohlbergian account. They identify three "schemas": personal interest, maintaining norms, and postconventional. In general, these three schemas correlate significantly with Kohlberg's three basic levels of moral development. However, they maintain that a much higher percentage of adults meet their postconventional criteria than Kohlberg suggests meet his postconventional level. See Rest et al., *Postconventional Moral Thinking* (Mahwah, NJ: Erlbaum, 1999); and their essays in *Educational Psychology Review,* 11, no. 4, 1999.

71. This case was developed by P. Aarne Vesilind, Department of Civil and Environmental Engineering at Duke University.

72. This account is based on Loren Graham's *The Ghost of the Executed Engineer: Technology and the Fall of the Soviet Union* (Cambridge, MA: Harvard University Press, 1993).

73. Ibid., p. 106.

74. Information for this case is based on a case study prepared by Manuel Velasquez, "The Ford Motor Car," in Manuel Velasquez, *Business Ethics: Concepts and Cases,* 3rd ed. (Englewood Cliffs, NJ: Prentice-Hall, 1992), pp. 110–113.

75. *Grimshaw v. Ford Motor Co.,* app., 174 Cal. Rptr. 348, p. 360.

76. This is reported in Ralph Drayton, "One Manufacturer's Approach to Automobile Safety Standards," *CTLA News,* VIII, no. 2 (February 1968), p. 11.

77. Mark Dowie, "Pinto Madness," *Mother Jones,* September/October 1977, p. 28.

78. Amy Docker Marcus, "MIT Students, Lured to New Tech Firms, Get Caught in a Bind," *The Wall Street Journal,* June 24, 1999, pp. A1, A6.

79. Ibid., p. A6.

80. Ibid.

81. Ibid.

82. John Markoff, "Odyssey of a Hacker: From Outlaw to Consultant," *New York Times,* January 29, 2001.

83. David Lorge Parnas, "SDI: A Violation of Professional Responsibility," in Deborah Johnson, ed., *Ethical Issues in Engineering* (Englewood Cliffs, NJ: Prentice-Hall, 1991), pp. 15–25. This case is based on Pritchard's discussion in "Computer

Ethics: The Responsible Professional," in James A. Jaksa and Michael S. Pritchard, eds., *Responsible Communication: Ethical Issues in Business, Industry, and the Professions* (Cresskill, NJ: Hampton Press, 1996), pp. 146–148.

84. Ibid., p. 17.

85. Ibid., p. 15.

86. Ibid., p. 25.

87. Parnas was convinced that the public, when informed, would agree with his conclusions about the SDI program. For a contrary view, see the debate between David Parnas and Danny Cohen, "Ethics and Military Technology: Star Wars," in Kristen Shrader-Frechette and Laura Westra, eds., *Technology and Values* (New York: Rowman & Littlefield, 1997), pp. 327–353.

88. Ibid.

89. This account is based on the authors' conversations with Ed Turner as well as information available at http://www.responsiblecharge.com.

90. This account is based on G. P. E. Meese, "The Sealed Beam Case," *Business & Professional Ethics*, 1, no. 3, Spring 1982, pp. 1–20.

91. H. H. Magsdick, "Some Engineering Aspects of Headlighting," *Illuminating Engineering*, June 1940, p. 533, cited in Meese, p. 17.

92. Much of this case is adapted from Michael S. Pritchard, "Service-Learning and Engineering Ethics," *Science and Engineering Ethics*, 6, 2000, pp. 413–422. An earlier version of this article is available at the Online Ethics Center (http://www.onlineethics.org/CMS/edu/resources/servicelearning.aspx).

93. Accreditation Board for Engineering and Technology, *Fifty-Third Annual Report*, 1985, p. 98.

94. Campus Compact supports the development of service learning programs throughout the country. For an early statement of its efforts, see Timothy Stanton, *Integrating Public Service with Academic Study* (Providence, RI: Campus Compact, Brown University, 1989).

95. Edmund Tsang, "Why Service Learning? And How to Integrate It into a Course in Engineering," in Kathryn Ritter-Smith and John Saltmarsh, eds., *When Community Enters the Equation: Enhancing Science, Mathematics and Engineering Education through Service-Learning* (Providence, RI: Campus Compact, Brown University, 1998). Currently at Western Michigan University as associate dean in the College of Engineering & Applied Sciences, Tsang continues

his work in service learning. He has also edited *Projects that Matter: Concepts and Models for Service-Learning in Engineering*, Vol. 14 (Washington, DC: American Association for Higher Education, 2000). Service learning is appropriate throughout both undergraduate and graduate programs in engineering. This is well illustrated by Kevin Passino, Professor of Electrical and Computing Engineering at Ohio State University. In addition to founding the student-centered Engineers for Community Service at his university, he is developing international service learning projects for PhD students. See his "Educating the Volunteer Engineer," available on his website (http://www.ece.osu.edu/~passino/professionalism.html) and forthcoming in *Science and Engineering Ethics*.

96. This account is based on a conversation with Tom Talley and Dave Wylie's, "AVIT Team Helps Disabled Children," *Currents* (Texas A & M University), Summer 1993, p. 6.

97. *Research Agenda for Combining Service and Learning in the 1990s* (Raleigh, NC: National Society for Internships and Experiential Education, 1991), p. 7.

98. CESG brochure.

99. CESG Strategic Plan Draft: 1997–2000, pp. 1–2.

100. This case is based on Glenn Collins, "What Smoke? New Device Keeps Cigarettes in a 'Box'," *New York Times*, October 23, 1997, pp. A1, C8.

101. Case presented by Pritchard in "Computer Ethics: The Responsible Professional," pp. 144–145.

102. Joe Gertner, "The Future Is Drying Up," *New York Times Magazine*, October 21, 2007.

103. One is environmental engineer Bradley Udall, son of U.S. Congressman Morris Udall and nephew of Stewart Udall, Secretary of the Interior under Presidents John F. Kennedy and Lyndon Johnson.

104. From Michael S. Pritchard, "Professional Responsibility: Focusing on the Exemplary," *Science and Engineering Ethics*, 4, 1998, p. 224.

105. See the May 1997 report by the Biomass Energy Design Project Team, "Design and Feasibility Study of a Biomass Energy Farm at Lafayette College as a Fuel Source for the Campus Steam Plant."

106. Accreditation Board for Engineering and Technology, *Engineering Criteria 2000*, 3rd ed. (Baltimore: Author, 1997). See also http://www.abet.org/eac2000html.

107. This case is based on Claudia H. Deutsch, "A Threat So Big, Academics Try Collaboration," *New York Times,* December 25, 2007 (http://www.nytimes.com/2007/12/25/business/25sustain.html?8br).

108. Ibid.

109. Ibid.

110. Ibid.

111. Joshua M. Pearce, "Service Learning in Engineering and Science for Sustainable Development," *International Journal for Service Learning in Engineering,* 1, No. 1, Spring 2006.

112. Karim Al-Khafaji and Margaret Catherine Morse, "Learning Sustainable Design through Service," *International Journal for Service Learning in Engineering,* 1, No. 1, Spring 2006.

113. Ibid., quoted from the article.

114. John Erik Anderson, Helena Meryman, and Kimberly Porsche, "Sustainable Building Materials in French Polynesia," *International Journal for Service Learning in Engineering,* 2, No. 2, Fall 2007.

115. From Michael S. Pritchard, "Professional Responsibility: Focusing on the Exemplary," *Science and Professional Ethics,* 4, 1998, pp. 225–226. This is based on Donald J. Giffels's commentary on Pritchard's speech, "Education for Responsibility: A Challenge to Engineers and Other Professionals," presented at the Third Annual Lecture in Ethics in Engineering, Center for Academic Ethics, Wayne State University, April 19, 1995.

116. This case is presented in greater detail, complete with an instructor's guide and student handouts, in R. W. Flumerfelt, C. E. Harris, M. J. Rabins, and C. H. Samson, eds, *Introducing Ethics Case Studies into Required Undergraduate Engineering Courses,* final report to NSF on grant no. DIR-9012252, November 1992, pp. 231–261. The case is available at the Texas A & M Engineering Ethics website (http://ethics.tamu.edu/).

117. This case is based on Molly Galvin, "Unlicensed Engineer Receives Stiff Sentence," *Engineering Times,* 16, no. 10, October 1994, pp. 1 and 6.

118. This discussion was researched and authored by Peggy DesAutels, a philosopher at University of Dayton who has special interests in gender and engineering issues.

119. 2001 statistics from the National Science Foundation, http://www.nsf.gov/statistics/wmpd.

120. 2003 statistic reported in *Beyond Bias and Barriers: Fulfilling the Potential of Women in Academic Science and Engineering* (Washington, DC: National Academies Press, 2006), pp. 14–17. This report was produced by the Committee on Maximizing the Potential of Women in Academic Science and Engineering and the Committee on Science, Engineering, and Public Policy, National Academy of Sciences, National Academy of Engineering, and Institute of Medicine of the National Academies.

121. 2003 statistic reported in *Beyond Bias and Barriers: Fulfilling the Potential of Women in Academic Science and Engineering,* pp. 14–17.

122. This table is modified from a table in *Beyond Bias and Barriers: Fulfilling the Potential of Women in Academic Science and Engineering,* pp. 5–6.

123. *Beyond Bias and Barriers: Fulfilling the Potential of Women in Academic Science and Engineering,* pp. 158–159.

124. Virginia Valian, "Beyond Gender Schemas: Improving the Advancement of Women in Academia," *Hypatia* 20, no. 3, Summer 2005, pp. 198–213.

125. F. Trix and C. Psenka, "Exploring the Color of Glass: Letters of Recommendation for Female and Male Medical Faculty," *Discourse and Society,* 14, 2003, pp. 191–220.

126. Valian, p. 202.

127. NSF ADVANCE Project at the University of Michigan (http://www.umich.edu/~advproj).

128. This case is supplied by an engineering colleague who was an expert witness in the case. We have given the company the fictitious name of "XYZ." For a more complete account, see R. W. Flumerfelt, C. E. Harris, M. J. Rabins, and C. H. Samson, *Introducing Ethics Case Studies into Required Undergraduate Engineering Courses,* pp. 287–312.

Codes of Ethics

IN THIS APPENDIX, the code of the National Society of Professional Engineers (NSPE) is printed, and web sources for most of the other major engineering codes are provided, together with a few comments on features of the codes that are worth particular notice. The NSPE code has been selected for inclusion for two primary reasons. First, membership in the NSPE is open to all professional engineers, regardless of their particular engineering discipline, such as electrical, mechanical, or civil engineering. For this reason, the code is in principle applicable to all engineers. This feature distinguishes the NSPE code from the codes of those professional societies that are open only to members of a particular engineering discipline. Electrical engineers, for example, might not be especially interested in the code of mechanical or civil engineering, but they should be interested in the provisions of the NSPE code since they are potential members of this organization. Second, the NSPE code is a very complete code and in general is representative of the other codes. Codes do, however, address the ethical problems that arise in their particular branch of engineering, and there may be some differences in the codes because of this. Codes may also differ because of the special "culture" of the professional societies.

Because the NSPE code is printed here in full and is in general representative of engineering codes of ethics, several features of the code deserve mention:

- The highest ethical obligation of engineers is to the "safety, health, and welfare of the public." Virtually every engineering code contains similar wording and makes it clear that the obligation to the public takes priority over obligations to clients or employers.
- Engineers must also act for clients or employers as "faithful agents or trustees," with the implicit understanding that this obligation is subordinate to the obligation to the public.
- Engineers must practice only in their areas of competence.
- Engineers must act objectively, truthfully, and in a way that avoids deception and misrepresentation, especially to the public. This includes avoiding bribes or other actions that might compromise an engineer's professional integrity.

- Engineers are encouraged (not required) to participate in civic affairs, such as career guidance for youth, and not only to promote or "work for the advancement of the safety, health, and well-being of their community."
- Engineers are encouraged (not required) to adhere to the principles of sustainable development in order to protect the environment for future generations. In an endnote, sustainable development is defined as "meeting human needs...while conserving and protecting environmental quality and the natural resource base essential for human development." Increasingly, codes are making reference to the concept of sustainable development as well as the obligation to protect the environment.
- Finally, engineers have an obligation to other engineers and to the engineering profession. The obligation to other engineers requires them to refrain from such activities as untruthfully criticizing the work of other engineers and to give credit to other engineers when appropriate. The obligation to the engineering profession requires them to conduct their work (and their advertising) with dignity as well as according to ethical standards.

NSPE CODE OF ETHICS FOR ENGINEERS[1]

Preamble

Engineering is an important and learned profession. As members of this profession, engineers are expected to exhibit the highest standards of honesty and integrity. Engineering has a direct and vital impact on the quality of life for all people. Accordingly, the services provided by engineers require honesty, impartiality, fairness, and equity, and must be dedicated to the protection of the public health, safety, and welfare. Engineers must perform under a standard of professional behavior that requires adherence to the highest principles of ethical conduct.

I. Fundamental Canons

Engineers, in the fulfillment of their professional duties, shall:

1. Hold paramount the safety, health, and welfare of the public.
2. Perform services only in areas of their competence.
3. Issue public statements only in an objective and truthful manner.
4. Act for each employer or client as faithful agents or trustees.
5. Avoid deceptive acts.
6. Conduct themselves honorably, responsibly, ethically, and lawfully so as to enhance the honor, reputation, and usefulness of the profession.

II. Rules of Practice

1. Engineers shall hold paramount the safety, health, and welfare of the public.

   a. If engineers' judgment is overruled under circumstances that endanger life or property, they shall notify their employer or client and such other authority as may be appropriate.

   b. Engineers shall approve only those engineering documents that are in conformity with applicable standards.

    c. Engineers shall not reveal facts, data, or information without the prior consent of the client or employer except as authorized or required by law or this Code.

    d. Engineers shall not permit the use of their name or associate in business ventures with any person or firm that they believe is engaged in fraudulent or dishonest enterprise.

    e. Engineers shall not aid or abet the unlawful practice of engineering by a person or firm.

    f. Engineers having knowledge of any alleged violation of this Code shall report thereon to appropriate professional bodies and, when relevant, also to public authorities, and cooperate with the proper authorities in furnishing such information or assistance as may be required.

2. Engineers shall perform services only in the areas of their competence.

    a. Engineers shall undertake assignments only when qualified by education or experience in the specific technical fields involved.

    b. Engineers shall not affix their signatures to any plans or documents dealing with subject matter in which they lack competence, nor to any plan or document not prepared under their direction and control.

    c. Engineers may accept assignments and assume responsibility for coordination of an entire project and sign and seal the engineering documents for the entire project, provided that each technical segment is signed and sealed only by the qualified engineers who prepared the segment.

3. Engineers shall issue public statements only in an objective and truthful manner.

    a. Engineers shall be objective and truthful in professional reports, statements, or testimony. They shall include all relevant and pertinent information in such reports, statements, or testimony, which should bear the date indicating when it was current.

    b. Engineers may express publicly technical opinions that are founded upon knowledge of the facts and competence in the subject matter.

    c. Engineers shall issue no statements, criticisms, or arguments on technical matters that are inspired or paid for by interested parties, unless they have prefaced their comments by explicitly identifying the interested parties on whose behalf they are speaking, and by revealing the existence of any interest the engineers may have in the matters.

4. Engineers shall act for each employer or client as faithful agents or trustees.

    a. Engineers shall disclose all known or potential conflicts of interest that could influence or appear to influence their judgment or the quality of their services.

    b. Engineers shall not accept compensation, financial or otherwise, from more than one party for services on the same project, or for services pertaining to the same project, unless the circumstances are fully disclosed and agreed to by all interested parties.

    c. Engineers shall not solicit or accept financial or other valuable consideration, directly or indirectly, from outside agents in connection with the work for which they are responsible.

d. Engineers in public service as members, advisors, or employees of a governmental or quasi-governmental body or department shall not participate in decisions with respect to services solicited or provided by them or their organizations in private or public engineering practice.

e. Engineers shall not solicit or accept a contract from a governmental body on which a principal or officer of their organization serves as a member.

5. Engineers shall avoid deceptive acts.

a. Engineers shall not falsify their qualifications or permit misrepresentation of their or their associates' qualifications. They shall not misrepresent or exaggerate their responsibility in or for the subject matter of prior assignments. Brochures or other presentations incident to the solicitation of employment shall not misrepresent pertinent facts concerning employers, employees, associates, joint venturers, or past accomplishments.

b. Engineers shall not offer, give, solicit, or receive, either directly or indirectly, any contribution to influence the award of a contract by public authority, or which may be reasonably construed by the public as having the effect or intent of influencing the awarding of a contract. They shall not offer any gift or other valuable consideration in order to secure work. They shall not pay a commission, percentage, or brokerage fee in order to secure work, except to a bona fide employee or bona fide established commercial or marketing agencies retained by them.

III. Professional Obligations

1. Engineers shall be guided in all their relations by the highest standards of honesty and integrity.

a. Engineers shall acknowledge their errors and shall not distort or alter the facts.

b. Engineers shall advise their clients or employers when they believe a project will not be successful.

c. Engineers shall not accept outside employment to the detriment of their regular work or interest. Before accepting any outside engineering employment, they will notify their employers.

d. Engineers shall not attempt to attract an engineer from another employer by false or misleading pretenses.

e. Engineers shall not promote their own interest at the expense of the dignity and integrity of the profession.

2. Engineers shall at all times strive to serve the public interest.

a. Engineers are encouraged to participate in civic affairs; career guidance for youths; and work for the advancement of the safety, health, and well-being of their community.

b. Engineers shall not complete, sign, or seal plans and/or specifications that are not in conformity with applicable engineering standards. If the client or employer insists on such unprofessional conduct, they shall notify the proper authorities and withdraw from further service on the project.

c. Engineers are encouraged to extend public knowledge and appreciation of engineering and its achievements.

d. Engineers are encouraged to adhere to the principles of sustainable development* in order to protect the environment for future generations.

3. Engineers shall avoid all conduct or practice that deceives the public.

   a. Engineers shall avoid the use of statements containing a material misrepresentation of fact or omitting a material fact.

   b. Consistent with the foregoing, engineers may advertise for recruitment of personnel.

   c. Consistent with the foregoing, engineers may prepare articles for the lay or technical press, but such articles shall not imply credit to the author for work performed by others.

4. Engineers shall not disclose, without consent, confidential information concerning the business affairs or technical processes of any present or former client or employer, or public body on which they serve.

   a. Engineers shall not, without the consent of all interested parties, promote or arrange for new employment or practice in connection with a specific project for which the engineer has gained particular and specialized knowledge.

   b. Engineers shall not, without the consent of all interested parties, participate in or represent an adversary interest in connection with a specific project or proceeding in which the engineer has gained particular specialized knowledge on behalf of a former client or employer.

5. Engineers shall not be influenced in their professional duties by conflicting interests.

   a. Engineers shall not accept financial or other considerations, including free engineering designs, from material or equipment suppliers for specifying their product.

   b. Engineers shall not accept commissions or allowances, directly or indirectly, from contractors or other parties dealing with clients or employers of the engineer in connection with work for which the engineer is responsible.

6. Engineers shall not attempt to obtain employment or advancement or professional engagements by untruthfully criticizing other engineers, or by other improper or questionable methods.

   a. Engineers shall not request, propose, or accept a commission on a contingent basis under circumstances in which their judgment may be compromised.

   b. Engineers in salaried positions shall accept part-time engineering work only to the extent consistent with policies of the employer and in accordance with ethical considerations.

   c. Engineers shall not, without consent, use equipment, supplies, laboratory, or office facilities of an employer to carry on outside private practice.

---

*"Sustainable development" is the challenge of meeting human needs for natural resources, industrial products, energy, food, transportation, shelter, and effective waste management while conserving and protecting environmental quality and the natural resource base essential for future development.
—As Revised July 2007

7. Engineers shall not attempt to injure, maliciously or falsely, directly or indirectly, the professional reputation, prospects, practice, or employment of other engineers. Engineers who believe others are guilty of unethical or illegal practice shall present such information to the proper authority for action.

   a. Engineers in private practice shall not review the work of another engineer for the same client, except with the knowledge of such engineer, or unless the connection of such engineer with the work has been terminated.
   b. Engineers in governmental, industrial, or educational employ are entitled to review and evaluate the work of other engineers when so required by their employment duties.
   c. Engineers in sales or industrial employ are entitled to make engineering comparisons of represented products with products of other suppliers.

8. Engineers shall accept personal responsibility for their professional activities, provided, however, that engineers may seek indemnification for services arising out of their practice for other than gross negligence, where the engineer's interests cannot otherwise be protected.

   a. Engineers shall conform with state registration laws in the practice of engineering.
   b. Engineers shall not use association with a nonengineer, a corporation, or partnership as a "cloak" for unethical acts.

9. Engineers shall give credit for engineering work to those to whom credit is due, and will recognize the proprietary interests of others.

   a. Engineers shall, whenever possible, name the person or persons who may be individually responsible for designs, inventions, writings, or other accomplishments.
   b. Engineers using designs supplied by a client recognize that the designs remain the property of the client and may not be duplicated by the engineer for others without express permission.
   c. Engineers, before undertaking work for others in connection with which the engineer may make improvements, plans, designs, inventions, or other records that may justify copyrights or patents, should enter into a positive agreement regarding ownership.
   d. Engineers' designs, data, records, and notes referring exclusively to an employer's work are the employer's property. The employer should indemnify the engineer for use of the information for any purpose other than the original purpose.
   e. Engineers shall continue their professional development throughout their careers and should keep current in their specialty fields by engaging in professional practice, participating in continuing education courses, reading in the technical literature, and attending professional meetings and seminars.

"By order of the United States District Court for the District of Columbia, former Section 11(c) of the NSPE Code of Ethics prohibiting competitive bidding, and all policy statements, opinions, rulings or other guidelines interpreting its scope, have been rescinded as unlawfully interfering with the legal right of engineers, protected under the antitrust laws, to provide price information to prospective clients; accordingly, nothing contained in the NSPE Code of Ethics, policy statements, opinions, rulings or other guidelines prohibits the submission of price quotations or competitive bids for engineering services at any time or in any amount."

Statement by NSPE Executive Committee

In order to correct misunderstandings which have been indicated in some instances since the issuance of the Supreme Court decision and the entry of the Final Judgment, it is noted that in its decision of April 25, 1978, the Supreme Court of the United States declared: "The Sherman Act does not require competitive bidding." It is further noted that as made clear in the Supreme Court decision:

1. Engineers and firms may individually refuse to bid for engineering services.
2. Clients are not required to seek bids for engineering services.
3. Federal, state, and local laws governing procedures to procure engineering services are not affected, and remain in full force and effect.
4. State societies and local chapters are free to actively and aggressively seek legislation for professional selection and negotiation procedures by public agencies.
5. State registration board rules of professional conduct, including rules prohibiting competitive bidding for engineering services, are not affected and remain in full force and effect. State registration boards with authority to adopt rules of professional conduct may adopt rules governing procedures to obtain engineering services.
6. As noted by the Supreme Court, "nothing in the judgment prevents NSPE and its members from attempting to influence governmental action..."

NOTE: In regard to the question of application of the Code to corporations vis-à-vis real persons, business form or type should not negate nor influence conformance of individuals to the Code. The Code deals with professional services, which services must be performed by real persons. Real persons in turn establish and implement policies within business structures. The Code is clearly written to apply to the Engineer, and it is incumbent on members of NSPE to endeavor to live up to its provisions. This applies to all pertinent sections of the Code.

1420 King Street
Alexandria, Virginia 22314-2794
703/684-2800 • Fax:703/836-4875
www.nspe.org

Publication date as revised: July 2007 Publication #1102

AMERICAN INSTITUTE OF CHEMICAL ENGINEERS (AIChE)

www.aiche.org/About/Code.asps
    The AIChE code requires members to "never tolerate harassment" and to "treat fairly all colleagues and co-workers." It states that members "shall" pursue the positive goal of "using their knowledge and skill for the enhancement of human welfare." Also, members "shall" protect the environment.

AMERICAN SOCIETY OF CIVIL ENGINEERS (ASCE)

www.asce.org/inside/codeofethics.cfm

The ASCE code contains a number of statements about obligations to protect the environment and to adhere to the principles of sustainable development. These obligations are characterized as something engineers "should" (not "shall") adhere to in their professional work.

AMERICAN SOCIETY OF MECHANICAL ENGINEERS, ASME INTERNATIONAL

www.asme.org/NewsPublicPolicy/Ethics/Ethics_Center.cfm

The ASME code is divided into two parts. The Fundamental Principles and Fundamental Canons are in one document, and the ASME Criteria for Interpretation of the Canons are in another document. The first of the three Fundamental Principles states that engineers "use their knowledge and skills for the enhancement of human welfare."

ASSOCIATION FOR COMPUTING MACHINERY (ACM)

Short version: www.acm.org/about/se-code#short
Full version: www.acm.org/about/se-code#full

The ACM code for "software engineering" has a more informal tone than the other codes and tends to use a different vocabulary from the other codes. According to the code, the "public interest" takes priority over the interests of the employer. Software "shall" not only be safe but also should "not diminish quality of life, diminish privacy, or harm the environment." The "ultimate effect" of work in software engineering should be "the public good." When appropriate, software engineers "shall" also "identify, document, and report significant issues of social concern."

INSTITUTE OF ELECTRICAL AND ELECTRONICS ENGINEERS (IEEE)

www.ieee.org/web/membership/ethics/code_ethics.html

According to the code, members recognize "the importance of our technologies in affecting the quality of life throughout the world." Members agree to "accept responsibility in making decisions consistent with the safety, health, and welfare of the public, and to disclose promptly factors that might endanger the public or the environment." They also agree to "improve the understanding of technology, its appropriate application, and potential consequences." Finally, members agree to "treat fairly all persons regardless of such factors as race, religion, gender, disability, age, or national origin."

INSTITUTE OF INDUSTRIAL ENGINEERS (IIE)

www.iienet2.org/Details.aspx?id=299

In addition to providing Fundamental Principles and Fundamental Canons of its own, the IIE also endorses the Canon of Ethics provided by the Accreditation Board for Engineering and Technology. The Fundamental Principles state that engineers uphold and advance the integrity, honor, and dignity of the engineering profession by (among other things) "using their knowledge and skill for the enhancement of human welfare." The Fundamental Principles and Fundamental Canons make no mention of the environment.

NOTE

1. Reprinted by permission of the National Society of Professional Engineers (NSPE), www.nspe.org.

BIBLIOGRAPHY

Books, Articles, and Monographs

Alger, P. L., Christensen, N. A., and Olmstead, S. P. *Ethical Problems in Engineering* (New York: Wiley, 1965).

Allen, A. L. "Genetic Privacy: Emerging Concepts and Values," in M. Rothstein, ed., *Genetic Secrets* (New Haven, CT: Yale University Press, 1997), pp. 36–59.

———. "Privacy," in H. LaFollette, ed., *Oxford Handbook of Practical Ethics* (Oxford: Oxford University Press, 2003), pp. 485–513.

Alpern, K. D. "Moral Responsibilities for Engineers," *Business and Professional Ethics Journal*, 2, no. 2, 1983, pp. 39–48.

Anand, S., and Sen, A. *Development as Freedom* (New York: Anchor Books, 1999).

———. "The Income Component of the Human Development Index," *Journal of Human Development*, 1, no. 1, 2000, pp. 83–106.

Anderson, R. M., Perrucci, R., Schendel, D. E., and Trachtman, L. E. *Divided Loyalties: Whistle-Blowing at BART* (West Lafayette, IN: Purdue Research Foundation, 1980).

Anderson, S. *The Man Who Tried to Save the World* (New York: Doubleday, 1999).

Baase, S. *A Gift of Fire: Social, Legal and Ethical Issues in Computers and the Internet* (Hoboken, NJ: Wiley, 2004).

Baier, K. *The Moral Point of View* (Ithaca, NY: Cornell University Press, 1958).

Bailey, M. J. *Reducing Risks to Life: Measurement of the Benefits* (Washington, DC: American Enterprise Institute for Public Policy Research, 1980).

Baker, D. "Social Mechanics for Controlling Engineers' Performance," in Albert Flores, ed., *Designing for Safety: Engineering Ethics in Organizational Contexts* (Troy, NY: Rensselaer Polytechnic Institute, 1982).

Baram, M. S. "Regulation of Environmental Carcinogens: Why Cost–Benefit Analysis May Be Harmful to Your Health," *Technology Review*, 78, July–August 1976.

Baron, M. *The Moral Status of Loyalty* (Dubuque, IA: Center for the Study of Ethics in the Professions and Kendall/Hunt, 1984).

Baum, R. J. "Engineers and the Public: Sharing Responsibilities," in D. E. Wueste, ed., *Professional Ethics and Social Responsibility* (Lanham, MD: Rowman & Littlefield, 1994).

———. *Ethics and Engineering* (Hastings-on-Hudson, NY: Hastings Center, 1980).

———, and Flores, A., eds. *Ethical Problems in Engineering*, vols. 1 and 2 (Troy, NY: Center for the Study of the Human Dimensions of Science and Technology, Rensselaer Polytechnic Institute, 1978).

Baxter, W. F. *People or Penguins: The Case for Optimal Pollution* (New York: Columbia University Press, 1974).

Bayles, M. D. *Professional Ethics*, 2nd ed. (Belmont, CA: Wadsworth, 1989).

Bazelon, D. L. "Risk and Responsibility," *Science*, 205, July 20, 1979, pp. 277–280.

Beauchamp, T. L. *Case Studies in Business, Society and Ethics*, 2nd ed. (Englewood Cliffs, NJ: Prentice-Hall, 1989).

Bellah, R., Madsen, R., Sullivan, W. M., Swidler, A., and Tipton, S. M. *Habits of the Heart: Individualism and Commitment in American Life* (New York: Harper & Row, 1985).

Belmont Report: Ethical Principles and Guidelines for Protection of Human Subjects of Biomedical and Behavioral Research, publication no. OS 78-00f12 (Washington, DC: DHEW, 1978).

Benham, L. "The Effects of Advertising on the Price of Eyeglasses," *Journal of Law and Economics,* 15, 1972, pp. 337–352.

Benjamin, M. *Splitting the Difference: Compromise in Ethics and Politics* (Lawrence: University Press of Kansas, 1990).

Black, B. "Evolving Legal Standards for the Admissibility of Scientific Evidence," *Science,* 239, 1987, pp. 1510–1512.

Blackstone, W. T. "On Rights and Responsibilities Pertaining to Toxic Substances and Trade Secrecy," *Southern Journal of Philosophy,* 16, 1978, pp. 589–603.

Blinn, K. W. *Legal and Ethical Concepts in Engineering* (Englewood Cliffs, NJ: Prentice-Hall, 1989).

Board of Ethical Review, NSPE. *Opinions of the Board of Ethical Review,* vols. I–VII (Arlington, VA: NSPE Publications, National Society of Professional Engineers, various dates).

Boeyink, D. "Casuistry: A Case-Based Method for Journalists," *Journal of Mass Media Ethics,* Summer 1992, pp. 107–120.

Bok, S. *Common Values* (Columbia: University of Missouri Press, 1995).

————. *Lying: Moral Choice in Public and Private Life* (New York: Vintage Books, 1979).

Borgmann, A. *Technology and the Character of Contemporary Life: A Philosophical Inquiry* (Chicago: University of Chicago Press, 1984).

Bowyer, K., ed. *Ethics and Computing,* 2nd ed. (New York: IEEE Press, 2001).

Broad, W., and Wade, N. *Betrayers of the Truth* (New York: Simon & Schuster, 1982).

Bucciarelli, L. L. *Designing Engineers* (Cambridge, MA: MIT Press, 1994).

Buchanan, R. A. *The Engineers: A History of the Engineering Profession in Britain, 1750–1914* (London: Jessica Kingsley Publishers, 1989).

Cady, J. F. *Restricted Advertising and Competition: The Case of Retail Drugs* (Washington, DC: American Enterprise Institute, 1976).

Callahan, D., and Bok, S. *Ethics Teaching in Higher Education* (New York: Plenum Press, 1980).

Callahan, J. C., ed. *Ethical Issues in Professional Life* (New York: Oxford University Press, 1988).

Callon, M., and Law, J. "Agency and the Hybrid Collectif," *South Atlantic Quarterly,* 94, 1995, 481–507.

Cameron, R., and Millard, A. J. *Technology Assessment: A Historical Approach* (Dubuque, IA: Center for the Study of Ethics in the Professions and Kendall/Hunt, 1985).

Carson, T. L. "Bribery, Extortion, and the 'Foreign Corrupt Practices Act,'" *Philosophy and Public Affairs,* 14, no. 1, 1985, pp. 66–90.

Chadwick, R., ed. *Ethics and the Professions* (Aldershot, UK: Avebury, 1994).

Chalk, R., Frankel, M., and Chafer, S. B. *AAAS Professional Ethics Project: Professional Ethics Activities of the Scientific and Engineering Societies* (Washington, DC: American Association for the Advancement of Science, 1980).

Childress, J. F., and Macquarrie, J., eds. *The Westminster Dictionary of the Christian Church* (Philadelphia: Westminster Press, 1986).

Cohen, R. M., and Witcover, J. *A Heartbeat Away: The Investigation and Resignation of Vice President Spiro T. Agnew* (New York: Viking Press, 1974).

Columbia Accident Investigation Board (CAIB). *The CAIB Report,* vols. I–VII. Available at www.caib.us/.

Cranor, C. F. "The Problem of Joint Causes for Workplace Health Protections [1]," *IEEE Technology and Society Magazine,* September 1986, pp. 10–12.

————. *Regulating Toxic Substances: A Philosophy of Science and the Law* (New York: Oxford University Press, 1993).

Curd, M., and May, L. *Professional Responsibility for Harmful Actions* (Dubuque, IA: Center for the Study of Ethics in the Professions and Kendall/Hunt, 1984).

Davis, M. "Avoiding the Tragedy of Whistleblowing," *Business and Professional Ethics Journal,* 8, no. 4, 1989, pp. 3–19.

————. "Better Communication between Engineers and Managers: Some Ways to Prevent Many Ethically Hard Choices," *Science and Engineering Ethics,* 3, 1997, pp. 184–193.

————. "Conflict of Interest," *Business and Professional Ethics Journal,* Summer 1982, pp. 17–27.

————. "Explaining Wrongdoing," *Journal of Social Philosophy,* 20, Spring–Fall 1988, pp. 74–90.

————. "Is There a Profession of Engineering?" *Science and Engineering Ethics,* 3, no. 4, 1997, pp. 407–428.

————. *Profession, Code and Ethics* (Burlington, VT: Ashgate, 2002).

————. "Thinking Like an Engineer: The Place of a Code of Ethics in the Practice of a Profession,"

Philosophy and Public Affairs, 20, no. 2, Spring 1991, pp. 150–167.

———. *Thinking Like an Engineer* (New York: Oxford University Press, 1998).

———, Pritchard, M. S., and Werhane, P. "Case Study in Engineering Ethics: 'Doing the Minimum,'" *Science and Engineering Ethics,* 7, no. 2, April 2001, pp. 286–302.

———, and Stark, A., eds. *Conflicts of Interest in the Professions* (New York: Oxford University Press, 2001).

De George, R. T. "Ethical Responsibilities of Engineers in Large Organizations: The Pinto Case," *Business and Professional Ethics Journal,* 1, no. 1, Fall 1981, pp. 1–14.

Donaldson, T., and Dunfee, T. W. *Ties that Bind: A Social Contracts Approach to Business Ethics* (Boston: Harvard Business School Press, 1999).

———. "Toward Unified Conception of Business Ethics: Integrative Social Contract Theory," *Academy of Management Review,* 19, no. 2, 1994, pp. 152–184.

Douglas, M., and Wildavsky, A. *Risk and Culture* (Berkeley: University of California Press, 1982).

Dusek, V. *Philosophy of Technology: An Introduction* (Malden, MA: Blackwell, 2006).

Eddy, E., Potter, E., and Page, B. *Destination Disaster: From the Tri-Motor to the DC-10* (New York: Quadrangle Press, 1976).

Elbaz, S. W. *Professional Ethics and Engineering: A Resource Guide* (Arlington, VA: National Institute for Engineering Ethics, 1990).

Engineering Times (NSPE). "AAES Strives towards Being Unified" and "U.S. Engineer: Unity Elusive," 15, no. 11, November 1993.

Ermann, M. P., Williams, M. B., and Shauf, M. S. *Computers, Ethics, and Society,* 2nd ed. (New York: Oxford University Press, 1997).

Ethics Resource Center and Behavior Resource Center. *Ethics Policies and Programs in American Business* (Washington, DC: Ethics Resource Center, 1990).

Evan, W., and Manion, M. *Minding the Machines* (Upper Saddle River, NJ: Prentice-Hall, 2002).

Faden, R. R., and Beauchamp, T. L. *A History and Theory of Informed Consent* (New York: Oxford University Press, 1986).

Fadiman, J. A. "A Traveler's Guide to Gifts and Bribes," *Harvard Business Review,* July–August 1986, pp. 122–126, 130–136.

Feenberg, A. *Questioning Technology* (New York: Routledge, 1999).

Feinberg, J. "Duties, Rights and Claims," *American Philosophical Quarterly,* 3, no. 2, 1966, pp. 137–144.

Feliv, A. G. "The Role of the Law in Protecting Scientific and Technical Dissent," *IEEE Technology and Society Magazine,* June 1985, pp. 3–9.

Fielder, J. "Organizational Loyalty," *Business and Professional Ethics Journal,* 11, no. 1, 1991, pp. 71–90.

———. "Tough Break for Goodrich," *Journal of Business and Professional Ethics,* 19, no. 3, 1986.

———, and Birsch, D., eds. *The DC-10* (New York: State of New York Press, 1992).

Firmage, D. A. *Modern Engineering Practice: Ethical, Professional and Legal Aspects* (New York: Garland STPM, 1980).

Flores, A., ed. *Designing for Safety* (Troy, NY: Rensselaer Polytechnic Institute, 1982).

———. *Ethics and Risk Management in Engineering* (Boulder, CO: Westview Press, 1988).

———. *Professional Ideals* (Belmont, CA: Wadsworth, 1988).

———, and Johnson, D. G. "Collective Responsibility and Professional Roles," *Ethics,* 93, April 1983, pp. 537–545.

Florman, S. C. *Blaming Technology: The Irrational Search for Scapegoats* (New York: St. Martin's, 1981).

———. *The Civilized Engineer* (New York: St. Martin's, 1987).

———. *The Existential Pleasures of Engineering* (New York: St. Martin's, 1976).

———. "Moral Blueprints," *Harper's Magazine,* 257, no. 1541, October 1978, pp. 30–33.

Flumerfelt, R. W., Harris, C. E., Jr., Rabins, M. J., and Samson, C. H., Jr. *Introducing Ethics Case Studies into Required Undergraduate Engineering Courses,* Report on NSF Grant DIR-9012252, November 1992.

Ford, D. F. *Three Mile Island: Thirty Minutes to Meltdown* (New York: Viking Press, 1982).

Frankel, M., ed. *Science, Engineering, and Ethics: State of the Art and Future Directions,* Report of an American Association for the Advancement of Science Workshop and Symposium, February 1988.

Fredrich, A. J. *Sons of Martha: Civil Engineering Readings in Modern Literature* (New York: American Society of Civil Engineers, 1989).

French, P. A. *Collective and Corporate Responsibility* (New York: Columbia University Press, 1984).

Friedman, M. "The Social Responsibility of Business Is to Increase Its Profits," *New York Times Magazine,* September 13, 1970.

Garrett, T. M., et al. *Cases in Business Ethics* (New York: Appleton Century Crofts, 1968).

General Dynamics Corporation. *The General Dynamics Ethics Program Update* (St. Louis: Author, 1988).

Gert, B. *Common Morality* (New York: Oxford University Press, 2004).

———. "Moral Theory, and Applied and Professional Ethics," *Professional Ethics,* 1, nos. 1 and 2, Spring–Summer 1992, pp. 1–25.

Gewirth, A. *Reason and Morality* (Chicago: University of Chicago Press, 1978).

Glantz, P., and Lipton, E. *City in the Sky ... The Rise and Fall of the World Trade Centers* (New York: Times Books/Holt, 2003).

———. "The Height of Ambition," *New York Times Magazine,* September 8, 2002, section 6, p. 32.

Glazer, M. "Ten Whistleblowers and How They Fared," *Hastings Center Report,* 13, no. 6, 1983, pp. 33–41.

———. *The Whistleblowers: Exposing Corruption in Government and Industry* (New York: Basic Books, 1989).

Glickman, T. S., and Gough, R. *Readings in Risk* (Washington, DC: Resources for the Future, 1990).

Goldberg, D. T. "Turning in to Whistle Blowing," *Business and Professional Ethics Journal,* 7, 1988, pp. 85–99.

Goldman, A. H. *The Moral Foundations of Professional Ethics* (Totowa, NJ: Rowman & Littlefield, 1979).

Goodin, R. E. *Protecting the Vulnerable* (Chicago: University of Chicago Press, 1989).

Gorlin, R. A., ed. *Codes of Professional Responsibility,* 2nd ed. (Washington, DC: Bureau of National Affairs, 1990).

Gorman, M. E., Mehalik, M. M., and Werhane, P. *Ethical and Environmental Challenges to Engineering* (Upper Saddle River, NJ: Prentice-Hall, 2000).

Graham, L. *The Ghost of an Executed Engineer* (Cambridge, MA: Harvard University Press, 1993).

Gray, M., and Rosen, I. *The Warning: Accident at Three Mile Island* (New York: Norton, 1982).

Greenwood, E. "Attributes of a Profession," *Social Work,* July 1957, pp. 45–55.

Gunn, A. S., and Vesilind, P. A. *Environmental Ethics for Engineers* (Chelsea, MI: Lewis, 1986).

Harris, C. E. *Applying Moral Theories,* 5th ed. (Belmont, CA: Wadsworth, 2006).

———. "Engineering Responsibilities in Lesser-Developed Nations: The Welfare Requirement," *Science and Engineering Ethics,* 4, no. 3, July 1998, pp. 321–331.

———, Pritchard, M. S., and Rabins, M. J. *Practicing Engineering Ethics* (New York: Institute of Electrical and Electronic Engineers, 1997).

Heilbroner, R., ed. *In the Name of Profit* (Garden City, NY: Doubleday, 1972).

Herkert, J. "Future Directions in Engineering Ethics Research: Microethics, Macroethics and the Role of Professional Societies," *Science and Engineering Ethics,* 7, no. 3, July 2001, pp. 403–414.

Hick, J. *Disputed Questions in Theology and the Philosophy of Religion* (New Haven, CT: Yale University Press, 1986).

Howard, J. L. "Current Developments in Whistleblower Protection," *Labor Law Journal,* 39, no. 2, February 1988, pp. 67–80.

Hunter, T. "Engineers Face Risks as Expert Witnesses," *Rochester Engineer,* December 1992.

Hynes, H. P. "Women Working: A Field Report," *Technology Review,* November–December 1984.

Jackall, R. "The Bureaucratic Ethos and Dissent," *IEEE Technology and Society Magazine,* June 1985, pp. 21–30.

———. *Moral Mazes: The World of Corporate Managers* (New York: Oxford University Press, 1988).

Jackson, I. *Honor in Science* (New Haven, CT: Sigma Xi, 1986).

Jaksa, J. A., and Pritchard, M. S. *Communication Ethics: Methods of Analysis,* 2nd ed. (Belmont, CA: Wadsworth, 1994).

James, G. G. "Whistle Blowing: Its Moral Justification," in W. M. Hoffman and R. E. Frederick, eds., *Business Ethics,* 3rd ed. (New York: McGraw-Hill, 1995), pp. 290–301.

Jamshidi, M., Shahinpoor, M., and Mullins, J. H., eds. *Environmentally Conscious Manufacturing: Recent Advances* (Albuquerque, NM: ECM Press, 1991).

Janis, I. *Groupthink,* 2nd ed. (Boston: Houghton Mifflin, 1982).

Johnson, D. G. *Computer Ethics,* 3rd ed. (Upper Saddle River, NJ: Prentice-Hall, 2001).

———. *Ethical Issues in Engineering* (Englewood Cliffs, NJ: Prentice-Hall, 1991).

———, and Nissenbaum, H. *Computer Ethics and Social Policy* (Upper Saddle River, NJ: Prentice-Hall, 1995).

———, and Snapper, J. W., eds. *Ethical Issues in the Use of Computers* (Belmont, CA: Wadsworth, 1985).

Johnson, E. "Treating Dirt: Environmental Ethics and Moral Theory," in T. Regan, ed., *Earthbound: New Introductory Essays in Environmental Ethics* (New York: Random House, 1984).

304 Bibliography

Jonsen, A. L., and Toulmin, S. *The Abuse of Casuistry* (Berkeley: University of California Press, 1988).

Jurmu, J. L., and Pinodo, A. "The OSHA Benzene Case," in T. L. Beauchamp, ed., *Case Studies in Business, Society, and Ethics,* 2nd ed. (Englewood Cliffs, NJ: Prentice-Hall, 1989), pp. 203–211.

Kahn, S. "Economic Estimates of the Value of Life," *IEEE Technology and Society Magazine,* June 1986, pp. 24–31.

Kant, I. *Foundations of the Metaphysics of Morals, with Critical Essays* (R. P. Wolff, ed.) (Indianapolis: Bobbs-Merrill, 1969).

Kemper, J. D. *Engineers and Their Profession,* 3rd ed. (New York: Holt, Rinehart & Winston, 1982).

Kettler, G. J. "Against the Industry Exemption," in J. H. Shaub and K. Pavlovic, eds., *Engineering Professionalism and Ethics* (New York: Wiley–Interscience, 1983), pp. 529–532.

Kipnis, K. "Engineers Who Kill: Professional Ethics and the Paramountcy of Public Safety," *Business and Professional Ethics Journal,* 1, no. 1, 1981.

Kline, A. D. "On Complicity Theory," *Science and Engineering Ethics,* 12, 2006, pp. 257–264.

Kolhoff, M. J. "For the Industry Exemption...," in J. H. Shaub and K. Pavlovic, eds., *Engineering Professionalism and Ethics* (New York: Wiley–Interscience, 1983).

Kroes, P., and Bakker, M., eds. *Technological Development and Science in the Industrial Age* (Dordrecht, The Netherlands: Kluwer, 1992).

Kuhn, S. "When Worlds Collide: Engineering Students Encounter Social Aspects of Production," *Science and Engineering Ethics,* 1998, pp. 457–472.

Kultgen, J. *Ethics and Professionalism* (Philadelphia: University of Pennsylvania Press, 1988).

———. "Evaluating Codes of Professional Ethics," in W. L. Robison, M. S. Pritchard, and J. Ellin, eds., *Profits and Professions* (Clifton, NJ: Humana Press, 1983), pp. 225–264.

Ladd, J. "Bhopal: An Essay on Moral Responsibility and Civic Virtue," *Journal of Social Philosophy,* XXII, no. 1, Spring 1991.

———. "The Quest for a Code of Professional Ethics," in R. Chalk, M. S. Frankel, and S. B. Chafer, eds., *AAAS Professional Ethics Project: Professional Ethics Activities of the Scientific and Engineering Societies* (Washington, DC: American Association for the Advancement of Science, 1980).

Ladenson, R. F. "Freedom of Expression in the Corporate Workplace: A Philosophical Inquiry," in W. L. Robison, M. S. Pritchard, and J. Ellin, eds., *Profits and Professions* (Clifton, NJ: Humana Press, 1983), pp. 275–285.

———. "The Social Responsibilities of Engineers and Scientists: A Philosophical Approach," in D. L. Babcock and C. A. Smith, eds., *Values and the Public Works Professional* (Rolla: University of Missouri–Rolla, 1980).

———, Choromokos, J., d'Anjou, E., Pimsler, M., and Rosen, H. *A Selected Annotated Bibliography of Professional Ethics and Social Responsibility in Engineering* (Chicago: Center for the Study of Ethics in the Professions, Illinois Institute of Technology, 1980).

LaFollette, H. *Oxford Handbook of Practical Ethics* (Oxford: Oxford University Press, 2003).

———. *The Practice of Ethics* (Oxford: Blackwell, 2007).

Langewiesche, W. "Columbia's Last Flight," *The Atlantic,* 292, no. 4, November 2003, pp. 58–87.

Larson, M. S. *The Rise of Professionalism* (Berkeley: University of California Press, 1977).

Latour, B. *Science in Action: How to Follow Scientists and Engineers through Society* (Cambridge, MA: Harvard University Press, 1987).

Layton, E. T., Jr. *The Revolt of the Engineers: Social Responsibility and the American Engineering Profession* (Baltimore, MD: John Hopkins University Press, 1971, 1986).

Leopold, A. *A Sand County Almanac* (New York: Oxford University Press, 1966).

Lichtenberg, J. "What Are Codes of Ethics for?" in M. Coady and S. Bloch, eds., *Codes of Ethics and the Professions* (Melbourne, Australia: Melbourne University Press, 1995), pp. 13–27.

Litai, D. *A Risk Comparison Methodology for the Assessment of Acceptable Risk,* Ph.D. dissertation, Massachusetts Institute of Technology, Cambridge, MA, 1980.

Lockhart, T. W. "Safety Engineering and the Value of Life," *Technology and Society (IEEE),* 9, March 1981, pp. 3–5.

Lowrance, W. W. *Of Acceptable Risk* (Los Altos, CA: Kaufman, 1976).

Luebke, N. R. "Conflict of Interest as a Moral Category," *Business and Professional Ethics Journal,* 6, no. 1, 1987, pp. 66–81.

Luegenbiehl, H. C. "Codes of Ethics and the Moral Education of Engineers," *Business and Professional Ethics Journal,* 2, no. 4, 1983, pp. 41–61.

———. "Whistleblowing," in C. Mitchum, ed., *Encyclopedia of Science, Technology, and Ethics* (Detroit, MI: Thomson, 2005).

Lunch, M. F. "Supreme Court Rules on Advertising for Professions," *Professional Engineer,* 1, no. 8, August 1977, pp. 41–42.

Lynch, W. T., and Kline, R. "Engineering Practice and Engineering Ethics," *Science, Technology and Human Values,* 25, 2000, pp. 223–231.

MacIntyre, A. *After Virtue* (Notre Dame, IN: University of Notre Dame Press, 1984).

_____. "Regulation: A Substitute for Morality," *Hastings Center Report,* February 1980, pp. 31–41.

———. *A Short History of Ethics* (New York: Macmillan, 1966).

Magsdick, H. H. "Some Engineering Aspects of Headlighting," *Illuminating Engineering,* June 1940, p. 533.

Malin, M. H. "Protecting the Whistleblower from Retaliatory Discharge," *Journal of Law Reform,* 16, Winter 1983, pp. 277–318.

Mantell, M. I. *Ethics and Professionalism in Engineering* (New York: Macmillan, 1964).

Margolis, J. "Conflict of Interest and Conflicting Interests," in T. Beauchamp and N. Bowie, eds., *Ethical Theory and Business* (Englewood Cliffs, NJ: Prentice-Hall, 1979), pp. 361–372.

Marshall, E. "Feynman Issues His Own Shuttle Report Attacking NASA Risk Estimates," *Science,* 232, June 27, 1986, p. 1596.

Martin, D. *Three Mile Island: Prologue or Epilogue?* (Cambridge, MA: Ballinger, 1980).

Martin, M. W. *Everyday Morals* (Belmont, CA: Wadsworth, 1989).

_____. *Meaningful Work* (New York: Oxford University Press, 2000).

———. "Personal Meaning and Ethics in Engineering," *Science and Engineering Ethics,* 8, no. 4, October 2002, pp. 545–560.

———. "Professional Autonomy and Employers' Authority," in A. Flores, ed., *Ethical Problems in Engineering,* vol. 1 (Troy, NY: Rensselaer Polytechnic Institute, 1982), pp. 177–181.

———. "Rights and the Meta-Ethics of Professional Morality," and "Professional and Ordinary Morality: A Reply to Freedman," *Ethics,* 91, July 1981, pp. 619–625, 631–622.

———. *Self-Deception and Morality* (Lawrence: University Press of Kansas, 1986).

———, and Schinzinger, R. *Engineering Ethics,* 4th ed. (New York: McGraw-Hill 2005).

Mason, J. F. "The Technical Blow-by-Blow: An Account of the Three Mile Island Accident," *IEEE Spectrum,* 16, no. 11, November 1979, pp. 33–42.

May, W. F. "Professional Virtue and Self-Regulation," in J. L. Callahan, ed., *Ethical Issues in Professional Life* (New York: Oxford, 1988), pp. 408–411.

McCabe, D. "Classroom Cheating among Natural Science and Engineering Majors," *Science and Engineering Ethics,* 3, no. 4, 1997, pp. 433–445.

McIlwee, J. S., and Robinson, J. G. *Women in Engineering: Gender, Power, and Workplace Culture* (Albany: State University of New York Press, 1992).

Meese, G. P. E. "The Sealed Beam Case," *Business and Professional Ethics Journal,* 1, no. 3, Spring 1982, pp. 1–20.

Meyers, C. "Institutional Culture and Individual Behavior: Creating the Ethical Environment," *Science and Engineering Ethics,* 10, 2004, p. 271.

Milgram, S. *Obedience to Authority* (New York: Harper & Row, 1974).

Mill, J. S. *Utilitarianism* (G. Sher, ed.) (Indianapolis, IN: Hackett, 1979).

———. *Utilitarianism, with Critical Essays* (S. Gorovitz, ed.) (Indianapolis, IN: Bobbs-Merrill, 1971).

Millikan, R. A. "On the Elementary Electrical Charge and the Avogadro Constant," *Physical Review,* 2, 1913, pp. 109–143.

Morgenstern, J. "The Fifty-Nine Story Crisis," *The New Yorker,* May 29, 1995, pp. 45–53.

Morrison, C., and Hughes, P. *Professional Engineering Practice: Ethical Aspects,* 2nd ed. (Toronto: McGraw-Hill Ryerson, 1988).

Murdough Center for Engineering Professionalism. *Independent Study and Research Program in Engineering Ethics and Professionalism* (Lubbock: College of Engineering, Texas Technological University, October 1990).

Murphy, C. and Gardoni, P. "The Acceptability and the Tolerability of Societal Risks," *Science and Engineering Ethics,* 14, no. 12, March 2008, pp. 77–92.

_____. "Determining Public Policy and Resource Allocation Priorities for Mitigating Natural Hazards: A Capabilities-Based Approach," *Science & Engineering Ethics,* 13, no. 4, December 2007, pp. 489–504.

_____. "The Role of Society in Engineering Risk Analysis: A Capabilities Approach," *Risk Analysis,* 26, no. 4, 2006, pp. 1073–1083.

Nader, R. "Responsibility and the Professional Society," *Professional Engineer,* 41, May 1971, pp. 14–17.

———, Petkas, P. J., and Blackwell, K. *Whistle Blowing* (New York: Grossman, 1972).

National Academy of Science, Committee on the Conduct of Science. *On Being a Scientist* (Washington, DC: National Academy Press, 1989).

New York Times. "A Post–September 11 Laboratory in High Rise Safety," January 23, 2003, p. A1.

Noonan, J. T. *Bribery* (New York: Macmillan, 1984).

Nussbaum, M. *Women and Human Development: The Capabilities Approach* (New York: Cambridge University Press, 2000).

———, and Glover, J., eds. *Women, Culture, and Development* (Oxford: Clarendon, 1995).

Okrent, D., and Whipple, C. *An Approach to Societal Risk Assessment Criteria and Risk Management,* Report UCLA-Eng-7746 (Los Angeles: UCLA School of Engineering and Applied Sciences, 1977).

Oldenquist, A. "Commentary on Alpern's 'Moral Responsibility for Engineers,'" *Business and Professional Ethics Journal,* 2, no. 2, Winter 1983.

Otten, J. "Organizational Disobedience," in A. Flores, ed., *Ethical Problems in Engineering,* vol. 1 (Troy, NY: Center for the Study of the Human Dimensions of Science and Technology, Rensselaer Polytechnic Institute, 1978), pp. 182–186.

Patton-Hulce, V. R. *Environment and the Law: A Dictionary* (Santa Barbara, CA: ABC Clio, 1995).

Peterson, J. C., and Farrell, D. *Whistleblowing: Ethical and Legal Issues in Expressing Dissent* (Dubuque, IA: Center for the Study of Ethics in the Professions and Kendall/Hunt, 1986).

Petroski, H. *Beyond Engineering: Essays and Other Attempts to Figure without Equations* (New York: St. Martin's, 1985).

———. *To Engineer Is Human: The Role of Failure in Successful Design* (New York: St. Martin's, 1982).

Petty, T. "Use of Corpses in Auto-Crash Test Outrages Germans," *Time,* December 6, 1993, p. 70.

Philips, M. "Bribery," in Werhane, P., and D'Andrade, K., eds., *Profit and Responsibility* (New York: Edwin Mellon Press, 1985), pp. 197–220.

Pinkus, R. L. D., Shuman, L. J., Hummon, N. P., and Wolfe, H. *Engineering Ethics* (New York: Cambridge University Press, 1997).

Pletta, D. H. *The Engineering Profession: Its Heritage and Its Emerging Public Purpose* (Washington, DC: University Press of America, 1984).

Pritchard, M. S. "Beyond Disaster Ethics," *The Centennial Review,* XXXIV, no. 2, Spring 1990, pp. 295–318.

———. "Bribery: The Concept," *Science and Engineering Ethics,* 4, no. 3, 1998, pp. 281–286.

———. "Good Works," *Professional Ethics,* 1, nos. 1 and 2, Spring–Summer 1992, pp. 155–177.

———. *Professional Integrity: Thinking Ethically* (Lawrence: University Press of Kansas, 2006).

———. "Professional Responsibility: Focusing on the Exemplary," *Science and Engineering Ethics,* 4, no. 2, 1998, pp. 215–233.

———. "Responsible Engineering: The Importance of Character and Imagination," *Science and Engineering Ethics,* 7, no. 3, 2001, pp. 391–402.

———, ed. *Teaching Engineering Ethics: A Case Study Approach,* National Science Foundation grant no. DIR-8820837 (June 1992).

———, and Holtzapple, M. "Responsible Engineering: Gilbane Gold Revisited," *Science and Engineering Ethics,* 3, no. 2, April 1997, pp. 217–231.

Rabins, M. J. "Teaching Engineering Ethics to Undergraduates: Why? What? How?" *Science and Engineering Ethics,* 4, no. 3, July 1998, pp. 291–301.

Rachels, J. *The Elements of Moral Philosophy,* 4th ed. (New York: Random House, 2003).

Raelin, J. A. *The Clash of Cultures: Managers and Professionals* (Boston: Harvard Business School Press, 1985).

Rawls, J. *A Theory of Justice* (Cambridge, MA: Harvard University Press, 1971).

Relman, A. "Lessons from the Darsee Affair," *New England Journal of Medicine,* 308, 1983, pp. 1415–1417.

Richardson, H. "Specifying Norms," *Philosophy and Public Affairs,* 19, no. 4, 1990, pp. 279–310.

Ringleb, A. H., Meiners, R. E., and Edwards, F. L. *Managing in the Legal Environment* (St. Paul, MN: West, 1990).

Rogers Commission. *Report to the President by the Presidential Commission on the Space Shuttle Challenger Accident* (Washington, DC: Author, June 6, 1986).

Ross, W. D. *The Right and the Good* (Oxford: Oxford University Press, 1988).

Rostsen, G. H. "Wrongful Discharge Based on Public Policy Derived from Professional Ethics Codes," *American Law Reports,* 52, 5th 405.

Rothstein, M., ed. *Genetic Secrets* (New Haven, CT: Yale University Press, 1997).

Ruckelshaus, W. D. "Risk, Science, and Democracy," *Issues in Science and Technology,* 1, no. 3, Spring 1985, pp. 19–38.

Sagoff, M. "Where Ickes Went Right or Reason and Rationality in Environmental Law," *Ecology Law Quarterly,* 14, 1987, pp. 265–323.

Salzman, J., and Thompson, B. H., Jr. *Environmental Law and Policy* (New York: Foundation Press, 2003).

Scharf, R. C., and Dusek, V., eds. *Philosophy of Technology* (Malden, MA: Blackwell, 2003).

Schaub, J. H., and Pavlovic, K. *Engineering Professionalism and Ethics* (New York: Wiley–Interscience, 1983).

Schlossberger, E. *The Ethical Engineer* (Philadelphia: Temple University Press, 1993).

———. "The Responsibility of Engineers, Appropriate Technology, and Lesser Developed Nations," *Science and Engineering Ethics,* 3, no. 3, July 1997, pp. 317–325.

Schrader-Frechette, K. S. *Risk and Rationality* (Berkeley: University of California Press, 1991).

Schwing, R. C., and Albers, W. A., Jr., eds., *Societal Risk Assessment: How Safe Is Safe Enough?* (New York: Plenum Press, 1980).

Science and Engineering Ethics, Special Issue on Ethics for Science and Engineering-Based International Industries, 4, no. 3, July 1998, pp. 257–392.

Shapiro, S. "Degrees of Freedom: The Interaction of Standards of Practice and Engineering Judgment," *Science, Technology and Human Values,* 22, no. 3, Summer 1997.

Simon, H. A. *Administrative Behavior,* 3rd ed. (New York: Free Press, 1976).

Singer, M. G. *Generalization in Ethics* (New York: Knopf, 1961).

———, ed. *Morals and Values* (New York: Charles Scribner's Sons, 1977).

Singer, P. *Practical Ethics* (Cambridge, UK: Cambridge University Press, 1979).

Sismondo, S. *An Introduction to Science and Technology Studies* (Malden, MA: Blackwell, 2004).

Slovic, P., Fischoff, B., and Lichtenstein, S. "Rating the Risks," *Environment,* 21, no. 3, April 1969, pp. 14–39.

Solomon, R. C., and Hanson, K. R. *Above the Bottom Line: An Introduction to Business Ethics* (New York: Harcourt Brace Jovanovich, 1983).

Spinello, R. A. *Case Studies in Information and Computer Ethics* (Upper Saddle River, NJ: Prentice-Hall, 1997).

———, ed. *Cyber Ethics: Morality and Law in Cyberspace* (New York: Jones & Bartlett, 2003).

———. *Regulating Cyberspace* (Westport, CT: Quorum Books, 2002).

———, and Tavani, H. T., eds. *Readings in Cyber Ethics* (New York: Jones & Bartlett, 2001).

Starry, C. "Social Benefits versus Technological Risk," *Science,* 165, September 19, 1969, pp. 1232–1238.

Stone, C. *Where the Law Ends* (Prospect Heights, IL: Waveland Press, 1991).

Strand, P. N., and Golden, K. C. "Consulting Scientist and Engineer Liability," *Science and Engineering Ethics,* 3, no. 4, October 1997, pp. 347–394.

Sullivan, T. F. P., ed. *Environmental Law Handbook* (Rockdale, MD: Government Institutes, 1997).

Taeusch, C. F. *Professional and Business Ethics* (New York: Holt, 1926).

Tavani, H. T. *Ethics and Technology: Ethical Issues in Information and Communication Technology* (Hoboken, NJ: Wiley, 2004).

Taylor, P. W. "The Ethics of Respect for Nature," *Environmental Ethics,* 3, no. 3, Fall 1981, pp. 197–218.

———. *Principles of Ethics: An Introduction* (Encino, CA: Dickenson, 1975).

Thompson, P. "The Ethics of Truth-Telling and the Problem of Risk," *Science and Engineering Ethics,* 5, 1999, pp. 489–510.

Toffler, A. *Tough Choices: Managers Talk Ethics* (New York: Wiley, 1986).

Travis, L. A. *Power and Responsibility: Multinational Managers and Developing Country Concerns* (Notre Dame, IN: University of Notre Dame Press, 1997).

Unger, S. H. *Controlling Technology: Ethics and the Responsible Engineer,* 2nd ed. (New York: Holt, Rinehart & Winston, 1994).

———. "Would Helping Ethical Professionals Get Professional Societies into Trouble?" *IEEE Technology and Society Magazine,* 6, no. 3, September 1987, pp. 17–21.

Urmson, J. O. "Hare on Intuitive Moral Thinking," in S. Douglass and N. Fotion, eds., *Hare and Critics* (Oxford: Clarendon, 1988), pp. 161–169.

———. "Saints and Heroes," in A. I. Meldon, ed., *Essays in Moral Philosophy* (Seattle: University of Washington Press, 1958), pp. 198–216.

Vallero, P. A. and Vesilind, P. A. *Socially Responsible Engineering: Justice in Risk Management* (Hoboken, NJ: John Wiley & Sons, Inc., 2007).

Van de Poel, I., and van Gorp, A. C. "The Need for Ethical Reflection in Engineering Design," *Science, Technology and Human Values,* 31, no. 3, 2006, pp. 333–360.

Vandivier, R. "What? Me Be a Martyr?" *Harper's Magazine,* July 1975, pp. 36–44.

Vaughn, D. *The Challenger Launch Decision* (Chicago: University of Chicago Press, 1996).

Vaughn, R. C. *Legal Aspects of Engineering* (Dubuque, IA: Kendall/Hunt, 1977).

Velasquez, M. *Business Ethics,* 3rd ed. (Englewood Cliffs, NJ: Prentice-Hall, 1992).

———. "Why Corporations Are Not Responsible for Anything They Do," *Business and Professional Ethics Journal,* 2, no. 3, Spring 1983, pp. 1–18.

Vesilind, P. A. "Environmental Ethics and Civil Engineering," *The Environmental Professional,* 9, 1987, pp. 336–342.

———. *Peace Engineering: When Personal Values and Engineering Careers Converge* (Woodsville, NH: Lakeshore Press, 2005).

———, and Gunn, A. *Engineering, Ethics, and the Environment* (New York: Cambridge University Press, 1998).

Vogel, D. A. *A Survey of Ethical and Legal Issues in Engineering Curricula in the United States* (Palo Alto, CA: Stanford Law School, Winter 1991).

Wall Street Journal. "Executives Apply Stiffer Standards Than Public to Ethical Dilemmas," November 3, 1983.

Weil, V., ed. *Beyond Whistleblowing: Defining Engineers' Responsibilities,* Proceedings of the Second National Conference on Ethics in Engineering, March 1982.

———. *Moral Issues in Engineering: Selected Readings* (Chicago: Illinois Institute of Technology, 1988).

———. "Professional Standards: Can They Shape Practice in an International Context?" *Science and Engineering Ethics,* 4, no. 3, 1998, pp. 303–314.

Weisskoph, M. "The Aberdeen Mess," *Washington Post Magazine,* January 15, 1989.

Wells, P., Jones, H., and Davis, M. *Conflicts of Interest in Engineering* (Dubuque, IA: Center for the Study of Ethics in the Professions and Kendall/Hunt, 1986).

Werhane, P. *Moral Imagination and Management Decision Making* (New York: Oxford University Press, 1999).

Westin, A. F. *Individual Rights in the Corporation: A Reader on Employee Rights* (New York: Random House, 1980).

———. *Whistle Blowing: Loyalty and Dissent in the Corporation* (New York: McGraw-Hill, 1981).

Whitbeck, C. *Ethics in Engineering Practice and Research* (New York: Cambridge University Press, 1998).

———. "The Trouble with Dilemmas: Rethinking Applied Ethics," *Professional Ethics,* 1, nos. 1 and 2, Spring–Summer 1992, pp. 119–142.

Wilcox, J. R., and Theodore, L., eds. *Engineering and Environmental Ethics* (New York: Wiley, 1998).

Williams, B., and Smart, J. J. C. *Utilitarianism: For and Against* (New York: Cambridge University Press, 1973).

Wong, D. *Moral Relativity* (Berkeley: University of California Press, 1984).